Carbon Capture Science & Technology (CCST)

Carbon Capture Science & Technology (CCST)

Editors

Zilong Liu
Meixia Shan
Yakang Jin

Basel • Beijing • Wuhan • Barcelona • Belgrade • Novi Sad • Cluj • Manchester

Editors

Zilong Liu
Faculty of Science
China University of
Petroleum (Beijing)
Beijing
China

Meixia Shan
School of Chemical
Engineering
Zhengzhou University
Zhengzhou
China

Yakang Jin
School of Physics
University of Electronic
Science and Technology
of China
Chengdu
China

Editorial Office
MDPI AG
Grosspeteranlage 5
4052 Basel, Switzerland

This is a reprint of articles from the Topic published online in the open access journals *Energies* (ISSN 1996-1073), *Molecules* (ISSN 1420-3049), *Nanomaterials* (ISSN 2079-4991), *Separations* (ISSN 2297-8739), and *Sustainability* (ISSN 2071-1050) (available at: https://www.mdpi.com/topics/Carbon_Capture).

For citation purposes, cite each article independently as indicated on the article page online and as indicated below:

Lastname, A.A.; Lastname, B.B. Article Title. *Journal Name* **Year**, *Volume Number*, Page Range.

ISBN 978-3-7258-1551-7 (Hbk)
ISBN 978-3-7258-1552-4 (PDF)
doi.org/10.3390/books978-3-7258-1552-4

© 2024 by the authors. Articles in this book are Open Access and distributed under the Creative Commons Attribution (CC BY) license. The book as a whole is distributed by MDPI under the terms and conditions of the Creative Commons Attribution-NonCommercial-NoDerivs (CC BY-NC-ND) license.

Contents

About the Editors vii

Xiaohui Wang, Qihong Zhang, Shiwei Liang and Songqing Zhao
Systematic Review of Solubility, Thickening Properties and Mechanisms of Thickener for Supercritical Carbon Dioxide
Reprinted from: *Nanomaterials* **2024**, *14*, 996, doi:10.3390/nano14120996 1

Xiaohui Wang, Shiwei Liang, Qihong Zhang, Tianjiao Wang and Xiao Zhang
Molecular Dynamics Simulation on Thickening and Solubility Properties of Novel Thickener in Supercritical Carbon Dioxide
Reprinted from: *Molecules* **2024**, *29*, 2529, doi:10.3390/molecules29112529 17

Florian M. Chimani, Aditya Anil Bhandari, Andreas Wallmüller, Gerhard Schöny, Stefan Müller and Josef Fuchs
Evaluation of CO_2/H_2O Co-Adsorption Models for the Anion Exchange Resin Lewatit VPOC 1065 under Direct Air Capture Conditions Using a Novel Lab Setup
Reprinted from: *Separations* **2024**, *11*, 160, doi:10.3390/separations11060160 28

Yan Ma, Kai Wang, Sikai Liang, Zhongqing Li, Zhiyuan Wang and Jun Shen
Investigation on YSZ- and SiO_2-Doped Mn-Fe Oxide Granules Based on Drop Technique for Thermochemical Energy Storage
Reprinted from: *Molecules* **2024**, *29*, 1946, doi:10.3390/molecules29091946 49

Heak Vannak, Yugo Osaka, Takuya Tsujiguchi and Akio Kodama
Air-Purge Regenerative Direct Air Capture Using an Externally Heated and Cooled Temperature-Swing Adsorber Packed with Solid Amine
Reprinted from: *Separations* **2023**, *10*, 415, doi:10.3390/separations10070415 70

Amiza Surmi, Azmi Mohd Shariff and Serene Sow Mun Lock
Modeling of Nitrogen Removal from Natural Gas in Rotating Packed Bed Using Artificial Neural Networks
Reprinted from: *Molecules* **2023**, *28*, 5333, doi:10.3390/molecules28145333 89

Ruzhan Bai, Na Li, Quansheng Liu, Shenna Chen, Qi Liu and Xing Zhou
Effect of Steam on Carbonation of CaO in Ca-Looping
Reprinted from: *Molecules* **2023**, *28*, 4910, doi:10.3390/molecules28134910 103

Aleksey Nazarov, Sergey Chetverikov, Darya Chetverikova, Iren Tuktarova, Ruslan Ivanov, Ruslan Urazgildin, et al.
Microbial Preparations Combined with Humic Substances Improve the Quality of Tree Planting Material Needed for Reforestation to Increase Carbon Sequestration
Reprinted from: *Sustainability* **2023**, *15*, 7709, doi:10.3390/su15097709 118

Marta Pacheco, Patrícia Moura and Carla Silva
A Systematic Review of Syngas Bioconversion to Value-Added Products from 2012 to 2022
Reprinted from: *Energies* **2023**, *16*, 3241, doi:10.3390/en16073241 130

Qian Zhang, Yunjia Wang and Lu Liu
Carbon Tax or Low-Carbon Subsidy? Carbon Reduction Policy Options under CCUS Investment
Reprinted from: *Sustainability* **2023**, *15*, 5301, doi:10.3390/su15065301 154

Seungpil Jung
Expansion of Geological CO$_2$ Storage Capacity in a Closed Aquifer by Simultaneous Brine Production with CO$_2$ Injection
Reprinted from: *Sustainability* **2023**, *15*, 3499, doi:10.3390/su15043499 180

José Ramón Fernández
An Overview of Advances in CO$_2$ Capture Technologies
Reprinted from: *Energies* **2023**, *16*, 1413, doi:10.3390/en16031413 197

Xiaofeng Xu, Dongdong He, Tao Wang, Xiangyu Chen and Yichen Zhou
Technological Innovation Efficiency of Listed Carbon Capture Companies in China: Based on the Dual Dimensions of Legal Policy and Technology
Reprinted from: *Energies* **2023**, *16*, 1118, doi:10.3390/en16031118 201

Hong Chen, Haowen Zhu, Tianchen Sun, Xiangyu Chen, Tao Wang and Wenhong Li
Does Environmental Regulation Promote Corporate Green Innovation? Empirical Evidence from Chinese Carbon Capture Companies
Reprinted from: *Sustainability* **2023**, *15*, 1640, doi:10.3390/su15021640 217

**Karolina Kiełbasa, Şahin Bayar, Esin Apaydin Varol, Joanna Sreńscek-Nazzal,
Monika Bosacka, Piotr Miądlicki, et al.**
Carbon Dioxide Adsorption over Activated Carbons Produced from Molasses Using H$_2$SO$_4$, H$_3$PO$_4$, HCl, NaOH, and KOH as Activating Agents
Reprinted from: *Molecules* **2022**, *27*, 7467, doi:10.3390/molecules27217467 241

About the Editors

Zilong Liu

Dr. Zilong Liu is a young talent at the China University of Petroleum-Beijing (CUPB) and has been since Oct. 2020. Before joining CUPB, He completed two years of postdoctoral research supported by Shell Global Solutions at Delft University of Technology (2018–2020). He obtained his Ph.D. from the University of Copenhagen (2018) and completed a half-year Ph.D. exchange at the Hong Kong University of Science and Technology (HKUST). He received his M.Sc and B.Sc. from China University of Petroleum (Qingdao) in 2015 and 2012, respectively. Dr. Liu is interested in applying theoretical and laboratory methodologies to explore surface and interface-related phenomena, such as gas adsorption and separation, hydrate mining, colloidal self-assembly, and solid–liquid interactions.

Meixia Shan

Dr. Meixia Shan obtained her PhD from Delft University of Technology in 2018. She completed one year of postdoctoral research at the same university and then joined the School of Chemical Engineering at Zhengzhou University as an Associate Professor. Dr. Shan's main research area focuses on designing and synthesizing porous organic framework membranes, metal–organic framework membranes, and mixed matrix membranes for gas separation and wastewater treatment.

Yakang Jin

Dr. Yakang Jin was awarded his Ph.D. in Mechanical Engineering from the Department of Mechanical and Aerospace Engineering at the Hong Kong University of Science and Technology (HKUST) in 2021. Then, they joined the School of Physics, University of Electronic Science and Technology of China (UESTC) as an Assistant Professor via an International Postdoctoral Exchange Fellowship Program. Dr. Jin's main research interest lies in micro/nano-mechanics and nanofluidics and their applications in energy and environmental science.

Review

Systematic Review of Solubility, Thickening Properties and Mechanisms of Thickener for Supercritical Carbon Dioxide

Xiaohui Wang [1,2], Qihong Zhang [1], Shiwei Liang [1] and Songqing Zhao [3,*]

[1] Beijing Key Laboratory of Optical Detection Technology for Oil and Gas, China University of Petroleum (Beijing), Beijing 102249, China; wangxiaohui@cup.edu.cn (X.W.); 2021211324@student.cup.edu.cn (Q.Z.); 2023211325@student.cup.edu.cn (S.L.)
[2] National Key Laboratory of Petroleum Resources and Engineering, China University of Petroleum (Beijing), Beijing 102249, China
[3] State Key Laboratory of Heavy Oil Processing, China University of Petroleum (Beijing), Beijing 102249, China
* Correspondence: zhaosongqing@aliyun.com

Abstract: Supercritical carbon dioxide (CO_2) has extremely important applications in the extraction of unconventional oil and gas, especially in fracturing and enhanced oil recovery (EOR) technologies. It can not only relieve water resource wastage and environmental pollution caused by traditional mining methods, but also effectively store CO_2 and mitigate the greenhouse effect. However, the low viscosity nature of supercritical CO_2 gives rise to challenges such as viscosity fingering, limited sand–carrying capacity, high filtration loss, low oil and gas recovery efficiency, and potential rock adsorption. To overcome these challenges, low–rock–adsorption thickeners are required to enhance the viscosity of supercritical CO_2. Through research into the literature, this article reviews the solubility and thickening characteristics of four types of polymer thickeners, namely surfactants, hydrocarbons, fluorinated polymers, and silicone polymers in supercritical CO_2. The thickening mechanisms of polymer thickeners were also analyzed, including intermolecular interactions, LA–LB interactions, hydrogen bonding, and functionalized polymers, and so on.

Keywords: supercritical CO_2; thickener; intermolecular interaction; functionalized polymer

1. Introduction

In recent years, carbon dioxide (CO_2) has become the focus of carbon emission reduction. As a major greenhouse gas, the utilization and storage of CO_2 (as a buffer gas when producing hydrogen) are required to be environmentally friendly and economically sustainable [1–6]. For the chemical properties of CO_2, the carbon atoms in the CO_2 molecule are hybridized in *sp* mode, and the electrons form two mutually perpendicular π bonds. The bond length of the carbon–oxygen double bond (O=C=O) is shorter than that of the carbonyl bond (C=O). Thus, the CO_2 structure is stable and the chemical properties are not very reactive [7], and the phase diagram of CO_2 can be seen in Figure 3 of Ref. [8]. The critical points of CO_2 are at 31 °C and 74 bar [8]. When the temperature and pressure exceed the critical point, CO_2 undergoes a transition into the supercritical state, resulting in the formation of supercritical CO_2.

Supercritical CO_2 is a substance that exhibits properties intermediate between those of a gas and a liquid. It possesses advantageous characteristics such as high diffusivity, low viscosity, low surface tension, and controllable solubility. This unique nature of supercritical CO_2 finds extensive applications in oil displacement technology and fracturing technology [9], effectively addressing the limitations associated with hydraulic fracturing [10,11]. These limitations include excessive water consumption, clay swelling, reservoir damage caused by residual working fluids, and inadequate flowback leading to groundwater pollution [12,13]. However, the low viscosity characteristics of pure supercritical CO_2, with a viscosity of only 0.02–0.05 mPa·s, will cause a series of problems including viscous fingering, limited sand–loading capacity, and filter damage. The viscosity fingering problem stems from the viscosity difference between supercritical CO_2 and crude oil, which causes supercritical CO_2 to form finger–like flows in the reservoir, bypassing the oil layer and reducing recovery efficiency [14]. High filtration loss means that part of the fracturing fluid is adsorbed, retained, and permeated into the formation during the fracturing process. The problem of limited sand–carrying capacity is due to the low viscosity and high diffusivity of supercritical CO_2, which limits its ability to carry sand particles and affects the effective support of fractures [10,15–17].

Therefore, understanding how to increase the viscosity of supercritical CO_2 has become extremely important. The most direct and effective method is to add a thickener to supercritical CO_2. The ideal supercritical CO_2 thickeners should be effective in increasing viscosity at low doses. From a quantitative point of view, the thickening ratio is sufficient when it can thicken supercritical CO_2 200–300 times, which is achieved in fluorinated thickeners [18], yet fluorinated thickeners are toxic, limiting their on–site application. In principle, the higher the viscosity of supercritical CO_2, the better it would fit within the range to be achieved. In oilfield applications, the ideal efficiency of thickening depends on the actual application needs. For a broad range of applications, the increase of 5 to 10 times will serve the purpose. For oil extraction in the Middle East, a viscosity increase of 10 times is more than adequate. And in the context of supercritical CO_2 fracturing technology, the viscosity enhancement should range from 20 to 30 times the original value. In addition to an increase in viscosity, an ideal thickener should also be cheap, environmentally friendly, safe, high efficiency, and soluble in supercritical CO_2 but insoluble in water [19]. At present, supercritical CO_2 thickeners are generally divided into the following four categories: surfactants, hydrocarbon polymers, fluorinated polymers, and silicone polymers [20]. In particular, fluorine–containing thickeners have the best solubility and thickening effects, but their use is restricted as they are expensive and not environmentally friendly [21,22]. Furthermore, they are not effective at low concentrations and also adsorb to rock [23]. This article mainly reviews the research progress of the four types of thickeners in terms of solubility, thickening properties, and mechanisms.

2. Characterization Parameters of Solubility and Thickening Properties

Through much literature research, it can be seen that the ideal supercritical CO_2 thickener requires both efficient solubility and thickening properties in CO_2 without any additional cosolvents. The characterization parameter of solubility and thickening properties are summarized below.

2.1. Solubility Properties

In terms of solubility, the main influencing factors are the interaction between the polymer and CO_2 molecules, solvent CO_2 density, solute relative molecular mass and molecular polarity, and especially the density factor. Solubility increases exponentially with the density of a supercritical CO_2 system [24]. For thickening properties, it is mainly influenced by the spatial network structure formed by the interaction between thickener molecules. This structure effectively impedes the flow of CO_2 molecules, resulting in the thickening of supercritical CO_2.

The dissolving properties of thickeners in supercritical CO_2 can be described by the solubility parameter δ [25], which is equal to the arithmetic square root of the cohesive energy density. The closer the polymer solubility parameter is to the CO_2 solubility parameter, the better the solubility in CO_2. The addition of cosolvents can reduce differences in the solubility parameters [25].

According to Enick's research on thermodynamics, in order to be dissolved in supercritical CO_2, Gibbs free energy (G_{mix}) must be reduced, that is $\Delta G_{mix} < 0$ [26], and its related expression is as follows:

$$\Delta G_{mix} = \Delta H_{mix} - T\Delta S_{mix} \tag{1}$$

Here, ΔH_{mix}, ΔS_{mix}, and T are the mixing enthalpy change, mixing entropy change, and absolute temperature, respectively. Therefore, the above problem is transformed into the problem of how to improve ΔH_{mix} and ΔS_{mix}.

For H_{mix}, the main influencing factors are the density of the CO_2 mixed solution, the interaction between CO_2 molecules, between the polymer thickener molecules, and between the polymer thickener and CO_2 molecules. However, the interaction between the polymer thickener and the CO_2 molecules is critical in determining whether the thickener can be dissolved in liquid CO_2. This interaction effectively promotes enthalpy reduction, which in turn reduces the free energy, allowing the thickener to dissolve in supercritical CO_2 [27]. The strength of the interaction between molecules can be described by cohesive energy density [25]. For CO_2 molecules, the electron distribution is near oxygen atoms and the CO_2 molecule has a zero dipole moment; however, a large quadrupole moment and low polarizability [28] can cause weak interactions with non–polar covalently bonded fragments like C–C motifs, but reasonably strong interactions with non–hydrogen–bonded polar functional groups like esters, ethers, C–F groups, or aromatic structures. Therefore, in order to be dissolved in CO_2, it should be weakly polar and have low cohesive energy density or contain a certain number of CO_2–philic groups.

As for ΔS_{mix}, the solubility of the polymer can be improved by increasing entropy of the system. For example, increasing the free volume of the polymer, improving the flexibility of the polymer chain, and lowering the glass transition temperature can all help reduce the interaction between polymer molecules and thus promote dissolution. Additionally, increasing the degree of branching of the polymer thickener can have a similar effect [29].

The interaction between the polymer and CO_2 is one of the key factors determining the solubility of the polymer. If the interaction energy between the polymer and CO_2 is strong, it usually means that the solubility of the polymer in supercritical CO_2 is better because this strong interaction helps to overcome the attraction between the polymer molecules or between the CO_2 molecules such that the thickener can be dispersed into the solvent. The interaction energy between the polymer and the CO_2 molecules provides a measure of the strength of this interaction and can be calculated via Formula (2) [30].

$$E_{inter} = E_{polymer-CO_2} - \left(E_{polymer} + E_{CO_2}\right) \quad (2)$$

The total energy E_{CO2} of supercritical CO_2, the total energy $E_{polymer}$, and the total energy $E_{polymer-CO2}$ of the mixed system can be calculated through molecular dynamics (MD). The larger the absolute value of E_{inter}, the stronger the interaction. Hu et al. used MD methods to study poly(vinyl acetate–alt–maleate) copolymers. The results show that this type of polymer reduces the interaction energy between the polymer and CO_2 after copolymerizing vinyl acetate, but it remains higher than the interaction energy between PVAc and CO_2 [31].

In addition, the adsorption of polymers in supercritical CO_2 can be reflected by the potential of mean force (PMF), which is calculated as shown in (3) [32]. The change in PMF can reflect the positional preference of polymers in CO_2. If the PMF value is negative, it indicates that the polymer is more stable at a certain position in supercritical CO_2, which contributes to polymer solubilization.

$$E_{(r)} = -k_B T \ln g_{(r)} \quad (3)$$

where $E_{(r)}$ is the mean force potential, k_B is Boltzmann's constant, T is the absolute temperature, $g_{(r)}$ is the radial distribution function between the polymer and CO_2. The physical significance of $g_{(r)}$ can be expressed as the ratio of the local density of the B atom to the intrinsic density of the A atom at a distance r from the central atom A. This indicates that CO_2 molecules need to overcome a certain barrier to approach the thickener. Moreover, the solubility of polymers in CO_2 is also affected by polymer–polymer interactions to the same degree. The weaker the polymer–polymer interaction and the stronger the polymer–CO_2 interaction, the more favorable the solubility.

Among the interaction between the polymer and CO_2, Lewis acid–base (LA–LB) interaction can effectively promote dissolution of polymer [25]. In an LA–LB interaction, the Lewis acid acts as an electron pair acceptor and can accept electron pairs, while the Lewis base acts as an electron pair donor and can donate electron pairs. For example, Gong et al. [25] found that the LA–LB interaction between O atoms in PVAc and C atoms in CO_2 can enhance the solubility of PVAc in supercritical CO_2. This interaction helps more CO_2 molecules distribute around the carbonyl groups in the PVAc molecular chain, thereby increasing the solubility of PVAc in supercritical CO_2.

2.2. Thickening Properties

The viscosity of the mixed solution can be calculated by the following formula [33].

$$\eta = \frac{\tau_w}{\gamma_w} = \frac{D\Delta p/4L}{8v/D} \tag{4}$$

where η(Pa·s) is the viscosity, τ_w and γ_w are the wall shear stress, and apparent shear rate, relatively. D is the capillary diameter, Δp is the capillary pressure difference, L is the capillary length, and v is the flow rate of CO_2 in the thickened liquid. Research shows that the thickened supercritical CO_2 system is a non–Newtonian fluid and the viscosity has a non–linear relationship with the shear rate [33]. Moreover, the viscosity can also be acquired by fitting the following transverse current autocorrelation function below:

$$C_{\perp(k,t)} \sim e^{-\frac{\eta k^2}{\rho}t} \tag{5}$$

which can be obtained through MD simulation [32].

3. Supercritical CO_2 Thickeners

3.1. Surfactants

Surfactant thickeners are polymers composed of nonpolar hydrophilic groups and polar hydrophobic groups [34,35]. In supercritical CO_2, the hydrophilic groups of the surfactant undergo physical interactions with a small amount of water, while the hydrophobic groups are exposed to and interact with CO_2, forming a reverse micelle structure that further develops into a spatial network structure [36,37]. This network structure can restrict the mobility of CO_2 molecules, thereby serving a thickening function.

The solubility of the polymer ZCJ–01 (copolymers of styrene with modified sulfonated fluorinated acrylates), surfactant thickener APRF–2 (consists of sodium succinate (–2–ethyl) sulfonate, ethanol, and H_2O, etc.) and surfactant thickener SC–T–18 in supercritical CO_2 was investigated by Zhai et al. [38]. The main component of SC–T–18 is a comb copolymer with polydimethylsiloxane as the main chain and amino groups as side chains. The research results show that SC–T–18 has the highest solubility among them, and the experimental values are in good agreement with the theoretical values [38]. SC–T–18 and supercritical CO_2 form a single, stable, homogeneous emulsion micelle after sufficient mixing, which is due to the side–chain amino group in SC–T–18 effectively increasing its solubility in CO_2 through an LA–LB interaction. At 25 °C and 6.894 MPa, Enick [39] conducted a study on the use of a 1 wt% surfactant tributyltin fluoride thickener and a 40–45 wt% pentane cosolvent to increase the viscosity of the supercritical CO_2 system. The results showed that this combination can thicken the viscosity by 10–100 times. However, the flaw of this thickener is the necessity to use a considerable number of cosolvents to facilitate its dissolution in supercritical CO_2, which makes the process very inefficient. Considering that the addition of cosolvents will bring more serious environmental problems, Shi et al. [40–42] introduced a fluoroalkyl group into the trialkyltin fluoride polymer molecule to obtain a semi–fluorinated trialkyltin fluoride thickener. Research shows that this thickener has high solubility in supercritical CO_2 systems without the use of any cosolvents. Under 10–18 MPa,

4% mass fraction of semi–fluorinated trialkyltin fluoride can increase the system viscosity by up to 3.3 times. Enick et al. [43] also used perfluoropolyether glycol and fluorinated diisocyanate to react to synthesize a fluorinated polyurethane thickener. At 25 °C and 25 MPa, 4% fluorinated polyurethane can increase the viscosity of the system by 1.7 times.

Trickett et al. [44] designed a surfactant that can not only dissolve in CO_2, but also form rod–shaped micelles with enhanced viscosity when a small amount of water is added. They changed Na^+ in dialkyl sulfosuccinic acid into M^{2+} (Co^{2+} or Ni^{2+}), as shown in Figure 1, turning the spherical micelles into rod–shaped micelles, and then forming reverse micelles in supercritical CO_2. Then, through the interaction between these micelles, the viscosity of the CO_2 mixed system increased. Research results show that $Co(di–HCF_4)_2$ and $Ni(di–HCF_4)_2$ with 6–10 wt% can increase the viscosity of the CO_2 mixed system by 20–90 wt%. For some surfactant thickeners that are difficult to dissolve in supercritical CO_2, CO_2–philic groups can be introduced, such as fluorinated amine– and oxygen–containing surfactants [45]. Semi–fluorinated and fluorinated surfactant thickeners are found to be soluble in CO_2 liquids and can also increase CO_2 viscosity through the addition of a small amount of water [46]. By studying the solubility of oxygenated hydrocarbon surfactants in CO_2, it was found that this thickener has a similar level to that of fluorinated surfactants, and both show high solubility properties, indicating that the O atoms in oxygenated hydrocarbon surfactants can increase solubility [47].

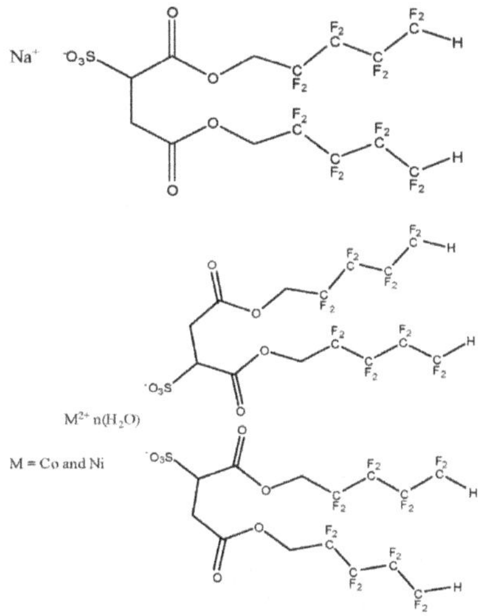

Figure 1. Structures of $Na^+(di–HCF_4)$ and $M^{2+}(di–HCF_4)$. Reprinted (adapted) with permission from [44]. Copyright© 2010, American Chemical Society.

The principle of surfactant thickening of CO_2 involves the formation of reverse micelles by surfactants. These reverse micelles overlap and entangle with each other, creating a spatial network structure. This structure restricts the flow of CO_2 molecules, resulting in the thickening effect of the system. In addition, hydrocarbons, polar, or ionic groups can also be introduced into surfactants to increase the viscosity of supercritical CO_2 systems [48]. The interaction between ion charges and water dipoles, as well as the Van der Waals force between alkyl chains, are also important factors in the formation of ionic surfactant micelles. The former is required to be stronger than the interaction between ions and CO_2, and as for the latter, specific functional groups can be introduced to enhance the Van der Waals forces. In order to enhance the solubility of thickener molecules in supercritical CO_2, CO_2–philic groups, such as fluoroalkyl groups, carbonyl groups, and oxygen atoms, can be introduced into surfactant thickener molecules.

3.2. Hydrocarbon Polymers

Generally, this type of thickener contains carbon (C), hydrogen (H) elements, and may contain oxygen (O) elements. Hydrocarbon polymer thickeners usually have low solubility and weak thickening properties. Heller et al. [49] conducted a study on commercially available hydrocarbon polymer thickeners and found that the viscosity of the supercritical CO_2 system did not significantly increase through the introduction of hydrocarbon polymer thickeners. They discovered that only a portion of the thickeners was soluble in supercritical CO_2. The solubility of this portion was attributed to the polymers' amorphous and irregular structure, which lacked a compact crystalline arrangement. This structural characteristic allowed for greater space, facilitating the penetration and solubilization of CO_2 molecules. Not only that, the solubility of this type of polymer in supercritical CO_2 was influenced by the interactions between the polymers, as well as between the polymers and CO_2. The solubility is higher when the interaction between the polymer and CO_2 is stronger.

Sarbu et al. [50] believed that polymers soluble in CO_2 should have monomer units of LA–LB interaction with CO_2 and monomer units with high free volume and high flexibility. Shen et al. [51] confirmed that among all the hydrocarbon thickeners, polyvinyl acetate (PVAc, Mw = 125,000) has the best solubility in supercritical CO_2. The reason is that PVAc contains acetic acid groups, which can effectively increase solubility, but its ability to thicken CO_2 is relatively weak. Because of this, PVAc has become an ideal supercritical CO_2 thickener design material. Shen et al. [52] used azobisisobutyronitrile (AIBN) as a catalyst to synthesize a polyvinyl acetate telomer through free radical reaction, and then polymerized it with styrene to form a binary copolymer, namely styrene vinyl acetate binary copolymer, and the reaction mechanism is shown in Ref. [52] (1), (2) and (3). This copolymer molecule has CO_2–philic groups and thickening groups, namely acetic acid groups and styrene groups, respectively, and is expected to become an economical and environmentally friendly thickener. Zhang Jian [53] used AIBN as an initiator to synthesize four–arm PVAc through the reversible addition–fragmentation–chain transfer (RAFT) polymerization method. In the case of adding cosolvent ethanol, at 35 °C and 15 MPa, adding four–arm PVAc with a concentration of 1 wt% and ethanol with a concentration of 5 wt% can increase the viscosity of the supercritical CO_2 system by 31–55%. However, this still does not meet the standards for actual use.

In recent years, MD simulations have been used to study the solubility of supercritical CO_2 and the thickening properties of thickeners. Xue et al. [32] used MD simulation methods to calculate the thickening mechanism of supercritical CO_2 thickener and found that polyvinyl acetate–copolyvinyl ether (PVAEE) molecular chains formed a spatial network structure by intertwining with each other. Due to the interaction between CO_2 molecules and polymers (including electrostatic interactions and Van der Waals interactions), the polymer restricted CO_2 molecules to the network structure formed by PVAEE (degree of polymerization of each chain was N = 50) molecular chains. This restriction reduced the supercritical CO_2 molecules flow, thereby increasing the viscosity of the supercritical CO_2 fluid. They also obtained PMF by calculating the radial distribution function (RDF) in the results. The contact minimum (CM) and the second solvent separation minimum (SSM) are 0.4 nm and 0.85 nm, respectively. Their corresponding energy values determine the binding stability of CO_2 and the polymer in the first and second solvent layers, respectively. A higher energy barrier exists between the CM and the SSM, which is the barrier of the solvent layer (BS). The equations for calculating the binding energy barrier and dissociation energy barrier between CO_2 and polymer groups are $\Delta E^+ = E_{BS} - E_{SSM}$ and $\Delta E^- = E_{BS} - E_{CM}$, respectively. The corresponding binding and dissociation energy barriers for the ether (ester) groups are 700 kJ/mol and 300 kJ/mol (540 kJ/mol and 190 kJ/mol), respectively. These values indicate that the ester group is more readily bonded to CO_2. Goicochea et al. [54] also used MD simulation to study the interaction between polymers and CO_2 molecules. Research shows that both intermolecular interactions and branching can improve the viscosity of supercritical CO_2. In particular, intermolecular π–π stacking plays a crucial role in the thickening effect of supercritical CO_2. These studies show that MD simulation is a very effective method to study supercritical CO_2 systems at the molecular level.

Double–chain polyether carbonate (TPA–PEC, Mw = 2168, N = 30), tri–chain polyether carbonate (TMA–PEC, Mw = 2211, N = 30), and four–chain polyether carbonate (TFA–PEC, Mw = 2254, N = 30) were synthesized as CO_2 thickeners by Chen et al. [55]. The respective dissolution properties were studied using the MD simulation method at 24.85 °C and 20 MPa. The results show that both TPA–PEC and TMA–PEC have better solubility than TFA–PEC due to their stronger interaction with the CO_2 molecules, but their thickening effect is poor. TFA–PEC has the highest viscosity and only needs an addition of 0.95 wt% to thicken the supercritical CO_2 viscosity by 11 times, while TMA–PEC needs to be added at 0.72 wt% to thicken the CO_2 viscosity by 3.9 times. Among the three, TPA–PEC has the worst CO_2 thickening ability. From the perspective of solubility and thickening properties, the multi–chain structure is beneficial to the thickening ability but not to solubility. Polyether carbonate is also a polymer thickener that is easily degraded under natural conditions and has the advantage of being environmentally friendly.

Afra et al. [23] investigated the supercritical CO_2 system of Poly–1–decene (P1D, Mw = 2950) in sandstone rock properties. It was shown that 1.5 wt% P1D increased the viscosity of the supercritical CO_2 system by a factor of six at 24.13 MPa and 35 °C. When the temperature was increased to 90 °C, the viscosity increased by a factor of 4.8. It was also shown that the large number of methyl groups in the P1D molecule contributes to its solubility in CO_2, while the branched structure of the molecule positively affects the thickening effect as well. In addition, poly–1–decene (an oligomer of about 20 repeating units) is not only very effective in CO_2 viscosification, but also reduces the remaining water saturation from 40 to about 27% at 24.13 MPa and 90 degrees celsius, which can improve the storage efficiency of CO_2.

Sun et al. [56] used AIBN as the initiator and synthesized a series of copolymers P(HFDA–co–MMA) and P(HFDA–co–EAL) using HFDA (1H,1H,2H,2H–perfluorodecyl acrylate), EAL (ethyl acrylate), and MMA (methyl methacrylate), as shown in Ref. [56] Scheme 1. The microstructure and intermolecular interactions in the supercritical CO_2 system were studied through MD simulations. According to their research, an increase in the content of the EAL group enhances the interaction between copolymer chains and reduces their flexibility, leading to a decrease in solubility. Moreover, the intermolecular association of the copolymer is strengthened, resulting in an increased thickening ability. At 35.05 °C and 30 MPa, the P(HFDA$_{0.19}$–co–EAL$_{0.81}$, Mw = 3576) copolymer with the highest EAL content increases the viscosity of supercritical CO_2 by 96 times at 5 wt% concentration and has the best thickening property among all copolymers. The solubility and thickening properties of P(HFDA$_{0.37}$–co–EAL$_{0.63}$) are higher than those of P(HFDA$_{0.39}$–co–MMA$_{0.61}$). P(HFDA$_{0.37}$–co–EAL$_{0.63}$, Mw = 3394) increases the viscosity of supercritical CO_2 by 70 times, while P(HFDA0.39–co–MMA0.61) only increases it by 40 times. It shows that although EAL and MMA are isomers, the differences in their structures and compositions make huge differences in the intermolecular interactions of copolymer–CO_2 and the association between copolymer chains. The presence of methyl groups in the main chain of P(HFDA–co–MMA) increases steric hindrance, which reduces intermolecular association, free volume, and chain flexibility.

Furthermore, at 344.3 K and 25–45 MPa, a coarse–grained molecular modeling study optimized by Kazuya [18,57] via the particle swarm optimization algorithm showed that branched hydrocarbon poly–1–decene oligomers (especially the model with six repeating units and Mw = 1000) showed a significant increase in solubility in supercritical CO_2 compared to straight–chained alkanes with the same molecular weight, up to a factor of 270 times. This increase is attributed to an increase in the number of branches in the molecular structure, especially structural edges (methyl groups), which have enhanced interactions with CO_2 and thus increase solubility. The branched structure of the thickener not only increases its solubility in CO_2 but also reduces the adsorption of the thickener to the rock as compared to the change in chemical composition [57]. These findings provide important molecular design principles for the development of thickeners with high solubility in supercritical CO_2. Ding et al. [58] conducted a study on the solubility and thickening properties of oligomers of 1–decene (O1D) with six repeat units and oligomers with branches of 1–dodecene and 1–hexadecene (O1D1H). The research confirmed that branches and methyl groups can promote solubility. At approximately 13.8 MPa and 35 °C, the solubility of O1D in supercritical CO_2 is 0.6 wt%. While at 24.1 MPa and 35 °C, the solubility of O1D1H is 0.3 wt%. However, in terms of relative viscosity, the 0.3 wt% concentration of O1D1H provides better viscosity performance than the 0.6 wt% concentration of O1D.

In general, the biggest problem with hydrocarbon polymer thickeners is their low solubility and difficulty in completely dissolving in supercritical CO_2. At present, the main way to quickly solve the problem of solubility is to add a large amount of cosolvent, but this also has the implication of environmental problems caused by the cosolvent, which is neither economically nor environmentally friendly. Therefore, it is still necessary to modify and design the hydrocarbon thickener molecules themselves to find thickeners with high solubility and high thickening properties.

3.3. Fluorinated Polymers

Compared with hydrocarbon polymer thickeners, fluorinated polymer thickeners obtained after fluorination have stronger CO_2–philic properties and can be effectively dissolved in liquid CO_2 without adding cosolvents. At the same time, the polymer has a better thickening effect.

DeSimone et al. [59] demonstrated for the first time that fluoropolymers can be strongly dissolved in supercritical CO_2 without the assistance of cosolvents and show good thickening properties. Research shows that at 50 °C and 300 bar, 3.7 wt% poly(1,1–dihydroperfluorooctyl acrylate) (PFOA, Mw = 1,400,000) in supercritical CO_2 can increase the viscosity of the system from 0.08 cP to 0.20–0.25 cP. However, PFOA is toxic to aquatic organisms, which may cause disruption to aquatic ecosystems.

Huang et al. [60] synthesized copolymers (PolyFAST) using a fluorinated acrylate and styrene copolymer. Among them, the fluorocarbon group is CO_2–philic and can improve the solubility of PolyFAST in supercritical CO_2, while the styrene group is CO_2–phobic and can thicken CO_2 but can also reduce the solubility of PolyFAST in supercritical CO_2. The ratio of fluorinated acrylate to styrene is crucial in determining the thickening effect. Through multiple experiments, it has been found that the most significant increase in viscosity occurs when using a ratio of 71 mol% fluorinated acrylate and 29 mol% styrene [60]. Additionally, incorporating 1–5 wt% PolyFAST in supercritical CO_2 can result in viscosity increasing up to 5–400 times. However, the production cost of this polymer is relatively high, and it is not environmentally friendly.

Heller et al. [49] studied telechelic polymer thickeners which have corresponding ionic groups at each end and form a network structure in the form of ion pair aggregation. Enick et al. [61] synthesized poly–sulfonated polyurethane. Fluorinated telechelic ionic polymers have good solubility in CO_2 and the addition of 4 wt% the polymer can increase the viscosity of CO_2 by 2.7 times at 25 °C and 25 MPa.

Shi et al. [42] synthesized a series of semi–fluorinated trialkyl tin fluorides. Among them, 4 wt% tris(1,1,2,2-tetrahydroperfluorohexyl)tin fluoride is soluble in CO_2 and can increase the thickening of the supercritical CO_2 system by 3.3 times. The mechanism is that the positively charged Sn atoms and the negatively charged F atoms form Sn–F bridges to create transient polymer chains.

Sun et al. [62] used an all–atom MD to simulate the molecular model of the polymer–CO_2 system and studied the solubility and thickening properties of the copolymer in supercritical CO_2. Research shows that 5 wt% of P(HFDA$_{0.49}$–co–VPc$_{0.51}$, Mw = 3023) can increase the viscosity of the CO_2 system by 62 times. The thickening performance of 1.5 wt% of P (HFDA$_{0.31}$–co–VAc$_{0.69}$, Mw = 3539) is higher than that of P (HFDA0.49–co–VAc0.51) under the same conditions, but it is not easily dissolved at higher concentrations. The main reason is the high concentration of VAc which increases the number of methyl groups in the polymer chain, resulting in a decrease in chain flexibility. Therefore, the length and composition of polymer side chains can greatly affect the thickening performance.

Huang zhou [27] synthesized a CO_2 thickener containing a fluoro–urea group. Research shows that DCT ([1,6–Bis(1,3–diperfluorooctanoic acid propyl–2–ureido)]heptane, double–chain thickener) begins to thermally degrade at around 200 °C, and SCT ((1,6–Difluorooctanoic acid ethyl ester urethyl)hexane, single–chain thickener) begins to thermally degrade at 170 °C. The optimal mass fraction obtained through single–factor experiments is 2 wt%, and DCT makes the viscosity of supercritical carbon dioxide is increased to 1.54 mPa·s, while the SCT is 1.46 mPa·s.

Kilic et al. [63] synthesized a series of aromatic acrylate–fluoroacrylate copolymer supercritical CO_2 thickeners, and studied their structure and mechanism of thickening supercritical CO_2. Research results show that the thickening property of this copolymer increases firstly and then decreases as the content of aromatic acrylate groups increases. The best solution is a 29% phenyl acrylate–71% fluoroacrylate copolymer. In the supercritical CO_2 system, the copolymer only needs 5 wt% to increase the viscosity of the system by 205 times at 21.85 °C and 41.4 MPa. At the same time, it was also found that 26% phenyl acrylate (PHA)–74% fluoroacrylate (FA) has a better thickening effect than 27% CHA (cyclohexyl acrylate)–74% FA (fluoroacrylate). This proves that π–π stacking between aromatic rings plays a crucial role in thickening supercritical CO_2.

In addition, Goicochea et al. [54], also used molecular simulation to study the thickening properties of polymer HFDA. Research shows that the thickening principle of fluorinated polymers mainly has two aspects. On the one hand, the fluorocarbon groups in the molecules can effectively enhance the CO_2–philic properties of the polymers; on the other hand, the coupling mechanism between polymer molecules, which is the π–π association between styrenes, is stronger than the intramolecular interaction, making it difficult for polymer molecules to diffuse and aggregate, hindering the flow of CO_2 molecules. This further enhances the thickening property of the supercritical CO_2 system.

At present, according to the above research results, it can be seen that fluorine–containing polymer thickeners have impressive characteristics in terms of CO_2–philic properties and CO_2–thickening properties. However, the economic cost of such fluorinated polymers is too high, and they cannot be metabolized by organisms in the ecosystem. At the same time, they can also cause varying degrees of damage to organisms, such as weakening germ cell activity, interfering with enzyme activity, and damaging cell membrane structures, and so on [22]. Nevertheless, the research on this type of polymer provides theoretical guidance for the future design of economical, environmentally friendly, and pollution–free supercritical CO_2 thickeners.

3.4. Silicone Polymer

Silicone polymers show reliable performance in thickening and are also pollution–free [64], and thus they can be an ideal potential supercritical CO_2 thickener.

Bae et al. [65,66] used polydimethylsiloxane (PDMS) as a thickener to thicken supercritical CO_2. Research shows that at 54 °C and 17.2 MPa, the viscosity of the 4% PDMS thickener + 20% toluene cosolvent + 76% liquid CO_2 system increases to a maximum of 1.2 mPa·s. Compared with pure supercritical CO_2, the viscosity increased by 30 times. But the disadvantage is that a large amount of cosolvent needs to be added. Zhao et al. [67] also used PDMS to thicken supercritical CO_2, and the difference was that kerosene was used as a cosolvent because kerosene has a better solubilizing effect than toluene. Research results show that at 51.85 °C, the viscosity of the 5% PDMS thickener + 5% kerosene cosolvent + 90% liquid CO_2 system increases to 4.67 mPa·s, which makes an increase of 54 times, while the amount of cosolvent is reduced.

Fink et al. [68] studied the feasibility of side–chain functionalization to improve the thickening properties of silicone polymers. The results show that silicone polymers with the appropriate amounts of side–chain functionalization act similarly to fluorinated polyether materials in supercritical CO_2. Kilic et al. [68,69] enhanced the solubility of PDMS in supercritical CO_2 through functionalization with propyldimethylamine. O'Brien et al. [70] synthesized a series of aromatic amidated polydimethylsiloxane (PDMS), as shown in Figure 5 of Ref. [70], and studied their solubility and thickening properties in supercritical CO_2. Research results show that PDMS with anthraquinone–2–carboxamide (AQCA) end groups can thicken supercritical CO_2 with hexane as a cosolvent, as shown in Figure 6 of Ref. [70]. The reason is that the content of CO_2–philic groups and benzene ring groups in PDMS containing AQCA is low, and hexane is needed to thicken supercritical CO_2.

Li et al. [71] synthesized a silicone terpolymer using 0.09 g tetramethylammonium hydroxide catalyst and a molar ratio of aminopropyltriethoxysilane and methyltriethoxysilane of 2:1. At 35 °C and 12 MPa, 3 wt% silicone terpolymer and 7 wt% toluene can thicken the viscosity of the supercritical CO_2 system by 5.7 times. The mechanisms of silicone terpolymer and toluene are shown in Figure 2. CO_2 interacts with amino groups. Specifically, N in the amino groups donates electrons to C in the CO_2 and CO_2 is located above N. Hydroxyl enhances the stability of the spatial network structure formed by siloxane and CO_2 molecules. The reason why this type of polymer can thicken supercritical CO_2 is that the hydroxyl group enhances the spatial network structure. Additionally, the chain structure generated by intermolecular interactions also plays a certain binding role, thereby increasing the flow resistance of CO_2.

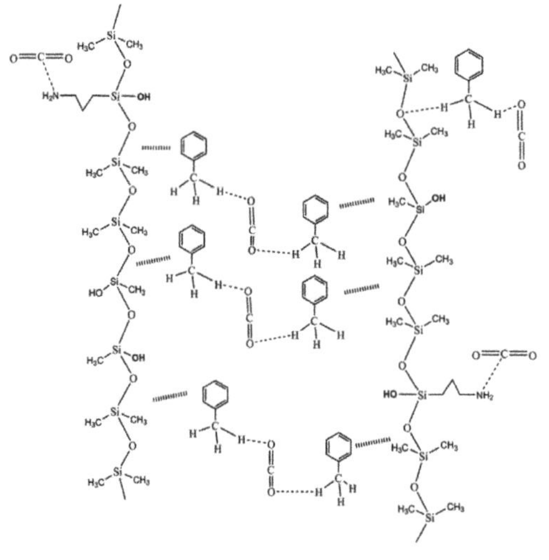

Figure 2. Thickening mechanism of silicone terpolymer [71].

Wang et al. [72] synthesized epoxy–terminated polydimethylsiloxane, as shown in Figure 1 of Ref. [72], and studied its thickening performance in supercritical CO_2. Research results show that when the shear rate increases, the polymer network structure will be destroyed by shear, and the viscosity of the system also decreases, that is, shear thinning. When the temperature rises, the activity and migration rate of various molecules in the system will be enhanced, which will weaken the intermolecular interaction, thereby destroying the network structure of the polymer and resulting in a decrease in the viscosity of the system. When the pressure increases in the range of 8–14 MPa, the degree of damage to the polymer's spatial network structure will decrease and the viscosity will increase.

Shen et al. [6] used benzoyl peroxide as the initiator and synthesized a graft copolymer of methylsilsesquioxane and vinyl acetate through graft polymerization, as shown in Figure 3 of Ref. [6]. The thickener does not contain fluorine. Studies have shown that the grafting of PVAc enhances the solubility of siloxane polymers in supercritical CO_2, and what plays a thickening role would be the network structure generated by polymethylsilsesquioxane. This research provides ideas for solving the solubility problem of polymers in supercritical CO_2.

4. Thickening Mechanism

To obtain an ideal thickener, it should have a certain amount of CO_2–philic groups (ether groups, carbonyl groups, acetate groups, acetyl groups, sugar ester groups, etc.) and CO_2–phobic groups in the molecule. CO_2–philic groups contribute to improve the solubility of the thickener, while CO_2–phobic groups enhance the viscosity of supercritical CO_2 through intermolecular association.

The introduced chain–like CO_2–philic groups should have good flexibility, low cohesive energy, and high free volume. CO_2–phobic groups can associate or their chains can cross and entangle with each other to form a spatial network structure to restrict the flow of CO_2 molecules [32]. According to the results of Sagisaka et al. [73], at a certain concentration, the surfactant self–assembles to form linear or rod–like micelles that would intertwine with each other, forming a network structure and increasing the viscosity of CO_2. It was also observed that rod–like reverse micelles, with different length–to–diameter ratios, exhibit varying thickening effects on supercritical CO_2 at the same temperature and pressure. For instance, at 45 °C and 350 bar, anisotropic reverse micelles of about 5 to 7 wt% with rod lengths of approximately 166 Å and 583 Å increase the viscosity by 24% and 200%, respectively. Meanwhile, these two groups cannot be too many or too few. If there are too few CO_2–phobic groups, the solubility of the thickener would be insufficient to achieve the desired thickening effect; while too many CO_2–phobic groups would also affect the solubility of the thickener [27]. Kilic et al. [63] showed that the thickening properties of aromatic acrylate–fluoroacrylate copolymers exhibited an increase and then a decrease with the content of aromatic acrylate groups. Copolymers containing 29% phenyl acrylate and 71% fluoroacrylate were found to be the most desirable. The addition of only 5 wt% of the copolymer could increase the viscosities of supercritical CO_2 by up to a factor of 205. Thus, research seeking an optimal ratio or dosage is still needed.

Generally, for surfactant thickeners, one end should be soluble in CO_2 and the other end should be soluble in water or organic solvents to reduce the surface tension of water or organic solvents in CO_2. At the same time, to form reverse micelles, two conditions should be satisfied, one is the multiple branched non–polar tail chain and low cohesive energy density, and the other one is a hydrogen–bonding interaction between the polar head group and water [74].

Hydrocarbon thickeners should meet two characteristics. On the one hand, they require large free volume, high chain flexibility, small steric resistance, weak interaction, low glass transition temperature and small steric hindrance, which help the polymer dissolve in CO_2. On the other hand, polymer chains can cross and entangle with each other to form a spatial network structure, which hinders the flow of CO_2 molecules and thicken CO_2 [27].

Fluorine–containing polymer thickeners are obtained by fluorination of hydrocarbon polymers. They are usually weakly polar and have dipole–quadrupole interactions with CO_2 molecules. At the same time, molecular chains can cross and entangle with each other to form a spatial network structure.

Silicone thickeners generally require cosolvents to improve solubility and thickening effects. The π–π stacking between phenyl groups produces intermolecular interactions, which has the thickening effect of supercritical CO_2.

For thickening properties of thicker in supercritical CO_2, in addition to the molecular structure, ratio of CO_2–philic groups and CO_2–phobic groups mentioned above, temperature, pressure, thickener molecular weight, and so on are also important influencing factors. The temperature and pressure conditions vary across different reservoir depths, leading to differences in the corresponding properties. The viscosity of supercritical CO_2 exhibits changes in response to temperature variations. Much research has been conducted [18,75] on the effects of temperature and pressure on CO_2 viscosity, and the results reveal a decreasing trend in CO_2 viscosity with increasing temperature, while an there is an increasing trend with increasing pressure.

5. Thickening Supercritical CO_2 in Porous Media

5.1. The Flow of CO_2 in Porous Media

The displacement process of CO_2 in heterogeneous porous media is one of the most important mechanisms [76]. Fluid physical parameters will cause phase flow instability during the CO_2 displacement process. Research shows that when CO_2 is injected into deep salt–water layers, it will displace the pores in a supercritical state. The dominant force in the displacement process is viscous force, and it will affect the form and distribution of fluid flow during the displacement process [77]. The simulation results show that under the condition of low–viscosity enhancement, the displacement process is obviously unstable, and the whole process has a relatively obvious fingering phenomenon; on the contrary, under the condition of higher viscosity enhancement, the displacement process is more stable, and no obvious fingering phenomenon occur [78].

5.2. Adsorption in Porous Media

During CO_2 fracturing, CO_2 thickeners may remain in the shale reservoir, and these chemicals may pollute the reservoir environment. Therefore, there is the need for the low adsorption of CO_2 thickeners in porous media in practical industrial applications. If the adsorption within the porous medium is excessive, it may lead to the blockage of the pores [58]. Afra et al. [23] conducted experiments on a variety of thickeners currently available on the market, and the results showed that several thickeners containing the element fluorine showed significant adsorption to the surface of porous media [23]. They also used MD simulations to study the adsorption problem and came up with the agreement between the simulation and experimental results. They proposed an effective theoretical approach to study the adsorption of thickeners in porous media [23]. Li et al. [79] modified the thickener and prepared a new type of PDMS, and then investigated its contact angle. The results showed that the contact angle of PDMS decreased from 138° to 99° with increasing temperature, with a significant decreasing trend, while the contact angle of the prepared novel PDMS decreased from 135° to 127° [79]. Compared with the two, the novel PDMS has less adsorption on the reservoir surface, which is more favorable to reduce the contamination of the thickener on the reservoir.

In conclusion, the thickener should not exhibit excessive adsorption on the surface of porous media, as it would lead to a poorer process, economic problems, and excessively large reductions in permeability due to wettability alteration. If the thickener is brine–soluble (which is unlikely, given the low mutual solubility of CO_2 and water), the thickener may separate out into the brine within the porous media. If the thickener is crude–oil–soluble, a portion of the thickener may ultimately contaminate the crude oil product and potentially cause problems in downstream processing equipment within refineries [23].

6. Summary and Outlook

This article provides a comprehensive review of four types of polymer thickeners, namely surfactants, hydrocarbons, fluorine–containing polymers, and silicones. We focused on analyzing their solubility and thickening characteristics in supercritical CO_2 systems, and also explained the thickening mechanisms. Furthermore, we discussed the flow and adsorption of thickeners in porous media.

For surfactants, the thickening property is adequate, while solubility is far from satisfactory [39,43]. For example, a 1 wt% surfactant tributyltin fluoride thickener and a 40–45 wt% pentane cosolvent can thicken the viscosity by 10–100 times [39]. However, the solubility of the thickener is poor, requiring a significant amount of cosolvent or CO_2–philic groups. In addition, silicones show similar solubility and thickening characteristics to surfactants, where 5% PDMS thickener with a small amount of cosolvent increases the viscosity to 4.67 mPa·s, which is an increase of up to 54 times [67]. The economic cost and environmental problems of cosolvents have become an urgent issue to be addressed.

Among the four thickeners, fluorinated thickeners have the most outstanding thickening properties, brought about by the interactions between the fluorine element and CO_2. According to our research, the addition of 5 wt% PolyFast could increase the viscosity of supercritical CO_2 by up to 400 times [67], which is the highest on record so far, according to our knowledge. Meanwhile, this copolymer also has fantastic solubility under reservoir conditions. However, this type of thickener is not commonly used, mainly because of its economic cost and biological toxicity. At present, most other polymer thickeners still require cosolvents to thicken liquid CO_2; however, it is not environment friendly.

It was found that PVAc is one of the optimal CO_2–philic hydrocarbon homopolymers because of the acetic acid group [73], yet its thickening properties are not ideal at present. However, PVAc is an ideal economical and environmentally friendly thickener, and an abundance of research has been conducted to improve the viscosity with it [32,52,53], such as forming a binary copolymer or spatial network structure, etc. This has made the PVAc–based system a mainstream thickener in work sites.

In recent years, the thickening mechanism and promotion of thickeners have been investigated through molecular modeling of polymer–CO_2 systems. Regarding the thickening mechanism, it has been recognized that CO_2–soluble polymers may have a moderately branched structure, high free volume, low solubility parameter, and contain Lewis acid–base groups. By introducing CO_2–philic groups, the interaction between the thickener molecules and CO_2 can be facilitated, thereby increasing the solubility of the thickener in supercritical CO_2. The polymers should also contain CO_2–phobic groups, which can combine with neighboring CO_2–phobic groups to form a viscosity–enhancing network structure. Furthermore, the thickeners should exhibit low adsorption onto rock to minimize the blockage of rock pores, maintain the fluidity of the fracturing fluid, and reducing pollution and damage to the rock environment. Therefore, further research may focus on these aspects, addressing economic and technological barriers, as well as environmental concerns. The development of efficient, environmentally friendly, and cost–controllable thickeners can help promote engineering site applications.

Author Contributions: Writing—review and editing, X.W. and Q.Z.; visualization, S.L.; supervision, S.Z.; funding acquisition, X.W. and S.Z. All authors have read and agreed to the published version of the manuscript.

Funding: This research was funded by the National Natural Science Foundation of China, grant number 11804028 and 12175023, and by the National Key Laboratory of Petroleum Resources and Engineering, China University of Petroleum (Beijing), grant number 2462024YJRC005.

Acknowledgments: The authors would like to acknowledge the National Key Laboratory of Petroleum Resources and Engineering, China University of Petroleum (Beijing), for supporting the research project.

Conflicts of Interest: The authors declare no conflicts of interest.

References

1. Muhammad, A.; Nurudeen, Y.; Nilanjan, P.; Alireza, K.; Stefan, I.; Hussein, H. Influence of pressure, temperature and organic surface concentration on hydrogen wettability of caprock; implications for hydrogen geo–storage. *Energy Rep.* **2021**, *7*, 5988–5996. [CrossRef]
2. Muhammad, A.; Nurudeen, Y.; Nilanjan, P.; Alireza, K.; Stefan, I.; Hussein, H. Influence of organic molecules on wetting characteristics of mica/H_2/brine systems: Implications for hydrogen structural trapping capacities. *J. Colloid Interface Sci.* **2022**, *608*, 1739–1749. [CrossRef] [PubMed]
3. Sherif, F.; Ahmed, E.-T.; Hesham, A.; Youssef, E.; Abdulmohsin, I. Increasing Oil Recovery from Unconventional Shale Reservoirs Using Cyclic Carbon Dioxide Injection. In Proceedings of the SPE Europec, EAGE Conference and Exhibition, Virtual, 1–3 December 2020. [CrossRef]
4. Sean, S.; Patricia, C.; Barbara, K.; Sittichai, N. Characterizing Pore–Scale Geochemical Alterations in Eagle Ford and Barnett Shale from Exposure to Hydraulic Fracturing Fluid and CO_2/H_2O. *Energy Fuels* **2020**, *35*, 583–598. [CrossRef]

5. Zhang, Q.; Zuo, L.; Wu, C.; Sun, C.; Zhu, X. Effects of crude oil characteristics on foaming and defoaming behavior at separator during CO_2 flooding. *Colloids Surf. A Physicochem. Eng. Asp.* **2020**, *608*, 125562. [CrossRef]
6. Aiguo, S.; Jinbo, L.; Yuehui, S.; Fuchang, S.; Zhengliang, W. Synthesis of the Copolymer of Vinylacete–Methylsilsesquioxane as a Potential Carbon Dioxide Thickener. *Polym. Mater. Sci. Eng.* **2011**, *27*, 157–159. (In Chinese) [CrossRef]
7. Zhao, Z.; Li, X.; Zhang, B.; Gan, B.; Li, G. Experimental Study on Supercritical CO_2 Fracturing. *Nat. Gas Explor. Dev.* **2016**, *39*, 58–63+14. (In Chinese)
8. Etienne, G.; Thierry, T.; Jean-Daniel, M.; Mathias, D. Structure–Property Relationships in CO_2–philic (Co)polymers: Phase Behavior, Self–Assembly, and Stabilization of Water/CO_2 Emulsions. *Chem. Rev.* **2016**, *116*, 4125–4169. [CrossRef] [PubMed]
9. XueSong, X.; Guo, Y.; Zhang, J.; Sun, N.; Shen, G.; Chang, X.; Yu, W.; Tang, Z.; Chen, W.; Wei, W.; et al. Fracturing with Carbon Dioxide: From Microscopic Mechanism to Reservoir Application. *Joule* **2019**, *3*, 1913–1926. [CrossRef]
10. Liu, W. Research advance in supercritical CO_2 thickeners. *Fault-Block Oil Gas Field* **2019**, *19*, 658–661. (In Chinese) [CrossRef]
11. Whorton, L.P.; Brownscombe, E.R.; Dyes, A.B. Method for producing oil by means of carbon dioxide. US. Patent US2623596A, 30 December 1952.
12. Birgit, C.G.; Ewers, U.; Fritz, H.F. Hydraulic fracturing: A toxicological threat for groundwater and drinking–water? *Environ. Earth Sci.* **2013**, *70*, 3875–3893. [CrossRef]
13. Min, L.; Mehdi, S.; Peyman, M. Impact of mineralogical heterogeneity on reactive transport modelling. *Comput. Geosci.* **2017**, *104*, 12–19. [CrossRef]
14. Zhang, Y. *Study on the Influencing Factors of Viscous Fingering of CO_2 in Oil Reservoirs*; China University of Petroleum: Beijing, China, 2010.
15. Shen, Z.; Wang, H.; Li, G. Numerical simulation of the cutting–carrying ability of supercritical carbon dioxide drilling at horizontal section. *Pet. Explor. Dev.* **2011**, *38*, 233–236. (In Chinese) [CrossRef]
16. Li, Q.; Chen, M.; Jin, Y.; Wang, M.; Jiang, H. Application of New Fracturing Technologies in Shale Gas Development. *Spec. Oil Gas Reserv.* **2012**, *19*, 1–7+141. (In Chinese) [CrossRef]
17. Wang, H.; Li, G.; Zhu, B.; Sepehrnoori, K.; Shi, L.; Zheng, Y.; Shi, X. Key problems and solutions in supercritical CO2 fracturing technology. *Front. Energy* **2019**, *13*, 667–672. [CrossRef]
18. Pal, N.; Zhang, X.; Ali, M.; Mandal, A.; Hoteit, H. Carbon dioxide thickening: A review of technological aspects, advances and challenges for oilfield application. *Fuel* **2022**, *315*, 122947. [CrossRef]
19. Xie, W.; Chen, S.; Wang, M.; Yu, Z.; Wang, H. Progress and Prospects of Supercritical CO2 Application in the Exploitation of Shale Gas Reservoirs. *Energy Fuels* **2021**, *35*, 18370–18384. [CrossRef]
20. LI, Q.; WANG, Y.; LI, Q.; WANG, F.; YUAN, L.; BAI, H. Thickening Performance and Thickening Mechanism of a Viscosifier for CO_2 Fracturing Fluid. *Drill. Fluid Complet. Fluid* **2019**, *36*, 102–108. (In Chinese) [CrossRef]
21. Xue, W.; Wenge, C.; Qiuyuan, Y.; Hongyun, N.; Qian, L.; Yun, L.; Ming, G.; Ming, X.; An, X.; Sijin, L.; et al. Preliminary investigation on cytotoxicity of fluorinated polymer nanoparticles. *J. Environ. Sci. China* **2018**, *69*, 217–226. [CrossRef] [PubMed]
22. Liu, Y.; Dong, W.; Ye, L.; Ren, P. Progress in Research on Pollution Status and Hazards of Perfluorinated Organic Compounds (Pfcs) in Surface Water. *Water Pollut. Control* **2015**, *33*, 43–47. (In Chinese) [CrossRef]
23. Salar, A.; Mohamed, H.A.; Abbas, F. Improvement in CO_2 geo–sequestration in saline aquifers by viscosification: From molecular scale to core scale. *Int. J. Greenh. Gas Control* **2023**, *125*, 103888. [CrossRef]
24. Wang, J.; Zhang, N.; Wu, Q.; Wang, S.; Zhang, P.; Zhao, J. Research progress on the solubility of supercritical CO_2. *Refin. Chem. Ind.* **2011**, *22*, 1–5+83. (In Chinese) [CrossRef]
25. Gong, H.; Zhang, H.; Xu, L.; Li, Y.; Dong, M. Effects of cosolvent on dissolution behaviors of PVAc in supercritical CO_2: A molecular dynamics study. *Chem. Eng. Sci.* **2019**, *206*, 22–30. [CrossRef]
26. Robert, M.E.; Eric, J.B.; Hamilton, A. *Novel CO_2–Thickeners for Improved Mobility Control*; UNT Digital Library: Denton, TX, USA, 2001. [CrossRef]
27. Huang, Z. *Preparation of CO_2 Thickening Agent and Evaluation of Its Fracturing Performance*; SouthWest Petroleum University: Chengdu, China, 2017.
28. Poovathinthodiyil, R.; Yutaka, I.; Scott, L.W. Polar Attributes of Supercritical Carbon Dioxide. *Acc. Chem. Res.* **2005**, *38*, 478–485. [CrossRef] [PubMed]
29. Gérard, C.; Romain, D.; Delmas, G. Thermodynamic properties of polyolefin solutions at high temperature: 2. Lower critical solubility temperatures for polybutene–1, polypentene–1 and poly(4–methylpentene–1) in hydrocarbon solvents and determination of the polymer–solvent interaction parameter for PB1 and one ethylene–propylene copolymer. *Polymer* **1981**, *22*, 1190–1198. [CrossRef]
30. Liu, B.; Wang, Y.; Liang, L.; Zeng, Y. Achieving solubility alteration with functionalized polydimethylsiloxane for improving the viscosity of supercritical CO_2 fracturing fluids. *RSC Adv.* **2021**, *11*, 17197–17205. [CrossRef]
31. Hu, D.; Sun, S.; Yuan, P.-Q.; Zhao, L.; Liu, T. Exploration of CO_2–Philicity of Poly(vinyl acetate–co–alkyl vinyl ether) through Molecular Modeling and Dissolution Behavior Measurement. *J. Phys. Chem. B* **2015**, *119*, 12490–12501. [CrossRef] [PubMed]
32. Xue, P.; Shi, J.; Cao, X.; Yuan, S. Molecular dynamics simulation of thickening mechanism of supercritical CO_2 thickener. *Chem. Phys. Lett.* **2018**, *706*, 658–664. [CrossRef]
33. Sui, S.; Xiao, P.; CUI, M. Effect of Modified Silicone CO_2 Thickener on Fluid Rheology and Oil Displacement Efficiency. *Oilfield Chem.* **2023**, *40*, 229–234. (In Chinese) [CrossRef]

34. Li, S.; Wang, J. Recent Developments of the Surfactants Using in Supercritical Carbon Dioxide. *Chem. World* **2007**, 496–499+510. (In Chinese) [CrossRef]
35. Liu, J.; Li, G.L.; Han, B. *Research on the Aggregation and Microenvironment of Fluorine-Free and Silicon-Free Nonionic Surfactants in Supercritical Carbon Dioxide*; Shandong University: Jinan, China, 2022.
36. Du, M. *Investigation on Supercritical Carbon Dioxide Fracturing Fluid System*; China University of Petroleum (East China): Qingdao, China, 2016.
37. Masanobu, S.; Yuuki, S.; Sajad, K.; Shirin, A.; Tretya, A.; Azmi, M.; Robert, M.E.; Sarah, E.R.; Christopher, H.; Julian, E. Thickening supercritical CO_2 at high temperatures with rod-like reverse micelles. *Colloids Surf. A Physicochem. Eng. Asp.* **2024**, *686*, 133302. [CrossRef]
38. Zhai, H.; Zhang, J.; Dong, J.; Wang, J.; Zhang, F.; Xiaofeng, L. Solubility Evaluation of Supercritical Carbon Dioxide Thickener. *Oilfield Chem.* **2021**, *38*, 422–426. (In Chinese) [CrossRef]
39. Enick, M.R. A Literature Review of Attempts to Increase the Viscosity of Dense Carbon Dioxide. 1998. Available online: https://www.semanticscholar.org/paper/A-Literature-Review-of-Attempts-to-Increase-the-of-Enick/899dccfc1e36981b94973f48d5bf62064ae56e14 (accessed on 11 March 2024).
40. Joel, F.; Naiping, H. The molecular basis of CO_2 interaction with polymers containing fluorinated groups: Computational chemistry of model compounds and molecular simulation of poly bis(2,2,2–trifluoroethoxy)phosphazene. *Polymer* **2003**, *44*, 4363–4372. [CrossRef]
41. Zhang, Y. *Design, Synthesis and Properties of Polyethers as Carbon Dioxide Thickening Agent*; Jilin University: Changchun City, China, 2017.
42. Chao, S.; Zhihua, H.; Eric, J.B.; Robert, M.E.; Sun-Young, K.; Dennis, P.C. Semi-Fluorinated Trialkyltin Fluorides and Fluorinated Telechelic Ionomers as Viscosity–Enhancing Agents for Carbon Dioxide. *Ind. Eng. Chem. Res.* **2001**, *40*, 908–913. [CrossRef]
43. Robert, M.E.; Eric, J.B.; Chen, S.; Eddy, K. Formation of fluoroether polyurethanes in CO_2. In Proceedings of the 4th International Symposium on Supercritical Fluids (ISSF 97), Sendai, Japan, 11–14 May 1998.
44. Kieran, T.; Dazun, X.; Robert, M.E.; Julian, E.; Martin, J.H.; Kevin, J.M.; Sarah, E.R.; Richard, K.H.; David, C.S. Rod-Like Micelles Thicken CO_2. *Langmuir* **2009**, *26*, 83–88. [CrossRef]
45. Hoefling, T.A.; Newman, D.A.; Robert, M.E.; Eric, J.B. Effect of structure on the cloud-point curves of silicone-based amphiphiles in supercritical carbon dioxide. *J. Supercrit. Fluids* **1993**, *6*, 165–171. [CrossRef]
46. Stephen, C.; Robert, M.E.; Sarah, E.R.; Richard, K.H.; Julian, E. Amphiphiles for supercritical CO_2. *Biochimie* **2012**, *94*, 94–100. [CrossRef]
47. Xin, F.; Vijay, P.; Michael, C.M.; Yan, W.; Juncheng, L.; Robert, M.E.; Andrew, D.H.; Christopher, B.R.; Johnson, J.K.; Eric, J.B. Oxygenated Hydrocarbon Ionic Surfactants Exhibit CO_2 Solubility. *J. Am. Chem. Soc.* **2005**, *127*, 11754–11762. [CrossRef] [PubMed]
48. Cui, W.; Wang, C.; Xu, J.; Duan, Y. Application and research of liquid carbon dioxide based gel fracturing fluid in tight gas reservoir. In Proceedings of the 2016 Natural Gas Academic Annual Conference, Yinchuan, China, 28 September 2016.
49. Heller, J.P.; Dandge, D.K.; Roger, J.C.; Donaruma, L.G. Direct Thickeners for Mobility Control of CO_2 Floods. *Soc. Pet. Eng. J.* **1985**, *25*, 679–686. [CrossRef]
50. Traian, S.; Thomas, S.; Eric, J.B. Non-fluorous polymers with very high solubility in supercritical CO_2 down to low pressures. *Nature* **2000**, *405*, 165–168. [CrossRef] [PubMed]
51. Shen, Z.; McHugh, M.; Xu, J.; Belardi, J.; Kilic, S.; Mesiano, A.; Bane, S.; Karnikas, C.; Beckman, E.; Enick, R. CO_2–solubility of oligomers and polymers that contain the carbonyl group. *Polymer* **2003**, *44*, 1491–1498. [CrossRef]
52. Shen, A.; Liu, J.; She, Y.; Shu, F.; Wang, Z. Design and Synthesis of Styrene Vinyl-acetate Copolymer as Potential CO_2 Thickener. *J. Oil Gas Technol.* **2011**, *33*, 131–134+168. (In Chinese)
53. Zhang, J. *Carbon Dioxide Thickening Using Poly(Vinyl Acetate) and Amphiphilic Surfactant*; East China University of Science and Technology: Shanghai, China, 2017.
54. Goicochea, A.G.; Abbas, F. CO_2 Viscosification by Functional Molecules from Mesoscale Simulations. *J. Phys. Chem. C* **2019**, *123*, 29461–29467. [CrossRef]
55. Chen, R.; Zheng, J.; Ma, Z.; Zhang, X.; Fan, H.; Bittencourt, C. Evaluation of CO_2–philicity and thickening capability of multichain poly(ether–carbonate) with assistance of molecular simulations. *J. Appl. Polym. Sci.* **2020**, *138*, 49700. [CrossRef]
56. Sun, W.; Wang, H.; Zha, Y.; Yu, J.; Zhang, J.; Ge, Y.; Sun, B.; Zhang, Y.; Gao, C. Experimental and microscopic investigations of the performance of copolymer thickeners in supercritical CO_2. *Chem. Eng. Sci.* **2020**, *226*, 115857. [CrossRef]
57. Kazuya, K.; Abbas, F. Branching in molecular structure enhancement of solubility in CO_2. *PNAS Nexus* **2023**, *2*, pgad393. [CrossRef] [PubMed]
58. Ding, B.; Kantzas, A.; Firoozabadi, A. Spatiotemporal X-ray Imaging of Neat and Viscosified CO_2 in Displacement of Brine-Saturated Porous Media. *SPE J.* **2024**, 1–16. [CrossRef]
59. Joseph, M.D.; Zihibin, G.; Elsbernd, C.S. Synthesis of Fluoropolymers in Supercritical Carbon Dioxide. *Science* **1992**, *257*, 945–947. [CrossRef] [PubMed]
60. Huang, Z.; Shi, C.; Xu, J.; Kilic, S.; Enick, R.M.; Beckman, E.J. Enhancement of the Viscosity of Carbon Dioxide Using Styrene/Fluoroacrylate Copolymers. *Macromolecules* **2000**, *33*, 5437–5442. [CrossRef]

61. Robert, M.E.; Eric, J.B.; Ali Vaziri, Y.; Val, J.K.; Hans, S.; Jon, L.H. Phase behavior of CO_2–perfluoropolyether oil mixtures and CO_2–perfluoropolyether chelating agent mixtures. *J. Supercrit. Fluids* **1998**, *13*, 121–126. [CrossRef]
62. Baojiang, S.; Wenchao, S.; Haige, W.; Ying, L.; Haiming, F.; Hao, L.; Xiuping, C. Molecular simulation aided design of copolymer thickeners for supercritical CO_2 as non-aqueous fracturing fluid. *J. CO2 Util.* **2018**, *28*, 107–116. [CrossRef]
63. Sevgi, K.; Robert, M.E.; Eric, J.B. Fluoroacrylate–aromatic acrylate copolymers for viscosity enhancement of carbon dioxide. *J. Supercrit. Fluids* **2019**, *146*, 38–46. [CrossRef]
64. Li, Q.; Wang, Y.; Li, Q.; Wang, F.; Li, Y.; Tang, L. Synthesis and performance evaluation of supercritical CO_2 thickener for fracturing. *Fault-Block Oil Gas Field* **2018**, *25*, 541–544. (In Chinese)
65. Jae-Heum, B.; Irani, C.A. A Laboratory Investigation of Viscosified CO_2 Process. *SPE Adv. Technol. Ser.* **1993**, *1*, 166–171. [CrossRef]
66. Jae-Heum, B. Viscosified CO_2 Process: Chemical Transport and Other Issues. In Proceedings of the SPE International Symposium on Oilfield Chemistry, San Antonio, TX, USA, 14–17 February 1995. [CrossRef]
67. Zhao, M.; Li, Y.; Gao, M.; Wang, T.; Dai, C.; Wang, X.; Guan, B.; Liu, P. Formulation and performance evaluation of polymer-thickened supercritical CO_2 fracturing fluid. *J. Pet. Sci. Eng.* **2021**, *201*, 108474. [CrossRef]
68. Fink, R.; Hancu, D.; Valentine, R.; Beckman, E.J. Toward the Development of "CO_2–philic" Hydrocarbons. 1. Use of Side–Chain Functionalization to Lower the Miscibility Pressure of Polydimethylsiloxanes in CO_2. *J. Phys. Chem. B* **1999**, *103*, 6441–6444. [CrossRef]
69. Sevgi, K.; Yang, W.; Johnson, J.K.; Eric, J.B.; Robert, M.E. Influence of tert–amine groups on the solubility of polymers in CO_2. *Polymer* **2009**, *50*, 2436–2444. [CrossRef]
70. Michael, J.O.B.; Robert, J.P.; Mark, D.D.; Jason, J.L.; Aman, D.; Eric, J.B.; Robert, M.E. Anthraquinone Siloxanes as Thickening Agents for Supercritical CO_2. *Energy Fuels* **2016**, *30*, 5990–5998. [CrossRef]
71. Li, Q.; Wang, Y.; Li, Q.; Foster, G.; Lei, C. Study on the optimization of silicone copolymer synthesis and the evaluation of its thickening performance. *RSC Adv.* **2018**, *8*, 8770–8778. [CrossRef] [PubMed]
72. Wang, Y.; Li, Q.; Dong, W.; Li, Q.; Wang, F.; Bai, H.; Zhang, R.; Owusu, A.B. Effect of different factors on the yield of epoxy-terminated polydimethylsiloxane and evaluation of CO_2 thickening. *RSC Adv.* **2018**, *8*, 39787–39796. [CrossRef] [PubMed]
73. Masanobu, S.; Shotaro, O.; Craig, J.; Atsushi, Y.; Azmi, M.; Frédéric, G.; Robert, M.E.; Sarah, E.R.; Adam, C.; Christopher, H.; et al. Anisotropic reversed micelles with fluorocarbon–hydrocarbon hybrid surfactants in supercritical CO_2. *Colloids Surf. B Biointerfaces* **2018**, *168*, 201–210. [CrossRef] [PubMed]
74. Hoefling, T.A.; David, S.; Reid, M.D.; Eric, J.B.; Robert, M.E. The incorporation of a fluorinated ether functionality into a polymer or surfactant to enhance CO_2–solubility. *J. Supercrit. Fluids* **1992**, *5*, 237–241. [CrossRef]
75. Liu, S.; Yuan, L.; Zhao, C.; Zhang, Y.; Song, Y. A review of research on the dispersion process and CO2 enhanced natural gas recovery in depleted gas reservoir. *J. Pet. Sci. Eng.* **2021**, *208*, 109682. [CrossRef]
76. Miri, R. *Effects of CO_2–Brine–Rock Interactions on CO_2 Injectivity–Implications for CCS*; University of Oslo: Oslo, Norway, 2015.
77. Lenormand, R.; Éric, T.; Zarcone, C. Numerical models and experiments on immiscible displacements in porous media. *J. Fluid Mech.* **1988**, *189*, 165–187. [CrossRef]
78. Li, Y. *The Numerical Simulation Study on CO_2–Water Two Phase Flow in Porous Media Based on CFD*; Dalian University of Technology: Dalian, China, 2015.
79. Li, Q.; Wang, Y.; Owusu, A.B. A modified Ester-branched thickener for rheology and wettability during CO_2 fracturing for improved fracturing property. *Environ. Sci. Pollut. Res.* **2019**, *26*, 20787–20797. [CrossRef] [PubMed]

Disclaimer/Publisher's Note: The statements, opinions and data contained in all publications are solely those of the individual author(s) and contributor(s) and not of MDPI and/or the editor(s). MDPI and/or the editor(s) disclaim responsibility for any injury to people or property resulting from any ideas, methods, instructions or products referred to in the content.

Article

Molecular Dynamics Simulation on Thickening and Solubility Properties of Novel Thickener in Supercritical Carbon Dioxide

Xiaohui Wang [1,2], Shiwei Liang [1], Qihong Zhang [1], Tianjiao Wang [1] and Xiao Zhang [1,3,*]

[1] Beijing Key Laboratory of Optical Detection Technology for Oil and Gas, China University of Petroleum-Beijing, Beijing 102249, China; wangxiaohui@cup.edu.cn (X.W.); 2023211325@student.cup.edu.cn (S.L.); 2021211324@student.cup.edu.cn (Q.Z.); 2023216161@student.cup.edu.cn (T.W.)

[2] National Key Laboratory of Petroleum Resources and Engineering, China University of Petroleum-Beijing, Beijing 102249, China

[3] College of Science, China University of Petroleum, Beijing 102249, China

* Correspondence: zhangxiao@cup.edu.cn

Abstract: Supercritical CO_2 has wide application in enhancing oil recovery, but the low viscosity of liquid CO_2 can lead to issues such as poor proppant-carrying ability and high filtration loss. Therefore, the addition of thickening agents to CO_2 is vital. Hydrocarbon polymers, as a class of green and sustainable materials, hold tremendous potential for acting as thickeners in supercritical CO_2 systems, and PVAc is one of the best-performing hydrocarbon thickeners. To further improve the viscosity enhancement and solubility of PVAc, here we designed a novel polymer structure, PVAO, by introducing CO_2-affine functional groups to PVAc. Molecular dynamics simulations were adopted to analyze viscosity and relevant solubility parameters systematically. We found that PVAO exhibits superior performance, with a viscosity enhancement of 1.5 times that of PVAc in supercritical CO_2. While in the meantime, PVAO maintains better solubility characteristics than PVAc. Our findings offer insights for the future design of other high-performance polymers.

Keywords: thickener; supercritical carbon dioxide; viscosity; molecular dynamics

1. Introduction

With the huge growth of energy demand in the world, enhanced oil recovery (EOR) has received considerable attention in the petroleum industry. Currently, most easily accessible oil reservoirs have been drilled, leading to an overall decline in oil production. Therefore, maximizing the utilization of known resources becomes more practical than exploring new oil wells; thus, EOR has become increasingly crucial.

In crude oil recovery, three processes are commonly employed. Primary oil recovery involves the extraction of oil under its own pressure and through gas expansion by dissolution, accounting for 5–20% of oil recovery. Secondary oil recovery employs water flooding to displace oil. In recent decades, hydraulic fracturing technology has emerged as an effective method for oil and gas production enhancement and has been widely utilized. However, hydraulic fracturing technology has its drawbacks, which include significant water resource consumption, potential damage to reservoirs, and the risk of groundwater contamination due to the addition of various chemical substances into the injected water. Additionally, even after primary and secondary oil recovery, over half of the oil remains trapped in the reservoir. To further increase production, the implementation of more advanced EOR techniques, commonly referred to as tertiary oil recovery, is necessary. EOR is a method of injecting displacing agents into reservoirs to improve the physical and chemical characteristics of the reservoir and its fluids, thereby enhancing oil displacement efficiency. With the implementation of tertiary oil recovery, the field's utilization rate can reach around 70%.

Among the various displacing agents, supercritical CO_2 (scCO_2) has garnered significant attention due to its excellent performance as an oil-displacing agent [1,2]. ScCO_2 exhibits intermediate properties between those of gas and liquid. It has obvious characteristics such as high diffusivity, low viscosity, low surface tension, and controllable solubility. In oil recovery, compared to traditional water flooding, scCO_2 has significant advantages. Firstly, scCO_2 possesses strong fracturing capability and is easily displaced [2], and it is also applicable to various types of reservoirs [3]. Secondly, with low raw material costs, the critical temperature and pressure of scCO_2 are 304.1 K and 7.38 MPa, respectively, which are lower than the temperature and pressure within the reservoir [4,5]. Therefore, carbon dioxide can be transformed into a supercritical liquid once entering the reservoir, saving the energy required for its conversion. Thirdly, scCO_2 can be converted into gas form and expelled from the reservoir after fracturing, causing no damage to the rock formation and preventing expansion; thus, it is non-toxic, non-polluting, non-flammable, and recyclable [6]. As an emerging oil and gas production method, scCO_2 fracturing technology exhibits significant advantages in terms of environmental friendliness, efficiency, and adaptability. It is expected to play an increasingly important role in future oil and gas development [7–9].

However, the low-viscosity nature of scCO_2 gives rise to challenges such as viscosity fingering, limited sand-carrying capacity, filter loss, and reduced efficiency in oil and gas recovery. To overcome these challenges, thickeners are required to enhance the viscosity of scCO_2. In recent years, researchers have been devoted to the design of novel CO_2-responsive polymers, and the investigation of CO_2 thickening agents has undergone several stages. Girard and Mertsch separately discovered that fluorine-containing polymers and silicon-based materials exhibit excellent solubility in CO_2, leading to a significant increase in viscosity [10–12]. However, the high costs of fluorine-containing and silicon-based polymers pose challenges for their large-scale industrial utilization as CO_2-responsive materials. Additionally, when fluorine-containing polymers are used in the field of oilfield chemistry, fluorine polymers are often discharged into the environment with wastewater, causing irreparable environmentally damage as they are non-biodegradable. As a result, the design of fluorine-free polymers composed solely of carbon (C), hydrogen (H), and oxygen (O) atoms, known as hydrocarbon polymers, has gained more attention. Extensive experimental studies have demonstrated the viability of several hydrocarbon polymers, such as poly (vinyl acetate) (PVAc), poly (vinyl ethyl ether) (PVEE), poly (propylene oxide) (PPO), and poly (vinyl methoxymethyl ether). Among these, PVAc is the most promising polymer due to its relatively high solubility in CO_2, which constitutes its main advantage over other materials [13].

Yet hydrocarbon polymers represented by PVAc still do not have the expected thickening performance compared with other thickeners, such as fluorine-containing polymers. To further improve the viscosity enhancement and solubility of PVAc, in this study, we designed a novel polymer structure, PVAO, by introducing CO_2-affine functional groups to PVAc. Molecular dynamics (MD) simulations were adopted to analyze viscosity and relevant solubility parameters systematically. Viscosity, radial distribution function, interaction energy, cohesive energy density, and solubility parameters are given. We found that PVAO exhibits superior performance both in viscosity enhancement and solubility characteristics than PVAc.

2. Results and Discussion

2.1. The Thickening Effect of PVAO

Predicting the viscosity of supercritical CO_2 remains a crucial task. Classical MD simulations have been employed for shear viscosity predictions. The primary methods utilized include non-equilibrium molecular dynamics (NEMD) and equilibrium molecular dynamics (EMD), with the Green–Kubo method based on EMD, which is the most widely applied approach [14]. Here, we present equilibrium MD calculations for the viscosity of pure scCO_2, PVAc/CO_2, and PVAO/CO_2 using the standard Green–Kubo method.

In the Green–Kubo theory, shear viscosity is calculated from the integral over time of the pressure tensor autocorrelation function [14], as follows:

$$\eta = \frac{V}{K_B T} \int_0^\infty \langle P_{\alpha\beta}(t) \cdot P_{\alpha\beta}(0) \rangle dt \tag{1}$$

where K_B is the Boltzmann constant, T is the absolute temperature, t is time, V is the volume of the simulation box, and $P_{\alpha\beta}$ denotes the element $\alpha\beta$ of the pressure tensor. The symmetry of the cubic simulation box implies that the three directions, x, y, and z, are equivalent. Theoretically, the autocorrelation function of the stress tensor should decay to zero as time progresses. Then, we can obtain a constant value that corresponds to the computed shear viscosity using Equation (1).

Yong Zhang et al. proposed a method for calculating shear viscosity by executing multiple independent trajectories and taking the average of the running time integrals. In order to calculate the viscosity, five independent trajectories were generated using different initial velocity distribution seeds, each with a length of 300 ps. Based on these trajectories, the average shear viscosity of each system at different temperatures was calculated using Equation (1) [15].

To validate the rationality of the methods, the shear viscosity of supercritical CO_2 was calculated using the proposed method under the conditions of 23 °C and 20 MPa. The simulation results demonstrated a viscosity value of 0.087 cp, which closely matched the experimental measurement of 0.094 cp [16]. This result serves as evidence supporting the feasibility of the employed methodology.

For the thickening effect of PVAO, the crucial aspect lies in determining whether the novel thickening agent exhibits superior viscosity enhancement compared to PVAc. Figure 1 presents a comparative analysis of the viscosity enhancement effects between PVAO and PVAc under identical conditions.

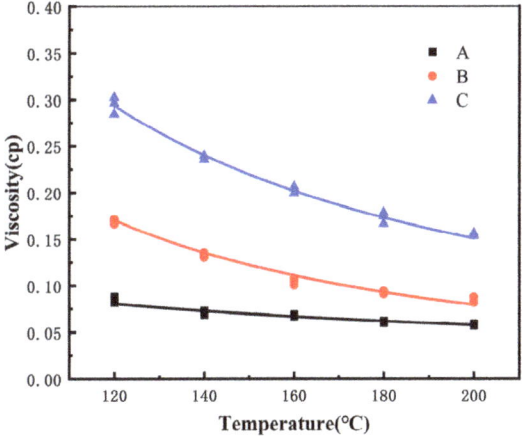

Figure 1. The viscosities of (**A**) $scCO_2$; (**B**) $scCO_2$ with one PVAc chain; and (**C**) $scCO_2$ with one PVAO chain with a pressure of 60 MPa and a temperature between 120 and 200 °C.

Figure 1 shows the viscosity of (A) $scCO_2$ with a pressure of 60 MPa and a temperature between 120 and 200 °C, (B) $scCO_2$ with one PVAc chain, and (C) $scCO_2$ with one PVAO chain. As shown in Figure 1, as the temperature increased, the viscosities of the three systems gradually decreased. Moreover, the viscosity-enhancing effect of PVAO was much stronger than that of PVAc. Under 120 °C and 60 MPa, the addition of a PVAO chain in $scCO_2$ fluid led to a significant increase in viscosity to 0.245 cp, approximately three times higher than that of pure CO_2 fluid, while under the same temperature and pressure conditions, the viscosity of scCO$_2$ with a PVAc chain was approximately twice that of pure scCO$_2$. PVAO is a promising hydrocarbon polymer with higher viscosity-enhancing efficiency. The reason for the significantly higher viscosity enhancement of PVAO compared to PVAc may be the fact that, at similar weight fractions, PVAO contains a greater number of key functional groups, which enhances the interactions between the C atoms in CO$_2$ and the O atoms in the branched chain of PVAO, thus contributing to its thickening effect.

2.2. Diffusivity

Mean square displacement (MSD) refers to the deviation of particle positions from a reference point with time. As the observation time approaches infinity, MSD becomes directly proportional to the observation time limit. In scCO$_2$ fluid systems, the MSD of CO$_2$ within a certain range of polymers exhibits a linear relationship with time evolution. Moreover, the slope of this relationship is related to the diffusion coefficient D, as expressed by the following formula [17]:

$$\text{MSD} = \left\langle |X_i(t_0+t) - X_i(t_0)|^2 \right\rangle \tag{2}$$

$$D = \frac{1}{6N} \lim_{t \to \infty} \frac{d}{dt} \sum \left\langle |X_i(t_0+t) - X_i(t_0)|^2 \right\rangle \tag{3}$$

The diffusion coefficient characterizes the extent of molecular diffusion in liquids, indicating the speed of molecular diffusion. The MSD curve can be obtained through MD simulations, which is shown in Figure 2. The slope of the curve can be determined by linear fitting. By comparing the magnitude of the diffusion coefficients, the strength of the interaction between polymers and CO$_2$ can be roughly estimated. The obtained curve clearly showed that the diffusion coefficients of the scCO$_2$ systems with PVAc and PVAO were smaller than those in an scCO$_2$ system without thickeners. The diffusion coefficients for the CO$_2$, PVAc/CO$_2$, and PVAO/CO$_2$ systems were denoted as $9.26 \pm 0.069 \times 10^{-7}$ cm^2/s, $3.88 \pm 0.073 \times 10^{-7}$ cm^2/s, and $4.58 \pm 0.058 \times 10^{-7}$ cm^2/s, respectively. This result indicated the ability of PVAO to bind CO$_2$ molecules, which reflected the thickening effects of PVAO and PVAc.

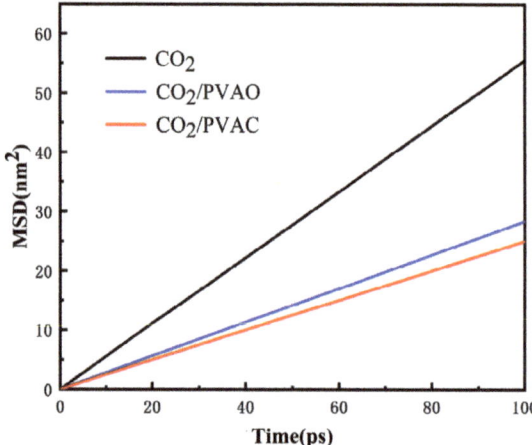

Figure 2. MSD–time curves of scCO$_2$ (black line), scCO$_2$ with PVAO (blue line), and scCO$_2$ with PVAc (red line).

2.3. Radial Distribution Function

The presence of lone pair electrons on O in CO_2 and the Lewis acid–Lewis base (LA–LB) interaction between the O atoms in PVAO and the C atoms in CO_2 are the primary influencing factors for the dissolution of ether-based and carbonyl-containing polymers in carbon dioxide [18,19], and the interaction between molecules or atoms can be described by the radial distribution function (RDF) [20,21].

RDF can be obtained by performing MD simulations using the *Forcite* module in Materials Studio, and it represents the relative local density of atom B with respect to the bulk density in a region around a central atom A, within a distance radius of r [22]. In essence, the RDF is a probability calculation that determines the likelihood of finding another atom at a distance of r from the reference atom. RDF can be denoted by $g(r, r')$. For small values of $|r - r'|$, $g(r, r')$ primarily characterizes the atomic packing and distances between bonds. For long-range situations, since the probability of finding an atom is approximately the same for a given distance, $g(r, r')$ becomes flat and ultimately approaches a constant value as $|r - r'|$ increases. Typically, when defining $g(r, r')$, it is normalized such that $g(r, r')$ approaches 1 as $|r - r'|$ tends to infinity. The formula for $g(r, r')$ is as follows [23]:

$$g(r) = \frac{dN}{4\rho\pi^2} \quad (4)$$

The integrated RDF between the oxygen atom in PVAO and the carbon atom in CO_2 was calculated and is illustrated in Figure 3b, and the specified oxygen atom in PVAO is labeled in Figure 3a. As shown in Figure 3b, both the O(a) and O(c) atoms in PVAO exhibited pronounced peaks in their RDF curves with respect to the C atoms in CO_2, whereas O(b) and O(d) displayed no significant peak features in the RDF curves. The results indicated the presence of LA–LB interactions between O(a) and O(c) in PVAO and the C atoms in CO_2, and that the LA–LB interactions between the carbonyl oxygen atom at the distal end of the PVAO side chain and the carbon atom in CO_2 were much stronger than the LA–LB interactions involving the oxygen atom at the proximal end. The insights derived from the RDF simulation results offer valuable guidance for the design of novel polymer structures. It was observed that the addition of carbonyl oxygen atoms at the distal end, as opposed to those in close proximity, was more likely to enhance the solubility of the polymer in CO_2.

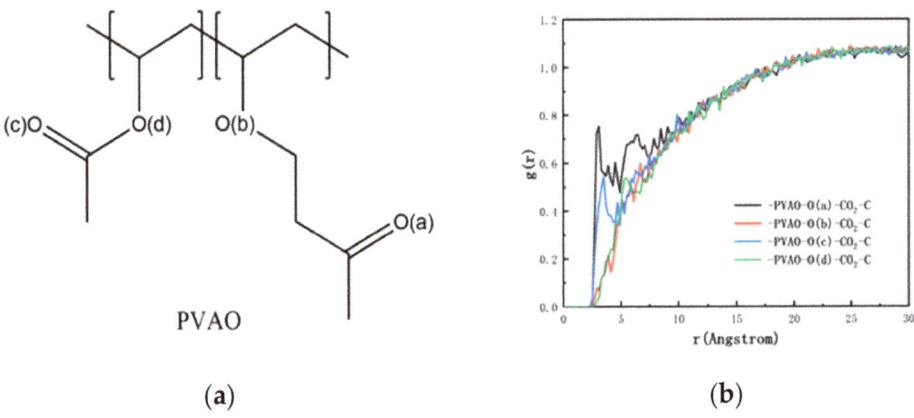

Figure 3. (**a**) Numbering diagram of O atoms in PVAO. (**b**) RDF between different O atoms in PVAO and C atoms in CO_2.

2.4. Interaction Energy

Interaction energy is the difference between the energy of the complex minus the energy of the isolated monomers in the complex. The lower the interaction energy, the more stable the structure. For the CO_2–polymer chain system, the interaction energy can be written as [20,21]:

$$E_{inter} = E_{CO_2-chain} - E_{CO_2} - E_{chain} \quad (5)$$

In the above equation, E_{inter} represents the interaction energy between CO_2 and the polymer chain, $E_{CO_2-chain}$ denotes the total energy of the CO_2–polymer system, E_{CO_2} and E_{chain} are the energies of CO_2 and the polymer chain, respectively.

In order to assess the polymer–CO_2 interactions, MD simulations were performed in the NPT ensemble. As shown in Table 1, the interaction energy of the PVAO-thickened CO_2 system was calculated to be -493.1 KJ/mol, while that of the PVAc-thickened CO_2 system was -474.6 KJ/mol, which was slightly higher than the former, indicating a lower stability compared to the PVAO-thickened CO_2 system. The standard deviations of the interaction energies for the two systems were separately computed over the last 10 frames. The standard deviations of PVAc/CO_2 and PVAO/CO_2 were 7.79 and 8.81 kJ/mol, respectively. The difference in the interaction energies can serve as a basis for evaluating the difference in solubility within the error bars. Thus, PVAO may have better compatibility with CO_2 and could be a more suitable polymer for CO_2 affinity. Moreover, these results indicated that the solubility of the PVAO chain in practical applications may not be weaker than that of PVAc.

Table 1. Interaction Energy between CO_2 and Polymer Single Chain at 120 °C and 60 MPa (Unit: kJ/mol).

Composition	$E_{CO_2-chain}$	E_{CO_2}	E_{chain}	E_{inter}
PVAc/CO_2	−3140.5	−1162.8	−1503.0	−474.6
PVAO/CO_2	−2273.9	−1772.0	−8.9	−493.1

However, it should be noted that, in addition to the interaction energy between CO_2 and the polymer chain, intermolecular interactions between polymer chains are also an important factor affecting solubility [20]. In order to achieve a more comprehensive prediction of solubility, it is necessary to incorporate cohesive energy density and solubility parameters.

2.5. Cohesive Energy Density and Solubility Parameters

Cohesive energy density (CED) and solubility parameters are quantitative measures used to characterize intermolecular interactions between molecules. CED is employed specifically for evaluating non-covalent bonding interactions quantitatively, which can be calculated by considering parameters such as partial charge distributions and atomic distances within a molecule. Accurate computation of CED (e_{coh}) holds significant importance in predicting molecular properties and reactivity, which can be utilized as descriptors to characterize both compatibility and solubility properties within the system under investigation [24]. Meanwhile, solubility parameters primarily describe solubility and compatibility. The expression for solubility parameter δ is given by the following equation: $\delta = \sqrt{e_{coh}}$. e_{coh}, and δ can serve as a basis for evaluating the molecular forces between polymer chains and also for evaluating the solubility of thickeners in scCO_2 [3,21]. The analysis of cohesive energy density required the utilization of the *Forcite* module within the MS software. The cohesive energy density was computed for the last 10 frames of the trajectory, and the average value was obtained. In order to investigate the internal interactions within polymer chains, the cohesive energy density and solubility parameters under 120 °C and 60 MPa conditions were calculated and are listed in Table 2. Each system consisted of three polymer chains, with the composition of each polymer chain outlined in Table 2.

Table 2. Cohesive energy density and solubility parameters under 120 °C and 60 MPa conditions.

Composition	e_{coh} (J/m^3)	δ ((J/m^3)$^{1/2}$)
CO_2	2.025×10^8	14.18
PVAc	2.561×10^8	15.98
PVAO	2.326×10^8	15.21

From Table 2, we can see that PVAc possessed the highest cohesive energy density and solubility parameters, indicating the strongest interactions between PVAc chains compared with the pure CO_2 and PVAO systems. The intensified interactions could result in increased interfacial tension, which might hinder the blending process between PVAc and CO_2. In addition, PVAO exhibited relatively lower cohesive energy density and milder interactions between internal polymer chains. The blending process between polymers is essentially a diffusion process between molecular chains, constrained by the interactions between the chains. The compatibility between different components can also be assessed by the difference in solubility parameter δ. When the δ values are closer, better compatibility can be observed, which follows the theory of similar dissolves mutually. The difference in solubility parameter ($|\Delta\delta|$) between the PVAO polymer and CO_2 was smaller than that between PVAc and CO_2, indicating that PVAO has merits over PVAc in terms of solubility.

3. Simulation Details and Methods

Qin et al. explored the structural and dynamic characteristics of scCO_2 fluids on hydroxylated and methylated amorphous silica surfaces using MD simulations [25]. Hu et al. investigated the interaction mechanisms between various functional groups and scCO_2 through MD simulations [20,21]. These studies undoubtedly demonstrate that MD simulations serve as powerful tools for investigating scCO_2 at the molecular level.

In our study, MD simulations were performed using Material Studio 8.0 developed by Accelrys [26]. The commonly used force fields for simulating polymers are AMBER [27], CHARMM [28], COMPASS [29,30], etc. However, the first two force fields are primarily employed for simulating biomolecules. The COMPASS force field is extensively utilized in covalent molecular systems, including a wide range of common organic molecules, small inorganic molecules, and polymers. COMPASS has been proven to be efficient in predicting the interactions of both organic and inorganic compounds [31,32]. The non-bonded interactions between atoms were described using long-range electrostatic interactions and short-range van der Waals (vdW) interactions. The electrostatic interactions were computed using Coulomb's law, while the vdW interactions were calculated using the Lennard–Jones potential. In our simulations, periodic boundary conditions were applied in all directions for each simulation cell. The atom-based method was employed to calculate van der Waals interactions, while the Ewald method was utilized to handle long-range electrostatic interactions [33]. The cutoff radius for non-bonded interactions was set at 1.25 nm and the buffer width was set at 0.05 nm.

The *Forcite* module was employed to perform structural optimization of the unit cell. The Smart Minimizer was utilized during the model structure optimization process. The lowest energy configuration was selected and annealed for 5 cycles within the temperature range of 300–500 K. Following annealing, NVT (constant number of particles, volume, and temperature) simulations for 300 ps and NPT (constant number of particles, pressure, and temperature) simulations for 300 ps were conducted, with a time step of 1 fs [31,34,35]. For both the NVT and NPT ensembles, we employed the Nose–Hoover method to implement the barostat for temperature and pressure control. The Q ratio was set to 0.01. The research conducted by D. J. Evans and B. L. Holian demonstrated that different thermostats have negligible effects on parameters such as shear viscosity and internal energy. The Nose–Hoover thermostat is commonly employed in both NVT and NPT ensembles to regulate the system temperature. The Nose–Hoover method, which strictly adheres to the canonical ensemble, is often utilized as a technique for equilibrium sampling [36]. Trajectories were saved at 5 ps intervals, and the configurations of the final 50 ps were used for data analysis. The parameters such as interaction energy and cohesive energy density were computed by averaging the values obtained from the last 10 frames. Subsequently, the thermodynamic parameters for the various systems were obtained [37,38]. The MD simulations were adopted to study PVAc and PVAO thickened $scCO_2$ systems, of which the newly designed structure PVAO was shown in Figure 4a, and the design principle was explained below. A snapshot of the MD simulation for the PVAO/CO_2 system is depicted in Figure 4b.

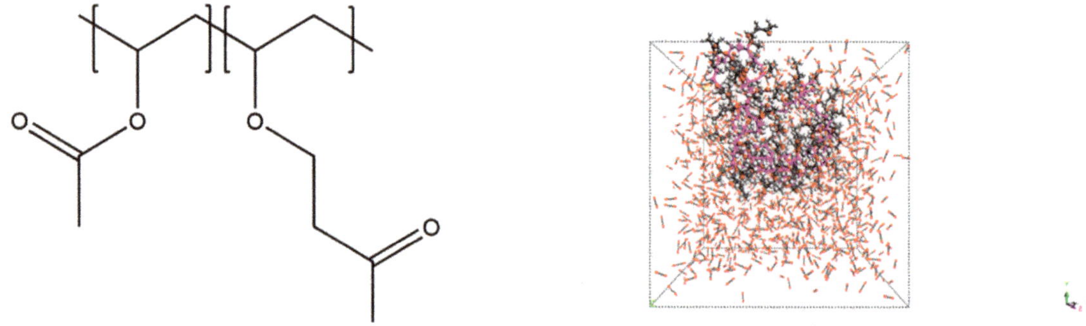

Figure 4. (a) The structural formula of polymer PVAO. (b) The snapshot of the simulation box.

4. Design

Kazarian et al. proposed that the dissolution of polymers in $scCO_2$ is primarily governed by the interactions between polymers and CO_2, including Lewis acid–base (LA–LB) interactions and weaker hydrogen bonding [39]. Beckman et al. confirmed that the O atoms of carbonyl groups can increase the solubility of polymers in CO_2 through LA–LB interactions, suggesting that the favorable dissolution behavior of PVAc may be attributed to the interactions between its carbonyl groups and CO_2 [18]. Raveendran et al. demonstrated that there are also interactions between the H atoms adjacent to the C atoms in polymer molecules and the O atoms in CO_2, which can be classified as hydrogen bonding [40]. Although such interactions are relatively weak, they can still enhance the interaction capability between polymer molecules and CO_2. The interactions between polymers and CO_2 primarily originate from functional groups, and identifying favorable functional groups is beneficial for designing new structures. According to the simulations conducted by Kilic et al., the interaction energies between the O atoms of ether groups and CO_2 are of the same order of magnitude as those between carbonyl groups and CO_2 [41]. Thus, ether groups are likely to play an active role in polymer–CO_2 interactions, providing insight for the design of new structures in this study. Here, we proposed a newly designed configuration which was found to exhibit superior performance in viscosity enhancement compared to PVAc under comparable weight percentages, while maintaining better solubility characteristics than PVAc. Figure 4a illustrates the newly designed configuration of poly [(vinyl acetate)-(4-vinyl ethoxy butan-2-one)], hereafter referred to as PVAO.

The research conducted by Hu et al. demonstrated that the simulation results of polymers are influenced by the number of repeating units. It is observed that when the number of repeating units exceeds 30–40, the thermodynamic parameters become insensitive to the molecular weight [20,21]. Then, we selected a PVAc chain with a degree of polymerization of n = 75 (Mn = 6452 g/mol) and constructed a similarly sized PVAO chain with a degree of polymerization of n = 33 (Mn = 7002 g/mol). Table 3 presents five systems, including pure $scCO_2$ with 1000 CO_2 molecules, a system containing 1 PVAc chain with 1000 CO_2 molecules, a system containing 1 PVAO chain with 1000 CO_2 molecules, and systems containing 3 PVAO chains and 3 PVAc chains. Considering the temperature and pressure of well sites, temperatures ranging from 120 to 200 °C and pressures ranging from 60 to 120 MPa were selected.

Table 3. Different polymer and CO_2 systems in MD simulations.

System	Composition	No. of Chains	Mn of Chain	No. of VAc Units	No. of VO Units	No. of CO_2	No. of Atoms	Size (nm)
1	CO_2					1000	3000	$4 \times 4 \times 4$
2	PVAc/CO_2	1	6452	75	0	1000	3902	$4.2 \times 4.2 \times 4.2$
3	PVAO/CO_2	1	7002	35	35	1000	4052	$4.2 \times 4.2 \times 4.2$
4	PVAc	3	6452	75	0	0	2706	$3 \times 3 \times 3$
5	PVAO	3	7002	35	35	0	3156	$3 \times 3 \times 3$

Furthermore, to validate the rationality of the parameters used in the MD simulations, an scCO_2 fluid model with 2.9 wt% PVAc content was constructed, and the relative viscosity was calculated to be 1.8 times that of pure scCO_2 fluid at 23 °C and 20 MPa, which closely matches the experimental data that shows a relative viscosity of 1.7 times [42]. All of the parameters were chosen to be consistent with the aforementioned system.

5. Conclusions

ScCO_2 has wide application in oil recovery, such as use as an oil-displacing agent in EOR. And improving its viscosity and solubility is one of the important research topics. Based on the structure of PVAc, a novel environmentally friendly polymer thickener was designed considering the interaction mechanisms of the functional groups in CO_2, aiming to find polymers with enhanced affinity for CO_2 through PVAc modifications.

Here, we proposed a newly designed polymer thickener named PVAO. MD simulations were conducted on the thickened scCO_2 system to investigate the viscosity enhancement effects and solution characteristics systematically. Under identical temperature and pressure conditions, and similar weight percentages, PVAO turned out to exhibit superior viscosity enhancement compared to PVAc. The viscosity of scCO_2 with a PVAO chain was approximately 1.5 times that of scCO_2 with PVAc. Further MD simulations were performed on PVAO to obtain its radial distribution function, identifying the functional groups that contributed to the crucial interactions. It was found that stronger LA–LB interactions were observed between the carbonyl oxygen atom at the distal end of the PVAO side chain and the carbon atom in CO_2, instead of the oxygen atom at the proximal end. The interaction energy, cohesive energy density, and solubility parameters of PVAO were obtained to analyze its dissolution capacity in CO_2. It was found that, under the simulated temperature and pressure conditions, PVAO exhibited better dissolution capacity than PVAc. Thus, PVAO is a novel CO_2-philicity polymer with higher viscosity-enhancement efficiency and better dissolution capacity than PVAc.

Comparatively, the viscosity enhancement and solubility of PVAO and PVAc were lower than fluorinated and siloxane-based polymers, and there is still a long way to go to improve the viscosity enhancement and solubility performance of polymers. Also, the synthesis pathway of PVAO still remains unclear. PVAO also faces potential challenges in oil production site application. Yet what counts is that they have merits over others considering environmental friendliness and economic efficiency aspects. Our findings offer insights for the design of other high-performance polymers and provide theoretical instruction for oil site applications.

Author Contributions: Simulations, writing and editing, X.W. and S.L.; data analysis and figure plotting, Q.Z. and T.W.; research conceptualization and supervision, X.Z. All authors have read and agreed to the published version of the manuscript.

Funding: This research was funded by the unveiling projects of the Department of Science and Technology of Shanxi Province, grant number 20201101004; the National Natural Science Foundation of China, grant numbers 11804028 and 12175023; and by the National Key Laboratory of Petroleum Resources and Engineering, China University of Petroleum (Beijing), grant numbers PRP/DX-2210 and PRE/DX-2409.

Institutional Review Board Statement: Not applicable.

Informed Consent Statement: Not applicable.

Data Availability Statement: Data is contained within the article.

Acknowledgments: The authors would like to acknowledge the National Key Laboratory of Petroleum Resources and Engineering, China University of Petroleum, for supporting the research project.

Conflicts of Interest: The authors declare no conflicts of interest.

References

1. Haizhu, W.; Gensheng, L.I.; Yong, Z.; Sepehrnoori, K.; Zhonghou, S.; Bing, Y.; Lujie, S. Research status and prospects of supercritical CO_2 fracturing technology. *Acta Pet. Sin.* **2020**, *41*, 116–126.
2. Joseph, M.D. Practical Approaches to Green Solvents. *Science* **2002**, *297*, 799–803. [CrossRef]
3. Eric, J.B. A challenge for green chemistry: Designing molecules that readily dissolve in carbon dioxide. *Chem. Commun.* **2004**, *17*, 1885–1888. [CrossRef] [PubMed]
4. Chatzis, G.; Samios, J. Binary mixtures of supercritical carbon dioxide with methanol. A molecular dynamics simulation study. *Chem. Phys. Lett.* **2003**, *374*, 187–193. [CrossRef]
5. Stubbs, J.M.; Siepmann, J.I. Binary phase behavior and aggregation of dilute methanol in supercritical carbon dioxide: A Monte Carlo simulation study. *J. Chem. Phys.* **2004**, *121*, 1525–1534. [CrossRef] [PubMed]
6. David, L.T.; Hongbo, L.; Dehua, L.; Xiangmin, H.; Maxwell, J.W.; Lee, L.J.; Kurt, W.K. A Review of CO_2 Applications in the Processing of Polymers. *Ind. Eng. Chem. Res.* **2003**, *42*, 6431–6456. [CrossRef]
7. Jason, L. Small Molecule Associative CO_2 Thickeners for Improved Mobility Control. Doctoral Dissertation, University of Pittsburgh, Pittsburgh, PA, USA, 2017.
8. Lee, J.J.; Stephen, C.; Aman, D.; Robert, M.E.; Eric, J.B.; Robert, J.P.; Michael, J.O.B.; Mark, D.D. Development of Small Molecule CO_2 Thickeners for EOR and Fracturing. In Proceedings of the SPE Improved Oil Recovery Conference, Tulsa, OK, USA, 12–16 April 2014.
9. Robert, M.E.; Olsen, D.K.; Ammer, J.R.; Schuller, W.A. Mobility and Conformance Control for CO_2 EOR via Thickeners, Foams, and Gels—A Literature Review of 40 Years of Research and Pilot Tests. In Proceedings of the SPE Improved Oil Recovery Conference, Tulsa, OK, USA, 14–18 April 2012.
10. Etienne, G.; Thierry, T.; Catherine, L.; Jean-Daniel, M.; Mathias, D. Distinctive Features of Solubility of RAFT/MADIX-Derived Partially Trifluoromethylated Poly(vinyl acetate) in Supercritical CO_2. *Macromolecules* **2012**, *45*, 9674–9681. [CrossRef]
11. Etienne, G.; Thierry, T.; Séverine, C.; Jean-Stéphane, C.; Jean-Daniel, M.; Mathias, D. Enhancement of Poly(vinyl ester) Solubility in Supercritical CO_2 by Partial Fluorination: The Key Role of Polymer–Polymer Interactions. *J. Am. Chem. Soc.* **2012**, *134*, 11920–11923.
12. Ruediger, M.; Bernhard, W. Solutions of Poly(dimethylsiloxane) in Supercritical CO_2: Viscometric and Volumetric Behavior. *Macromolecules* **1994**, *27*, 3289–3294. [CrossRef]
13. Jiarui, X. Carbon Dioxide Thickening Agents for Reduced CO_2 Mobility. Doctoral Dissertation, University of Pittsburgh, Pittsburgh, PA, USA, 2003.
14. Cui, S.T.; Cummings, P.T.; Cochran, H.D. The calculation of viscosity of liquid n-decane and n-hexadecane by the Green-Kubo method. *Mol. Phys.* **1998**, *93*, 117–122. [CrossRef]
15. Yong, Z.; Akihito, O.; Edward, J.M. Reliable Viscosity Calculation from Equilibrium Molecular Dynamics Simulations: A Time Decomposition Method. *J. Chem. Theory Comput.* **2015**, *11*, 3537–3546. [CrossRef] [PubMed]
16. Arno, L.; Chris, D.M. Reference Correlation for the Viscosity of Carbon Dioxide. *J. Phys. Chem. Ref. Data* **2017**, *46*, 013107. [CrossRef] [PubMed]
17. Van der Vegt, N.F.A. Temperature Dependence of Gas Transport in Polymer Melts: Molecular Dynamics Simulations of CO_2 in Polyethylene. *Macromolecules* **2000**, *33*, 3153–3160. [CrossRef]
18. Christian, D.; Eric, J.B. Phase behavior of polymers containing ether groups in carbon dioxide. *J. Supercrit. Fluids* **2002**, *22*, 103–110. [CrossRef]
19. Zhihao, S.; Mark, A.M.; Jiarui, X.; Belardi, J.; Sevgi, K.; Anita, J.M.; Bane, S.; Karnikas, C.; Eric, J.B.; Robert, M.E. CO_2-solubility of oligomers and polymers that contain the carbonyl group. *Polymer* **2003**, *44*, 1491–1498. [CrossRef]

20. Dongdong, H.; Shaojun, S.; Pei-Qing, Y.; Ling, Z.; Tao, L. Exploration of CO_2-Philicity of Poly(vinyl acetate-co-alkyl vinyl ether) through Molecular Modeling and Dissolution Behavior Measurement. *J. Phys. Chem. B* **2015**, *119*, 12490–12501. [CrossRef] [PubMed]
21. Dongdong, H.; Shaojun, S.; Pei-Qing, Y.; Ling, Z.; Tao, L. Evaluation of CO_2-Philicity of Poly(vinyl acetate) and Poly(vinyl acetate-alt-maleate) Copolymers through Molecular Modeling and Dissolution Behavior Measurement. *J. Phys. Chem. B* **2015**, *119*, 3194–3204. [CrossRef]
22. Sadegh, Y.-N.; Jaber, S.; Javad, K.-S.; Ali, N.; Elham, A. Determination of momentum accommodation coefficients and velocity distribution function for Noble gas-polymeric surface interactions using molecular dynamics simulation. *Appl. Surf. Sci.* **2019**, *493*, 766–778. [CrossRef]
23. Dai, X.; Chongtao, W.; Meng, W.; Ruying, M.; Yu, S.; Junjian, Z.; Xiaoqi, W.; Xuan, S.; Veerle, V. Interaction mechanism of supercritical CO_2 with shales and a new quantitative storage capacity evaluation method. *Energy* **2023**, *264*, 126424. [CrossRef]
24. Hojatollah, M.; Hedayat, A.; Parissa, K.P.; Nia, R. Supercritical Methanol and Ethanol Solubility Estimation by Using Molecular Dynamics Simulation. *Chem. Eng. Technol.* **2023**, *46*, 2167–2174. [CrossRef]
25. Yan, Q.; Xiaoning, Y.; Yupeng, Z.; Ping, J.L. Molecular Dynamics Simulation of Interaction between Supercritical CO_2 Fluid and Modified Silica Surfaces. *J. Phys. Chem. C* **2008**, *112*, 12815–12824. [CrossRef]
26. Yalin, L.; Xiaoxiao, D.; Yuejin, Z.; Peng, H.; Bing, L.; Jianlin, L. Effect of the Water Film Rupture on the Oil Displacement by Supercritical CO_2 in the Nanopore: Molecular Dynamics Simulations. *Energy Fuels* **2022**, *36*, 4348–4357. [CrossRef]
27. Alberto, P.; Ivan, M.; Daniel, S.; Jiří, Š.; Thomas, E.C.; Charles, A.L.; Modesto, O. Refinement of the AMBER Force Field for Nucleic Acids: Improving the Description of α/γ Conformers. *Biophys. J.* **2007**, *92*, 3817–3829. [CrossRef]
28. MacKerell, A.D., Jr.; Bashford, D.; Bellott, M.L.D.R.; Dunbrack, R.L., Jr.; Evanseck, J.D.; Field, M.J.; Fischer, S.; Gao, J.; Guo, H.; Ha, S.; et al. All-Atom Empirical Potential for Molecular Modeling and Dynamics Studies of Proteins. *J. Phys. Chem. B* **1998**, *102*, 3586–3616. [CrossRef] [PubMed]
29. David, L.R.; Huai, S.; Eichinger, B.E. Computer simulations of poly(ethylene oxide): Force field, pvt diagram and cyclization behaviour. *Polym. Int.* **1997**, *44*, 311–330. [CrossRef]
30. Huai, S. COMPASS: An ab Initio Force-Field Optimized for Condensed-Phase Applications Overview with Details on Alkane and Benzene Compounds. *J. Phys. Chem. B* **1998**, *102*, 7338–7364. [CrossRef]
31. Morteza, N.; Samane, M.; Yongan, G. Effects of Viscous and Capillary Forces on CO_2 Enhanced Oil Recovery under Reservoir Conditions. *Energy Fuels* **2007**, *21*, 3469–3476. [CrossRef]
32. Jie, Y.; Yi, R.; Anmin, T.; Huai, S. COMPASS Force Field for 14 Inorganic Molecules, He, Ne, Ar, Kr, Xe, H_2, O_2, N_2, NO, CO, CO_2, NO_2, CS_2, and SO_2, in Liquid Phases. *J. Phys. Chem. B* **2000**, *104*, 4951–4957.
33. Kaplun, A.B.; Meshalkin, A.B. Unified equation for calculating the viscosity coefficient of argon, nitrogen, and carbon dioxide. *High Temp.* **2016**, *54*, 808–814. [CrossRef]
34. Xiaoqi, W.; Yongan, G. Oil Recovery and Permeability Reduction of a Tight Sandstone Reservoir in Immiscible and Miscible CO_2 Flooding Processes. *Ind. Eng. Chem. Res.* **2011**, *50*, 2388–2399. [CrossRef]
35. Hongyu, G.; Hao, Z.; Long, X.; Yajun, L.; Mingzhe, D. Effects of cosolvent on dissolution behaviors of PVAc in supercritical CO_2: A molecular dynamics study. *Chem. Eng. Sci.* **2019**, *206*, 22–30. [CrossRef]
36. Denis, J.E.; Brad Lee, H. The Nose–Hoover thermostat. *J. Chem. Phys.* **1985**, *83*, 4069–4074. [CrossRef]
37. Eichinger, B.E.; David, L.R.; Judith, A.S. Cohesive properties of Ultem and related molecules from simulations. *Polymer* **2002**, *43*, 599–607. [CrossRef]
38. Huai, S.; David, L.R. Polysiloxanes: Ab initio force field and structural, conformational and thermophysical properties. *Spectrochim. Acta Part A Mol. Biomol. Spectrosc.* **1997**, *53*, 1301–1323. [CrossRef]
39. Sergei, G.K.; Michael, V.; Frank, V.B.; Charles, L.L.; Charles, A.E. Specific Intermolecular Interaction of Carbon Dioxide with Polymers. *J. Am. Chem. Soc.* **1996**, *118*, 1729–1736. [CrossRef]
40. Poovathinthodiyil, R.; Scott, L.W. Cooperative C−H···O Hydrogen Bonding in CO_2−Lewis Base Complexes: Implications for Solvation in Supercritical CO_2. *J. Am. Chem. Soc.* **2002**, *124*, 12590–12599. [CrossRef] [PubMed]
41. Sevgi, K.; Michalik, S.; Yang, W.; Johnson, J.K.; Robert, M.E.; Eric, J.B. Phase Behavior of Oxygen-Containing Polymers in CO_2. *Macromolecules* **2007**, *40*, 1332–1341.
42. Peter, L.; Adel, A.; Lee, J.J.; Eric, J.B.; Robert, M.E. Thickening CO_2 with Direct Thickeners, CO_2-in-Oil Emulsions, or Nanoparticle Dispersions: Literature Review and Experimental Validation. *Energy Fuels* **2021**, *35*, 8510–8540. [CrossRef]

Disclaimer/Publisher's Note: The statements, opinions and data contained in all publications are solely those of the individual author(s) and contributor(s) and not of MDPI and/or the editor(s). MDPI and/or the editor(s) disclaim responsibility for any injury to people or property resulting from any ideas, methods, instructions or products referred to in the content.

Article

Evaluation of CO_2/H_2O Co-Adsorption Models for the Anion Exchange Resin Lewatit VPOC 1065 under Direct Air Capture Conditions Using a Novel Lab Setup

Florian M. Chimani *, Aditya Anil Bhandari, Andreas Wallmüller, Gerhard Schöny, Stefan Müller and Josef Fuchs

Institute of Chemical, Environmental and Bioscience Engineering, TU Wien, Getreidemarkt 9/166, 1060 Wien, Austria
* Correspondence: florian.chimani@tuwien.ac.at

Abstract: This study aimed to develop a laboratory-scale direct air capture unit for evaluating and comparing amine-based adsorbents under temperature vacuum swing adsorption conditions. The experimental campaign conducted with the direct air capture unit allowed for the determination of equilibrium loading, CO_2 uptake capacity, and other main performance parameters of the investigated adsorbent Lewatit VP OC 1065®. The investigations also helped to understand the co-adsorption of CO_2 and H_2O on the tested material, which is crucial for improving temperature vacuum swing adsorption processes. This was achieved by obtaining pure component isotherms for CO_2 and H_2O and using three different co-adsorption isotherm models from the literature. It was found that the weighted average dual-site Toth model emerged as the most accurate and reliable model for simulating this co-adsorption behaviour. Its predictions closely align with the experimental data, particularly in capturing the adsorption equilibrium at various temperatures. It was also observed that this lab-scale unit offers advantages over thermogravimetric analysis when conducting adsorption experiments on the chosen amine. The final aim of this study is to provide a pathway to develop devices for testing and developing efficient and cost-effective adsorbents for direct air capture.

Keywords: direct air capture; CO_2 adsorption; co-adsorption; isotherm modelling; negative emissions technology

Citation: Chimani, F.M.; Bhandari, A.A.; Wallmüller, A.; Schöny, G.; Müller, S.; Fuchs, J. Evaluation of CO_2/H_2O Co-Adsorption Models for the Anion Exchange Resin Lewatit VPOC 1065 under Direct Air Capture Conditions Using a Novel Lab Setup. *Separations* **2024**, *11*, 160. https://doi.org/10.3390/separations11060160

Academic Editors: Zilong Liu, Meixia Shan and Yakang Jin

Received: 29 April 2024
Revised: 17 May 2024
Accepted: 20 May 2024
Published: 22 May 2024

Copyright: © 2024 by the authors. Licensee MDPI, Basel, Switzerland. This article is an open access article distributed under the terms and conditions of the Creative Commons Attribution (CC BY) license (https://creativecommons.org/licenses/by/4.0/).

1. Introduction

As the atmospheric CO_2 concentration continues to increase, there is a growing risk of severe and irreversible impacts due to the dramatic progression of climate change [1]. In response to these concerns, the United Nations Framework Convention on Climate Change facilitated the development of the Paris Agreement, a landmark international accord on climate change [2]. Under this agreement, participating countries submitted Nationally Determined Contributions (NDCs) outlining their commitments to reduce greenhouse gas emissions. However, analyses suggest that adopting the existing commitments will result in a temperature increase of around 2.4–2.6 °C by the end of the century, considering both conditional and unconditional NDCs [3]. Therefore, there is a growing recognition of the need for effective carbon dioxide removal (CDR) strategies to complement emission reduction efforts and achieve the desired climate goals. CDR encompasses various approaches that aim to reduce atmospheric CO_2 levels, including methods for direct extraction of CO_2 from the atmosphere and enhancing carbon sinks on land and in the oceans to enhance CO_2 removal [4].

Primarily, six technical CDR approaches have been identified for the removal and sequestration of carbon dioxide: coastal blue carbon, terrestrial carbon removal and sequestration, bioenergy with carbon capture and storage (BECCS), carbon mineralization, geological sequestration, and direct air capture (DAC) [5]. Each of the aforementioned technical methods presents its own set of advantages and disadvantages. Yet, factors such as the land area required, water demand, technology learning curve, scalability, and life-cycle considerations cause DAC to garner more attention than coastal blue carbon and BECCS as a method for carbon removal [6]. Furthermore, since DAC can also be used in the CO_2 utilization industry, which helps climate mitigation efforts, there is a growing interest in developing and scaling up DAC technology [7].

Despite this growing interest in DAC technology, the industry is still in its infancy, with a limited number of major players such as Carbon Engineering, Climeworks, and Global Thermostat and a plethora of upcoming startups. However, these companies have their own drawbacks, such as high costs and low scale of operations [8].

One way to drastically reduce the costs is to bring down the energy requirements of the CO_2 capture process. This, as indicated by several research papers, boils down to choosing the most suitable active material that captures CO_2 and the subsequent method of adsorption–desorption [9]. Several different functional materials are being used in the industry or investigated at research institutes, for example, aqueous alkali hydroxide solutions, solid amine-functionalized adsorbents, solid oxide-based adsorbents, and membrane-based filters [10]. Similarly, there are various methods proposed for adsorption–desorption, for example, temperature swing adsorption (TSA), temperature vacuum swing adsorption (TVSA), pressure vacuum swing adsorption (PVSA), moisture swing adsorption (MSA), electro-swing process, and electrolysis [9,11,12]

Amine-functionalized adsorbents that work on TVSA are often proposed as active materials for DAC CO_2 capture [13]. It is known that the characterization, comparison, and evaluation of solid amine-functionalized adsorbents are critical steps in determining their suitability for CO_2 capture applications. To characterize any adsorbent for CO_2 capture, two types of properties are typically measured: intrinsic properties and performance parameters. Intrinsic properties are textural features of the material that are determined after synthesis and depend on the material's structure. These properties include pore size, surface area, and pore volume, which play a critical role in the adsorption process. Performance parameters, on the other hand, describe the functional behaviour of the adsorbent, such as the CO_2 uptake capacity, selectivity, degradation over lifetime, and regeneration potential. These parameters are used to evaluate the adsorbent's effectiveness for CO_2 capture applications [14].

However, assessing adsorbents for CO_2 capture can be challenging because there are many different types of adsorbents available, and each has its own unique properties. Furthermore, different research studies may use varying test conditions, making it difficult to compare the adsorbent's performance parameters accurately. For example, the temperature, pressure, and gas composition used in the experiments can significantly affect the adsorbent's performance.

Hence, this study aimed to develop a suitable lab setup and a testing method that can help evaluate and compare amine-functionalized adsorbents under TVSA-DAC conditions for a range of performance parameters. Performance parameters such as temperature, relative humidity, CO_2 partial pressure, and adsorption kinetics are tested to assess the efficiency of the adsorbent. The collected data are then used to explore models that explain the multilayer adsorption of both CO_2 and water (H_2O) on these materials, which is crucial for improving TVSA processes [15]. The results will be compared with thermogravimetric analysis (TGA) for H_2O and CO_2 adsorption on amines to see if this lab unit offers advantages over the conventional methods of measuring adsorption kinetics. This comprehensive approach aids in developing more efficient materials for carbon capture and utilization applications.

The design of the test equipment is based on state-of-the-art knowledge of DAC technology and amine-functionalized adsorbent materials, as well as practical considerations such as ease of use and integrability. For this study, Lewatit VP OC 1065®, a solid amine-functionalized adsorbent, is used. The equipment, methods, and models developed in this work should also apply to other amine-functionalized adsorbents. The final goal of this research is to contribute to the scaling up DAC technology globally by providing a pathway to develop a suitable setup for testing and developing efficient and cost-effective adsorbents.

2. Material and Methods

A novel DAC unit on a laboratory scale at TU Wien was built and used for the experimental investigations of carbon capture potential of amine-functionalized sorbents under direct air capture conditions. The purpose of this facility was to create an environment in which an adsorbent can be loaded and unloaded with CO_2 under specific conditions, thus providing information on the optimum operating conditions for different adsorbents.

The material used in this study, Lewatit, consists of a polystyrene polymer cross-linked with divinylbenzene and functionalized with primary amine groups. This adsorbent was proposed in a variety of CO_2 capture processes, for example, by Veneman et al. [16] and Sonnleitner et al. [17] in a continuous TSA CO_2 capture process, or by Low et al. [18] in DAC applications, with the latter providing detailed BET analysis of the adsorbent. The key material data obtained from the material datasheet [19] are presented in Table 1.

Table 1. Material properties of the tested adsorbent.

	Lewatit
Composition	Cross-linked polystyrene functionalized with primary amines
Average particle diameter	0.52 mm
Average pore diameter	25 nm
Bulk density	630–710 kg m^{-3}

2.1. Laboratory Unit Setup

The experimental setup of the laboratory unit can be divided into two process modes: adsorption and desorption. This unit captures CO_2 in batches rather than continuously. For research purposes, this has the advantage that the mass and energy balance can be calculated for both process modes separately.

Figure 1 shows a picture of the DAC unit operating in its desorption mode. Figure 2a shows a schematic flow diagram of the adsorption setup, and Figure 2b shows the desorption setup. For the CO_2 loading of the adsorbent, a stream of dry compressed air containing 450 ± 30 ppm of CO_2 is used. To reduce or enhance the CO_2 loading of the gas stream, pure CO_2 or nitrogen can be added via a mass flow controller (MFC). Part of the humidity of the dry gas stream can be sent through a temperature-regulated humidifier to increase the humidity. Before the gas stream reaches the fixed adsorbent bed, it passes through a honeycomb heat exchanger. This enables precise control over the gas stream's temperature. The fixed bed containing the adsorbent is located between two cones and sealed off, ensuring the adsorbent is only in contact with the gas stream from the heat exchanger.

The adsorbent bed is mounted between two heating plates made of aluminum in the regeneration setup, as shown in Figure 2b. These heating plates feature fine perforations and an intricate gas channelling system. This design efficiently extracts released gas from the fixed bed while ensuring the uniform distribution of purge gas and preventing the formation of channels. Furthermore, the heating plates have additional channels allowing water or thermal oil to be pumped through, allowing the system to regulate the temperature precisely. It is essential to mention that these two channel systems are separated, and no molecular exchange is happening between the water-saturated CO_2 gas stream and the thermal oil. The released gas can be pumped off at the top side on the upper plate, while the purge gas can be introduced at the lower plate. During regeneration, the influence of this purge gas on the desorption process is to be investigated. For this purpose, the apparatus is equipped with a steam generator connected to a membrane pump. The released gas mixture is pumped off via a downstream vacuum pump. A heat exchanger for cooling is installed on the suction side to prevent condensation and overheating in the pump. This heat exchanger collects residual water in the gas stream via condensation. The nitrogen flow is selected so that the downstream CO_2 measurement operates in the intended measuring range (0–10,000 ppm). From the N_2 flow and CO_2 concentration, the absolute amount of CO_2 can finally be determined in the originally extracted gas flow.

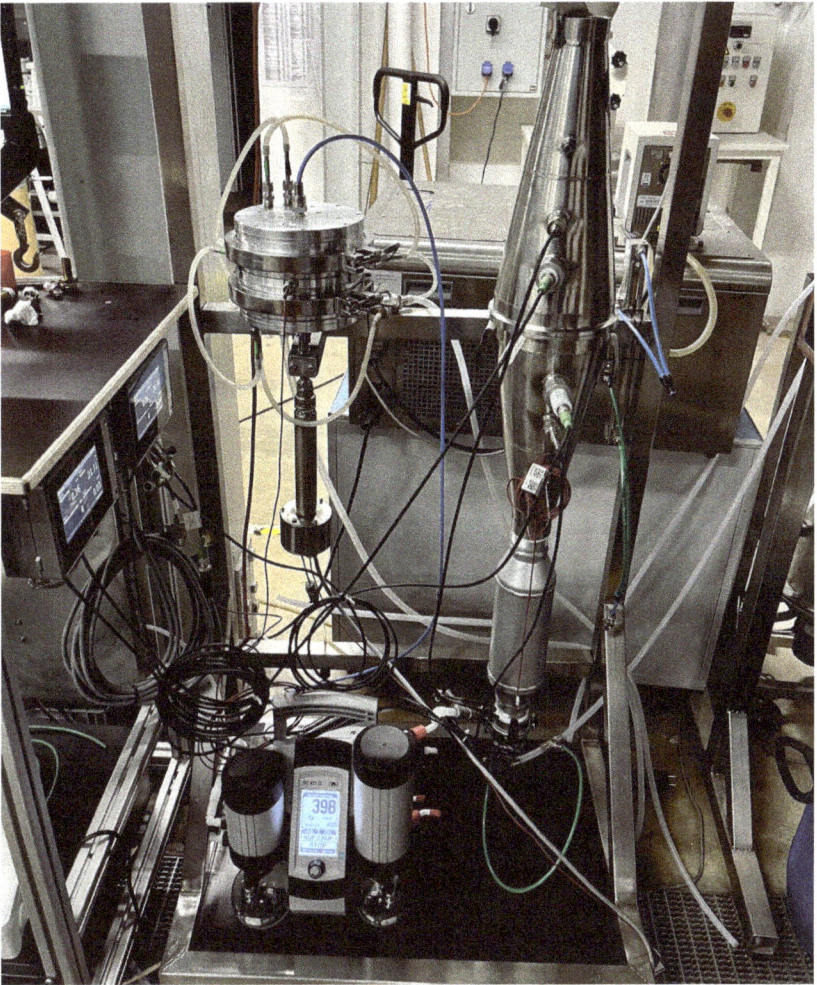

Figure 1. Experimental DAC plant set up in the Technikum at TU Wien. The system is shown in the desorption mode. In the lower part of the picture, the vacuum pump is operating at 398 mbar.

Pt100 sensors with a diameter of 3 mm are used to measure the temperature in the fixed bed as well as the temperature of the supply air after the heat exchanger. Due to the low inert mass of the sensors, they react immediately to temperature changes and provide reliable data. The humidity sensors, which are installed in the conical pipes, measure not only the relative humidity but also the temperature. Pressure sensors in the conical pipes monitor the pressure loss across the fixed bed. However, due to the low bed height and low gas velocity, the pressure loss is <1 mbar and can therefore be neglected. Table 2 shows a list of the sensors used for the investigations. This is sufficient for adsorption measurements before and after the fixed bed. However, since the gas stream from the vacuum pump is almost pure CO_2, it is mixed with a defined quantity of nitrogen. To obtain a homogeneous mixing of the gases and to avoid introducing counterpressure into the vacuum pump, the nitrogen is fed straight through a T-piece. At the same time, the CO_2 flows in laterally through a taper.

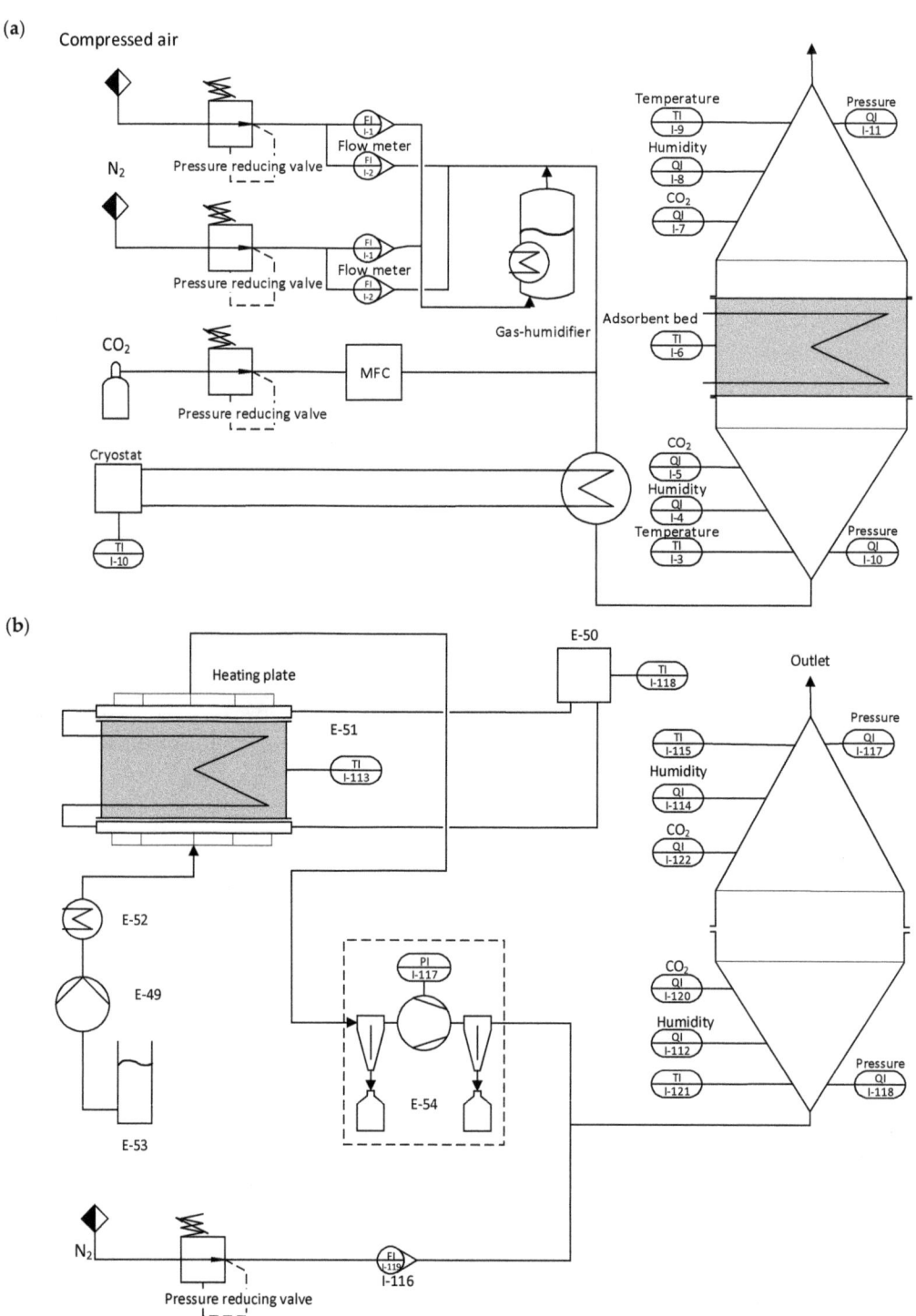

Figure 2. (**a**) Adsorption setup of the laboratory unit. (**b**) Desorption setup of the laboratory unit.

Table 2. Sensors used for gas analysis.

Sensor	Measurement Range	Measurement Error	Manufacturer	Designation
Pt100 temp. sensor	−30 to +180 °C	N/A	Pohltechnik	C-H100 × 3 sil 2 m 3 L
Humidity sensor	0–100% rel. H	±0.8% rel. H	Vaisala	HMP7
CO_2 sensor	0–10,000 ppm	±2%	Vaisala	GMP252
Pressure Transmitter	0–10 mbar	N/A	Kalinsky	TYPE DS 2-010

2.2. CO_2 Adsorption Experiments

All CO_2 adsorption measurements were conducted in an aluminum fixed bed column. Approximately 500 g of adsorbent inside the column was tightly packed and fixed in position with wire meshes. The adsorbent was fully desorbed after each adsorption step. All experiments were carried out using TVSA. The experiment consisted of the adsorbent loading and regeneration phase. The adsorbent loading or adsorption phase was initiated by introducing a CO_2/N_2- or H_2O/N_2-loaded air stream for the pure component adsorption data or a CO_2/H_2O-loaded air stream for the co-adsorption data. During each adsorption step, all operating parameters were kept constant and were recorded at 2 s intervals. Equilibrium was reached when the partial pressure of CO_2 measured by sensors before and after the adsorbent bed was consistent, indicating that no further adsorption or desorption was occurring. Various heating and cooling systems were used to ensure isothermal conditions during adsorption. The fluidizing humidifier (bubbler) was held at a constant temperature using a PID controller. The separated dry and humid gas streams are reunited before passing through a heat exchanger, which is temperature-controlled via a cryostat.

After the adsorption was completed, the adsorbent bed was placed onto the desorption apparatus, and the initial evacuation step was initiated. This purging step was crucial to removing residual air inside the adsorbent bed and all other connected tubes. This was realized by connecting the adsorbent bed to a vacuum pump and setting an absolute pressure of 50 mbar for one minute. Although some CO_2 can be desorbed through physisorption even during an isothermal evacuation, no measurable CO_2 was detected in this step. The adsorbent's regeneration started once the system's absolute pressure did not rise when the vacuum pump was turned off. This ensured that the system had no leaks and, thus, no false air was passing through the adsorbent bed.

The heating of the adsorbent during regeneration was realized in two different ways. Two aluminum heating plates transfer thermal energy to the fixed bed via heat conduction. Additionally, the fixed bed is heated from the inside via a heat exchanger in the form of bent 6 mm aluminum tubes. As a purging agent, nitrogen or steam can be used. For the latter, the membrane pump transports a defined amount (75 g/h in this study) of water through an evaporator and into the fixed bed. The vacuum pump was set to a specified pressure setting for the desorption duration to ensure a constant gas stream through the adsorbent bed. After fully regenerating, the adsorbent was cooled to approximately 55 °C to avoid sorbent degradation. The desorption conditions were kept the same for the entire duration of the experiment as presented in Table 3.

Table 3. Operating conditions during testing.

Process Step	Pressure (abs.)	Temperature	Duration
Adsorption	1.017 bar	15–30 °C	180–600 min
Evacuation	<0.05 bar	Isothermal, T_{ads}	1 min
Desorption	0.2–0.6 bar	70–90 °C	90 min
Cooling	0.1 bar	<55 °C	10 min

2.3. Applied Methods

2.3.1. Adsorption Capacity

The equilibrium loading is determined by balancing according to Equation (1). Adapted for CO_2, the formula is as follows.

$$X = \frac{m_{ads}}{m_{adsorbent}} = \frac{m_{in} - m_{out}}{m_{adsorbent}} \qquad (1)$$

where X is the equilibrium loading of CO_2, m_{in} and m_{out} are the mass of CO_2 entering and leaving the adsorbent bed during the adsorption period, m_{ads} is the total mass of CO_2 adsorbed, and $m_{adsorbent}$ is the total mass of the adsorbent inside the fixed bed when it is not loaded (fully desorbed). For the entire duration of the experiment, it is assumed that there is a constant mass of CO_2 flowing into the reactor, and thus the total mass of incoming CO_2 m_{in} can be calculated. This assumption is justified because no change is made to the reactor inlet stream during this time. To determine the incoming CO_2, the duration of adsorption is multiplied by the concentration measured upstream of the reactor after the experiment, using the ideal gas equation.

$$m_{in} = \int_{t_0}^{t} \frac{c\dot{V}pM}{RT} dt \qquad (2)$$

where c is the CO_2 concentration, $\dot{V}$ is the volumetric flow, p is the total pressure (1 atm.), R is the ideal gas constant, T is the temperature, and M is the molar mass of CO_2.

The same procedure can be followed for the mass of CO_2 leaving the reactor. Here, the total duration of the adsorption is divided into discrete time intervals with a length of 2 s. For these n time intervals, the concentration measured at the outlet is then used to determine an outgoing mass in this time interval. These partial masses are summed up as shown in Equation (3).

$$m_{out} = \sum_{i=1}^{n} \int_{t_{i-1}}^{t_i} \frac{c\dot{V}pM}{RT} dt \qquad (3)$$

The calculation of the adsorption capacity q_{CO_2} follows Equation (4):

$$q_{CO_2} = \frac{m_{ads}}{M m_{adsorbent}} \qquad (4)$$

2.3.2. Pure Component Adsorption Isotherms

The pure component CO_2 adsorption isotherms were modelled using the temperature-dependent Toth isotherm. For amine functionalized adsorbents, this model has proven accurate for higher CO_2 partial pressures [16,20,21], as well as in low partial pressure regions, as is the case for direct air capture [9,22]:

$$q_{CO_2}(T, p_{CO_2}) = n_s(T) \frac{b(T) p_{CO_2}}{[1 + (b(T) p_{CO_2})^{\tau(T)}]^{1/\tau(T)}} \qquad (5)$$

where n_s is the maximum adsorption capacity, b is the adsorption affinity, p_{CO_2} is the partial pressure of CO_2 in the gas phase, and τ is an exponential factor describing the heterogeneity of the adsorbent called the Toth constant. To obtain the temperature-dependent Toth equation, the aforementioned parameters must also be based on temperature-dependent calculations:

$$n_s(T) = n_{s,0} \exp\left[\chi\left(1 - \frac{T}{T_0}\right)\right] \qquad (6)$$

$$b(T) = b_0 \exp\left[\frac{\Delta H_0}{RT_0}\left(\frac{T_0}{T} - 1\right)\right] \qquad (7)$$

$$\tau(T) = \tau_0 + \alpha\left(1 - \frac{T_0}{T}\right) \qquad (8)$$

where $n_{s,0}$, b_0, and τ_0 represent the values of the Toth parameters at a reference temperature T_0. χ and α are dimensionless parameters and ΔH_0 is the isosteric heat of adsorption at zero fractional loading.

For CO_2 capture applications, water adsorption onto solid species is an essential field of study [23–25]. The Guggenheim–Anderson–de Boer (GAB) model, an extension of the Brunauer–Emmet–Teller (BET) model, has proven to be the most accurate regarding solid sorbent adsorption. The equation for the GAB model is:

$$q_{H_2O}(\varphi) = n_m \frac{c_g K_{ads} \varphi}{(1 - K_{ads}\varphi)(1 + (c_g - 1)K_{ads}\varphi)} \quad (9)$$

where q_{H_2O} is the adsorption capacity at a relative humidity φ and n_m is the monolayer adsorption capacity of H_2O. The parameters c_g and K_{ads} are temperature dependent and are calculated as follows:

$$c_g(T) = c_0 \exp(\frac{\Delta H_C}{RT}) \quad (10)$$

$$K_{ads}(T) = K_0 \exp(\frac{\Delta H_K}{RT}) \quad (11)$$

where ΔH_C and ΔH_k are the adsorption enthalpies of mono- and multilayer adsorption, and c_0 and K_0 are dimensionless parameters [26–29].

2.3.3. Co-Adsorption Isotherms

Previous research has shown that the presence of water enhances the CO_2 uptake of amine-based adsorbents [16,30–32], meaning the CO_2 isotherm becomes steeper, especially in low partial pressure ranges, and the overall maximum uptake increases. There are different approaches in the literature when it comes to describing the co-adsorption of CO_2 and H_2O. In the following paragraphs, three different models are compared to describe this phenomenon.

The first approach is empirical, where an enhancement factor (EF) is introduced to describe the adsorption capacity. Wurzbacher et al. [29] used a similar approach to describe binary CO_2 and H_2O adsorption onto their adsorbent. An enhancement factor β_{EF} is introduced based on previous adsorption data from Sonnleitner et al. [17]. This factor includes a constant k and the relative humidity φ:

$$q_{CO_2}(T, p_{CO_2}, \varphi) = \beta_{EF}(\varphi) q_{CO_2}(T, p_{CO_2}) \quad (12)$$

$$\beta_{EF}(\varphi) = 1 + \varphi * k \quad (13)$$

A different approach was followed by Stampi-Bombelli et al. [28], where a new isotherm model (SB model) based on the Toth isotherm was proposed. The model accounts for the water uptake dependency in the maximum uptake term n_s and the affinity coefficient b:

$$q_{CO_2}(T, p_{CO_2}, q_{H_2O}) = n_s(T, q_{H_2O}) \frac{b(T, q_{H_2O}) p_{CO_2}}{[1 + (b(T, q_{H_2O}) p_{CO_2})^{\tau(T)}]^{1/\tau(T)}} \quad (14)$$

$$n_s(T, q_{H_2O}) = n_s \left[\frac{1}{1 - y q_{H_2O}}\right] y > 0 \quad (15)$$

$$b(T, q_{H_2O}) = b(T)(1 + \beta q_{H_2O}) \beta > 0 \quad (16)$$

This model results in an increased maximum CO_2 uptake and isotherm affinity when water is present in the gas and it reduces to the single component Toth model when water is absent.

The third approach is the weighted average dual-site Toth (WADST) co-adsorption model introduced by Young et al. [27]. This approach is based on the different availability of water molecules at sites on the adsorbent where one site has a water molecule available, and the other does not. The probability that one site has a water molecule available is described via an Arrhenius-style equation described by a critical water loading parameter A.

$$q_{CO_2}(p_{CO_2}, q_{H_2O}, T) = \left(1 - e^{-\frac{A}{q_{H_2O}}}\right) \frac{n_{s,dry}(T)b_{dry}(T)p_{CO_2}}{[1 + (b_{dry}(T)p_{CO_2})^{\tau_{dry}(T)}]^{\frac{1}{\tau_{dry}(T)}}} + e^{-\frac{A}{q_{H_2O}}} \frac{n_{s,wet}(T)b_{wet}(T)p_{CO_2}}{[1 + (b_{wet}(T)p_{CO_2})^{\tau_{wet}(T)}]^{\frac{1}{\tau_{wet}(T)}}} \quad (17)$$

The first part of the equation simply describes the Toth model shown in Equation (5) including the fitted parameters from the pure component adsorption isotherms, while the wet site is defined by the same equations and fitted to co-adsorption experiments. While the same model for the adsorption capacity q_{H_2O} is used, Equation (9) describes the temperature dependency of c_g and K_{ads} according to Anderson's derivation [33] as follows:

$$c_g(T) = \exp\left(\frac{E_1 - E_{10+}}{RT}\right) \quad (18)$$

$$K_{ads}(T) = \exp\left(\frac{E_{2-9} - E_{10+}}{RT}\right) \quad (19)$$

$$E_1(T) = C - \exp(DT) \quad (20)$$

$$E_{2-9}(T) = F + GT \quad (21)$$

$$E_{10+}(T) = -44.38T + 57220 \quad (22)$$

where E_1 refers to the heat of adsorption for the 1st layer, E_{2-9} represents the heat of adsorption for the 2nd to 9th layers, and E_{10+} corresponds to the heat of adsorption for the 10th layer and beyond, which is comparable to the latent heat of water condensation. The unknown dependencies of temperature (C, D, F, G) on E_1, E_{2-9}, and E_{10+} were empirically fitted to experimental water isotherms for Lewatit.

3. Results

3.1. Adsorption Breakthrough Curves

Adsorption performance evaluation was implemented in three different test series. For this purpose, an elaborate test matrix was defined in which the supplied air temperature, the CO_2 concentration, and the relative humidity in the supplied airflow were changed (Table 4). This matrix involved a total of 24 adsorption tests followed by a constant regeneration of the adsorbent. A constant volume flow of 6 Nm^3/h was sent through the fixed bed over the entire duration of the experiment. The temperature variation between 15 °C and 30 °C represents typical direct air capture conditions. For this purpose, comparisons were made between the tests, which differed only in the parameter investigated under otherwise identical or averaged conditions. The regeneration of the adsorbent after every adsorption was carried out with a set vacuum pump pressure of 300 mbar(a), 75 mL/h/kg$_{Lewatit}$ membrane pump flow rate (conveying water into the evaporator), and 90 °C cryostat temperature for external heating. Table 3 shows a detailed summary of the desorption conditions.

Table 4. Adsorption matrix. Process conditions in the supplied airflow.

Temperature [°C]	CO_2 Concentration [ppm]	Relative Humidity [%]
15	400/700/1000/1300	35
20	400/700/1000/1300	35
25	400/700/1000/1300	35
30	400/700/1000/1300	35

The first experiments with 400 ppm CO_2 in the supplied air stream showed slower adsorption due to the low CO_2 concentration, as seen in Figure 3a. At 15 °C, equilibrium loading was reached after about 10 h. As expected, the lower temperature of the supply air stream positively affected the maximum CO_2 uptake of the adsorbent. In the first 200 min of the test, the data showed that practically no CO_2 was measured after the adsorbent bed. To gain valuable insights into separation efficiency and adsorption kinetics, it is recommended to increase the gas-to-solid (G/S) ratio and operate the experiment as an ideal fluidized bed. The initial kinetics and separation efficiency at zero loading can only be obtained this way. After a certain amount of CO_2 has passed through the adsorbent bed, the so-called breakthrough occurs where an increase in the CO_2 concentration after the bed is first visible. The expected trend in the breakthrough time starting at 30 °C with falling temperature down to 15 °C of supply air temperature was confirmed during the experiments. The measured carbon dioxide captured during subsequent desorption fitted well with the adsorption data. In accordance with the adsorption data, the desorbed CO_2 mass decreased with the increasing supply air temperature of the previous adsorption.

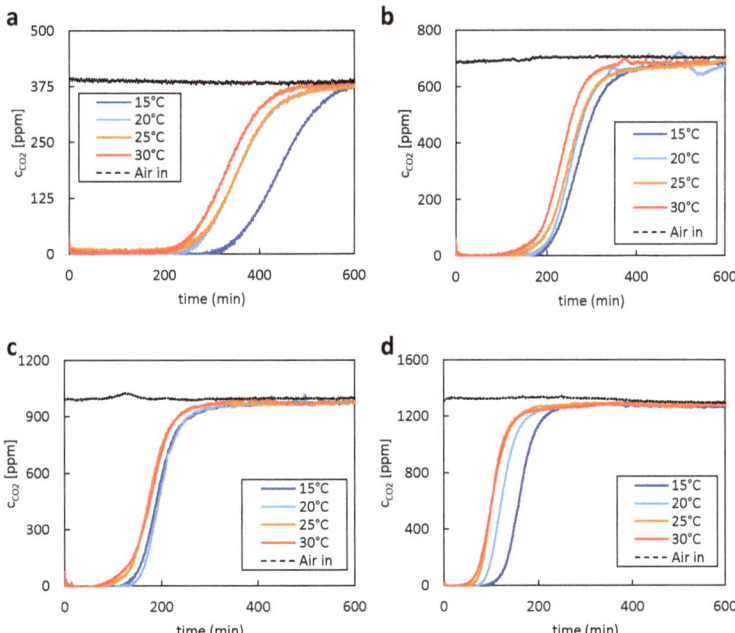

Figure 3. Experimental adsorption CO_2 breakthrough curves at a constant relative humidity of 35%, constant volume flow of 6 Nm^3/h, at 15 °C (black line), 20 °C (blue line), 25 °C (orange line), and 30 °C (red line) and at (**a**) 400 ppm, (**b**) 700 ppm, (**c**) 1000 ppm, (**d**) 1300 ppm of CO_2 respectively. The dotted lines describe the air stream measured before entering the adsorbent bed.

For the experiments with higher CO_2 concentrations than 400 ppm, CO_2 was added to the supply air stream using a mass flow controller. Therefore, it was possible to precisely adjust the CO_2 concentration and compensate for natural fluctuations of the CO_2 content of the ambient air. The same behaviour regarding equilibrium loading and breakthrough curves could be observed. With increasing temperature, the equilibrium loading and total adsorption time decreased. In case of the experiments with 700 ppm of CO_2, shown in Figure 3b, the overall desorption time was reduced to approximately 6.5 h. A similar reduction in desorption time could also be determined for the experiments with 1000 and 1300 ppm of CO_2. During these experiments, shown in Figure 3c,d, desorption times of 5 h and 3.5 h were achieved, respectively.

No increase in the fixed-bed temperature was observed over the entire duration of the tests. Although the adsorption of CO_2 is an exothermic reaction, the heat generated during the reaction had no relevant effect. This is due to the relatively large volume of air that flows around the adsorbent and immediately dissipates any heat. Overall, the data clearly indicated the significant impact of partial pressure on the overall adsorption time.

Veneman et al. [16] reported that the presence of CO_2 does not significantly affect the sorption capacity of Lewatit VP OC 1065 for H_2O. However, the adsorption capacity for CO_2 can dramatically increase when water is present. This phenomenon has been observed before for other amine-based adsorbents and is attributed to the interference of H_2O in the adsorption mechanism [34–36]. The presence of water influences the reaction stoichiometry during CO_2 adsorption because water can act as a free base. This facilitates the formation of bicarbonate ions, enabling one amine group to potentially react with one CO_2 molecule. On the other hand, under dry conditions, the absence of water leads to carbamate formation. Hence, the reaction requires two amine molecules to bind one molecule of CO_2. This discrepancy in stoichiometry highlights the significance of water in the adsorption process [27].

Figure 4 shows the impact of the variation of relative humidity in the supply air stream. Looking at the breakthrough curves with the corresponding CO_2 balances confirms the above-stated observations. Higher relative humidity in the supply air stream causes an increase in adsorption time; therefore, an improved CO_2 uptake can be confirmed. It is important to mention that some materials may experience a decrease in adsorption capacity due to pore blockage or competition for adsorption sites between water and CO_2.

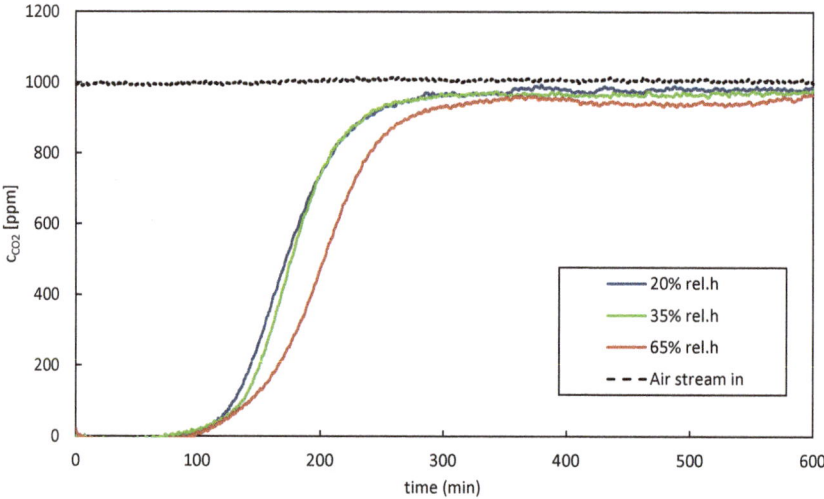

Figure 4. Experimental adsorption CO_2 breakthrough curves at a constant temperature of 20 °C, steady volume flow of 6 Nm^3/h, constant partial pressure of CO_2 at 1000 ppm, and 20% (blue line), 35% (green line), and 65% (red line) relative humidity. The dotted line represents the air stream measured before entering the adsorbent bed.

3.2. H_2O and CO_2 Adsorption Isotherms

Most research today uses TGA to obtain pure component CO_2 and H_2O isotherms, as it serves as a valuable tool for investigating adsorption behaviour [20,31,37–39]. However, the unique setup used in this study allows for precise measurements of pure component isotherms without the use of TGA. The mathematical model used for curve fitting was based on the GAB model Equations (9)–(11), where K_{ads}, c_g, and n_m represent the parameters of interest. This study utilized a least squares curve fitting approach in MATLAB, called the "lsqucurvefit" function. By fitting the model equation to the experimental data points using the least squares method, the MATLAB code iteratively adjusted the parameter values to minimize the sum of squared differences between the predicted and actual measured equilibrium loadings. Evaluating the accuracy of a model entails comparing its predictions or fitted values with experimental data. We utilized the coefficient of determination (R^2) as a metric to quantify the accuracy of our fit. The resulting parameter values provide valuable insights into the water sorption behaviour of Lewatit under different relative humidity conditions. The H_2O adsorption isotherm obtained coincides with the literature data for the same material [16,18].

The adsorption isotherms for pure CO_2 were obtained and fitted using the same methodology as employed for the H_2O isotherms. The isotherms were compared to previous adsorption experiments at TU Wien by Sonnleitner et al. [17] using a fluidizing bed reactor as well as TGA. The temperature-dependent Toth isotherm model described in Section 2.3.2 was used for these isotherms. In the presence of extremely low partial pressures of CO_2, the fitted data demonstrate equilibrium loadings comparable to the findings reported by Veneman et al. [16]. Moreover, the equilibrium loadings are slightly elevated compared to the results reported by Sonnleitner et al. [17] for Lewatit. Given that both models are optimized for higher equilibrium loadings, discrepancies between the experimental data and the models become evident at elevated partial pressures of CO_2.

The slight variations in the CO_2 adsorption capacity results obtained could be attributed to several factors. The particle size of the adsorbent material may not have been uniform across the samples used in the tests. Variations in particle size can affect the available surface area for adsorption and thus impact the results. Also, different batches of the adsorbent material obtained from production might exhibit slight variations in their chemical composition or physical properties, leading to differences in CO_2 adsorption capacities. Furthermore, variations in the amine loading, which refers to the amount of amine functional groups attached to the adsorbent material, can affect its adsorption capacity for CO_2. Other possible points of difference include testing methodology, sample preparation, measurement techniques, and material aging. These factors highlight the importance of controlling and standardizing these parameters to ensure consistent and comparable results in CO_2 adsorption studies.

The temperature-dependent Toth model parameters, as well as the GAB model parameters, are given in Table 5. The values for the heat of adsorption ΔH_0 are within reasonable boundaries for amine-functionalized solid adsorbents in low partial pressure ranges, ensuring credibility and justifying the adoption of these data in the models used [9,27,40]. Figure 5 illustrates the adsorption isotherm of water on Lewatit, while Figure 6 displays the CO_2 adsorption isotherms on the same material.

Table 5. Toth isotherm parameters.

Toth Parameters	This Work	Sonnleitner et al. [17]	Veneman et al. [16]	Unit
n_{s0}	1.81	3.13	3.4	mol kg^{-1}
χ	0	0	0	-
T_0	343	343	353	K
b_0	169.62	282	408.84	bar^{-1}
ΔH_0	64.02	106	86.7	kJ mol^{-1}
t_0	0.96	0.34	0.3	-
α	0.34	0.42	0.14	-
R^2	0.99	-	-	-
GAB parameters	Value	Unit		
c_G	3.68	-		
K_{ads}	0.73	-		
n_m	4.99	mol kg^{-1}		
R^2	0.99	-		

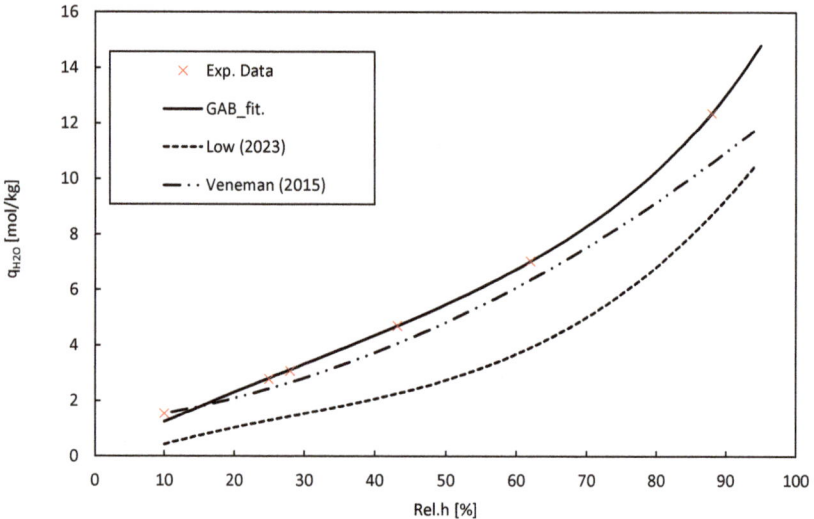

Figure 5. H$_2$O adsorption isotherms at 20 °C for Lewatit obtained from lab unit measurements and fitted with GAB model and compared to data from Veneman et al. [16] and Low et al. [18].

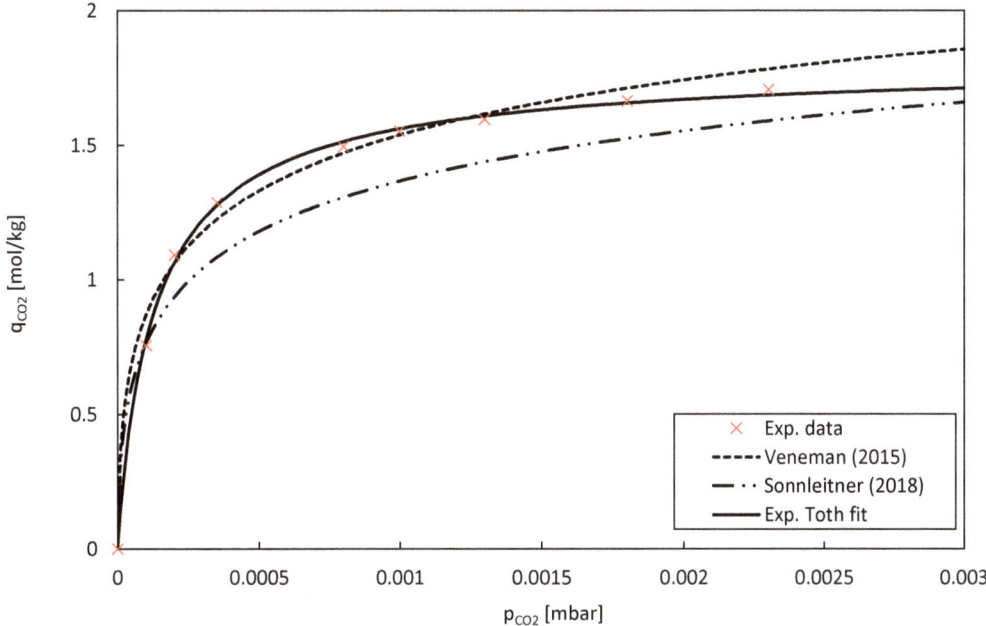

Figure 6. CO_2 adsorption isotherms on Lewatit obtained from lab unit measurements at 20 °C and fitted with Toth model. In comparison, TGA measurements from Sonnleitner et al. [17] and from Veneman et al. [16] fitted with the Toth model are displayed as well.

3.3. Co-Adsorption CO_2 Isotherms

Figure 7 presents the co-adsorption CO_2 isotherms in low partial pressure regions at a relative humidity of 35–38% and different temperatures. Since the pure component temperature-dependent Toth model lacks the effect of H_2O, the models based on Equations (12)–(22) were used. As reported by Young et al. [27], the co-adsorption enhancement effect is particularly noticeable at lower CO_2 concentrations, leading to increased adsorption capacity compared to dry conditions. This finding is crucial for DAC processes, which typically operate at partial CO_2 pressures around 0.4 mbar. As the partial pressure of CO_2 increases, the enhancement effect tends to become negligible. Nonetheless, this behaviour is advantageous for DAC, as desorption typically begins at significantly higher partial pressures than adsorption. If the enhancement effect of water remained high during desorption, it would hinder the desorption process by competing with CO_2 for adsorption sites and potentially impeding the release of CO_2 molecules. The parameters of the three models used in this study are shown in Table 6.

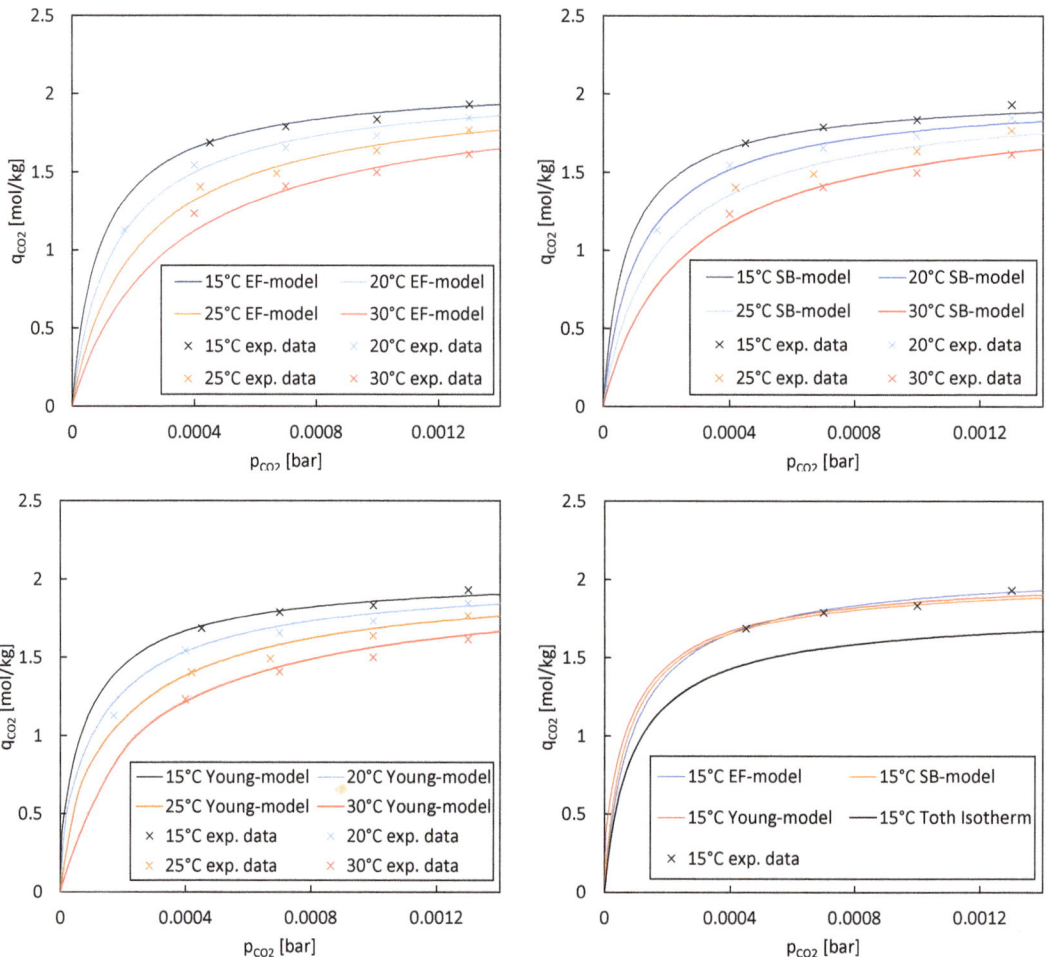

Figure 7. Experimental co-adsorption CO$_2$ uptake (markers) at temperatures ranging from 15 °C to 30 °C and a relative humidity of 35–38%, respectively. The CO$_2$ co-adsorption isotherms are fitted with experimental data from this work and expanded with the co-adsorption models from Young et al. [27], Stampi-Bombelli et al. [28], and the EF model.

Table 6. Co-adsorption parameters.

EF Model Parameters	Value	Unit
k	0.42	-
SB model parameters		
γ	0.027	-
β	0.061	-
Young model parameters		
A	14.72	mol kg^{-1}
b_{wet}	0.37	bar^{-1}
$n_{s,wet}$	12.64	mol kg^{-1}
τ_{wet}	8.56	-
χ_{wet}	0	-
α_{wet}	6.62×10^{-6}	-
ΔH_{wet}	203,687	J mol^{-1}

4. Discussion

In this study, we evaluated three different models, namely, the EF model, SB model, and Young model, to describe the adsorption isotherms of a particular system across various temperatures. We selected these three distinct co-adsorption models from the literature because each offers a fundamentally different approach. By choosing models with diverse theoretical foundations, we aimed to facilitate a comprehensive comparison and evaluation of their respective strengths and limitations. Our analysis focused on the coefficient of determination (R^2) as a measure of how well each model fitted the experimental data (see Table 7). Our findings revealed notable differences in the performance of the models across different temperatures. The EF model exhibited higher R^2 values at lower temperatures, such as 15 °C, indicating a better fit to the experimental data in these conditions. Conversely, the SB model and Young model demonstrated more consistent performance across a range of temperatures.

Table 7. Fit of the models for different temperatures and their overall fit.

Model	EF Model	SB Model	Young Model
[°C]	R^2	R^2	R^2
15	0.987	0.978	0.975
20	0.977	0.983	0.982
25	0.909	0.928	0.930
30	0.930	0.973	0.981
Overall fit	**0.951**	**0.965**	**0.967**

When considering the overall fit across all temperatures, the SB model and Young model emerged as strong contenders, with comparable R^2 values suggesting robust performance across varying temperature conditions. These models provided a good representation of the adsorption isotherms, capturing a significant proportion of the variance in the experimental data. These results underscore the importance of evaluating model performance across different temperature regimes. The temperature sensitivity observed in the EF model highlights the necessity of considering temperature effects when selecting an appropriate model for describing adsorption processes. The experimental data exhibited variations and potential sources of inaccuracies. Maintaining constant parameters like temperature, relative humidity, and CO_2 partial pressure while investigating the impact of a single parameter proved challenging. It has been found that experimental conditions have a more prolonged effect on adsorption results than just one adsorption and regeneration cycle. To ensure accuracy, multiple measurements were performed under controlled conditions, and the results were validated through comparison with known standards or the literature data. Overall, while experimental measurements may have inherent variability, efforts were made to minimize uncertainties and ensure the reliability of the equilibrium loading and CO_2 uptake capacity values reported in the study. However, additional investigations are required to assess the impact of relative humidity on the isotherms.

While the R^2 values offer valuable insights into model performance, it is essential to interpret them in the context of other factors. Factors such as the physical basis of the models, simplicity versus complexity, and their ability to generalize to new conditions should also be considered in model selection.

In conclusion, our study provides valuable insights into the suitability of different models for describing adsorption isotherms across varying temperature conditions. The findings contribute to a better understanding of adsorption processes and aid in the selection of appropriate models for predictive purposes in practical applications.

When evaluating the suitability of the laboratory unit compared to TGA, it is important to consider different aspects. The laboratory unit offers several advantages. First, it allows greater flexibility in adjusting parameters such as pressure, temperature, and gas composition. In comparison, TGA usually offers more limited adjustment options. Real-time analysis of dynamic processes cannot be achieved using standard TGA setups, which can be a limitation when studying fast reactions or phenomena. In addition, the laboratory unit can often simulate more realistic conditions specific to the system or process under study. This allows for more accurate modelling of actual adsorption conditions. Another vital advantage of the laboratory plant is that it can accommodate co-adsorption isotherms. In contrast, TGA usually focuses on the adsorption of a single gas. The laboratory system can provide a more realistic representation of complex adsorption processes in which multiple gas components are adsorbed simultaneously by recording co-adsorption isotherms. This is particularly relevant for applications such as gas purification or DAC.

The ability to record co-adsorption isotherms allows more accurate characterization of the interactions between the different gas components and the adsorbent. This allows, for example, a better understanding of synergistic effects or competitive phenomena during adsorption and desorption. This understanding is essential to improve the efficiency of adsorption processes and to determine optimal conditions for the adsorption of specific gas components. Additional conditions and measurements may vary depending on the study's specific objectives. For example, variations in temperature and pressure can be used to study the effects on adsorption capacity and other adsorption properties. Isotherm measurements at different partial pressures of the adsorbate molecule can be performed to determine adsorption isotherms and better understand adsorption behaviour. Desorption studies allow the investigation of desorption behaviour and provide information on the stability and reusability of the adsorbent.

5. Conclusions

This study aimed to develop a laboratory-scale DAC unit for evaluating and comparing amine-based adsorbents under TSA conditions. The unit was designed to provide information about the optimum operating conditions for different adsorbents and assess their effectiveness and efficiency in capturing CO_2 from air.

The experimental campaign conducted with the lab unit allowed for the determination of equilibrium loadings, CO_2 uptake capacities, and other performance parameters of the adsorbents. The Toth isotherm model was used to characterize the pure component CO_2 adsorption isotherms, while the Guggenheim–Anderson–de Boer model was applied to study water co-adsorption onto the adsorbents. Co-adsorption isotherms were also examined, considering the enhanced CO_2 uptake in the presence of water. Three approaches, namely, the EF model, SB model, and Young model, were compared to describe the co-adsorption phenomenon. The results and discussions provided insights into the adsorption performance of the amine-based adsorbents under various test conditions, including temperature, CO_2 concentration, and relative humidity. The experiments demonstrated the effect of these parameters on adsorption behaviour and maximum CO_2 uptake. In addition, the experimental data validated the SB and Young model to be used in simulation studies. Comparing the results of the lab unit with TGA analysis shows several advantages. The lab unit allows greater flexibility in adjusting performance parameters such as pressure, temperature, and gas composition which allows for more accurate modelling of real-life adsorption conditions. The biggest advantage of this lab setup is that it can accommodate co-adsorption isotherms.

The behaviour of adsorption capacity concerning relative humidity largely depends on the specific properties of the adsorbent material and the gases involved in the process. Overall, relative humidity can impact DAC processes' performance, energy requirements, and water management aspects. Therefore, it is an important factor to be taken into account when designing and optimizing DAC systems.

Overall, this study contributes to developing more efficient and cost-effective amine-based adsorbents for DAC applications. Providing a comprehensive evaluation and comparison of adsorbents under TSA-DAC conditions paves the way for scaling up the DAC industry globally. Further research can build upon this work by exploring additional adsorbent materials, optimizing operating conditions, and investigating the scalability of the developed lab unit. The continued advancement of DAC technology and the identification of effective adsorbents are crucial steps in mitigating climate change and reducing CO_2 levels in the atmosphere.

Author Contributions: Conceptualization, F.M.C. and A.A.B.; Methodology, F.M.C.; Validation, G.S.; Resources, S.M.; Data curation, F.M.C. and A.W.; Writing—original draft, F.M.C. and A.A.B.; Writing—review & editing, J.F.; Supervision, S.M. and J.F.; Project administration, J.F. All authors have read and agreed to the published version of the manuscript.

Funding: Open Access Funding by TU Wien. This study was conducted as part of the DAC project supported and funded by the Dharma Karma Foundation.

Data Availability Statement: Data available on request due to restrictions.

Acknowledgments: The authors acknowledge TU Wien Bibliothek for financial support through its Open Access Funding Programme.

Conflicts of Interest: Author Gerhard Schöny was employed by the company DACworx. Author Aditya Anil Bhandari was employed by the company Youweb Incubator. The remaining authors declare that the research was conducted in the absence of any commercial or financial relationships that could be construed as a potential conflict of interest.

Abbreviations

BECCS	Bioenergy with carbon capture and storage
BET	Brunauer–Emmet–Teller
CO_2	Carbon dioxide
CDR	Carbon dioxide removal
DAC	Direct air capture
EF	Enhancement factor
GAB	Guggenheim–Anderson–de Boer
H_2O	Water
MFC	Mass flow controller
MSA	Moisture swing adsorption
NDCs	Nationally Determined Contributions
N_2	Nitrogen
ppm	Parts per million
PID	Proportional–integral–derivative controller
PVSA	Pressure vacuum swing adsorption
Rel.H	Relative humidity
SB	Stampi-Bombelli
TSA	Temperature swing adsorption
TVSA	Temperature vacuum swing adsorption
TGA	Thermogravimetric analysis
WADST	Weighted average dual-site Toth model

List of Symbols

Symbol	Description	Units
X	Equilibrium loading	(-)
m_{in}/m_{out}	Mass in/out	(kg)
m_{ads}	Mass adsorbed	(kg)
$m_{adsorbent}$	Mass of adsorbent	(kg)
c	Concentration	(mol m^{-3})
$\dot{V}$	Volumetric flow	(m^3 h^{-1})
p	Pressure	(Pa)
R	Ideal gas constant	(J mol^{-1} K^{-1})
T	Temperature	(K)
T_0	Reference temperature	(K)
M	Molar mass	(kg mol^{-1})
t	Time	(s)
t_0	Time at reference point	(s)
q_{CO_2}	Loading of CO$_2$ on the adsorbent	(mol kg^{-1})
n_s	Max. adsorption capacity	(mol kg^{-1})
n_{s0}	Max. adsorption capacity at reference temperature	(mol kg^{-1})
b	Adsorption affinity	(Pa^{-1})
b_0	Adsorption affinity at reference temperature	(Pa^{-1})
p_{CO_2}	Partial pressure of CO$_2$	(Pa)
τ	exponential factor describing the heterogeneity of the adsorbent	(-)
α	Factor describing temperature dependency	(-)
χ	Factor describing temperature dependency	(-)
ΔH_0	Isosteric heat of adsorption at zero fractional loading	(J mol^{-1})
q_{H_2O}	Loading of water on the adsorbent	(mol kg^{-1})
n_m	Monolayer adsorption capacity	(mol kg^{-1})
c_g	Affinity parameter	(-)
c_0	Affinity parameter at reference temperature	(-)
K_{ads}	Affinity parameter	(-)
K_0	Affinity parameter at reference temperature	(-)
φ	Relative humidity	(-)
$\Delta H_C / \Delta H_k$	Adsorption enthalpies of mono and multilayer adsorption	(J mol^{-1})
β_{EF}	Enhancement factor	(-)
k	Constant describing enhancement factor	(-)
β	Modified Toth parameter	(-)
y	Modified Toth parameter	(-)
A	Critical water loading parameter	(-)
E_1, E_{2-9}, E_{10+}	Heat of adsorption for the 1st, 2nd to 9th and 10th layer and beyond	(J mol^{-1})
C, F	Constants in WADST model	(J mol^{-1})
D	Constant in WADST model	(K^{-1})
G	Constant in WADST model	(J mol^{-1} K^{-1})

References

1. Hoegh-Guldberg, O.; Jacob, D.; Taylor, M.; Bolaños, T.G.; Bindi, M.; Brown, S.; Camilloni, I.A.; Diedhiou, A.; Djalante, R.; Ebi, K.; et al. The human imperative of stabilizing global climate change at 1.5 °C. *Science* **2019**, *365*, aaw6974. [CrossRef] [PubMed]
2. Pauw, W.P.; Klein, R.J.T.; Mbeva, K.; Dzebo, A.; Cassanmagnago, D.; Rudloff, A. Beyond headline mitigation numbers: We need more transparent and comparable NDCs to achieve the Paris Agreement on climate change. *Clim. Change* **2017**, *147*, 23–29. [CrossRef]
3. Emissions Gap Report 2022: The Closing Window Climate Crisis Calls for Rapid Transformation of Societies. Nairobi. 2022. Available online: https://www.unep.org/emissions-gap-report-2022 (accessed on 5 June 2023).
4. Keller, D.P.; Lenton, A.; Littleton, E.W.; Oschlies, A.; Scott, V.; Vaughan, N.E. The Effects of Carbon Dioxide Removal on the Carbon Cycle. *Curr. Clim. Change Rep.* **2018**, *4*, 250–265. [CrossRef] [PubMed]
5. National Academies of Sciences, Engineering, and Medicine. *Negative Emissions Technologies and Reliable Sequestration: A Research Agenda*; The National Academic Press: Washington, DC, USA, 2019. [CrossRef]
6. Ozkan, M.; Nayak, S.P.; Ruiz, A.D.; Jiang, W. Current status and pillars of direct air capture technologies. *iScience* **2022**, *25*, 103990. [CrossRef] [PubMed]

7. Breyer, C.; Fasihi, M.; Bajamundi, C.; Creutzig, F. Direct Air Capture of CO_2: A Key Technology for Ambitious Climate Change Mitigation. *Joule* **2019**, *3*, 2053–2057. [CrossRef]
8. Ozkan, M. Direct air capture of CO_2: A response to meet the global climate targets. *MRS Energy Sustain.* **2021**, *8*, 51–56. [CrossRef] [PubMed]
9. Elfving, J.; Bajamundi, C.; Kauppinen, J.; Sainio, T. Modelling of equilibrium working capacity of PSA, TSA and TVSA processes for CO_2 adsorption under direct air capture conditions. *J. CO_2 Util.* **2017**, *22*, 270–277. [CrossRef]
10. Erans, M.; Sanz-Pérez, E.S.; Hanak, D.P.; Clulow, Z.; Reiner, D.M.; Mutch, G.A. Direct air capture: Process technology, techno-economic and socio-political challenges. *Energy Environ. Sci.* **2022**, *15*, 1360–1405. [CrossRef]
11. Wang, T.; Hou, C.; Ge, K.; Lackner, K.S.; Shi, X.; Liu, J.; Fang, M.; Luo, Z. Spontaneous Cooling Absorption of CO_2 by a Polymeric Ionic Liquid for Direct Air Capture. *J. Phys. Chem. Lett.* **2017**, *8*, 52. [CrossRef]
12. McQueen, N.; Gomes, K.V.; McCormick, C.; Blumanthal, K.; Pisciotta, M.; Wilcox, J. A review of direct air capture (DAC): Scaling up commercial technologies and innovating for the future. *Prog. Energy* **2021**, *3*, 032001. [CrossRef]
13. Gelles, T.; Lawson, S.; Rownaghi, A.A.; Rezaei, F. Recent advances in development of amine functionalized adsorbents for CO_2 capture. *Adsorption* **2020**, *1*, 5–50. [CrossRef]
14. Farmahini, A.H.; Krishnamurthy, S.; Friedrich, D.; Brandani, S.; Sarkisov, L. Performance-Based Screening of Porous Materials for Carbon Capture. *Chem. Rev.* **2021**, *121*, 10666–10741. [CrossRef] [PubMed]
15. Drechsler, C.; Agar, D.W. Investigation of water co-adsorption on the energy balance of solid sorbent based direct air capture processes. *Energy* **2020**, *192*, 116587. [CrossRef]
16. Veneman, R.; Frigka, N.; Zhao, W.; Li, Z.; Kersten, S.; Brilman, W. Adsorption of H_2O and CO_2 on supported amine sorbents. *Int. J. Greenh. Gas. Control* **2015**, *41*, 268–275. [CrossRef]
17. Sonnleitner, E.; Schöny, G.; Hofbauer, H. Assessment of zeolite 13X and Lewatit® VP OC 1065 for application in a continuous temperature swing adsorption process for biogas upgrading. *Biomass Convers. Biorefinery* **2018**, *8*, 379–395. [CrossRef]
18. Low, M.-Y.A.; Danaci, D.; Azzan, H.; Woodward, R.T.; Petit, C. Measurement of Physicochemical Properties and CO_2, N_2, Ar, O_2, and H_2O Unary Adsorption Isotherms of Purolite A110 and Lewatit VP OC 1065 for Application in Direct Air Capture. *J. Chem. Eng. Data* **2023**, *68*, 3511. [CrossRef] [PubMed]
19. Lanxess. Product Information Lewatit®VP OC 1065. Available online: https://lanxess.com/en/Products-and-Brands/Products/l/LEWATIT--VP-OC-1065 (accessed on 16 May 2024).
20. Veneman, R.; Zhao, W.; Li, Z.; Cai, N.; Brilman, D.W.F. Adsorption of CO_2 and H_2O on supported amine sorbents. In *Energy Procedia*; Elsevier Ltd.: Amsterdam, The Netherlands, 2014; pp. 2336–2345. [CrossRef]
21. Serna-Guerrero, R.; Belmabkhout, Y.; Sayari, A. Modeling CO_2 adsorption on amine-functionalized mesoporous silica: 1. A semi-empirical equilibrium model. *Chem. Eng. J.* **2010**, *161*, 173–181. [CrossRef]
22. Buijs, W.; De Flart, S. Direct Air Capture of CO_2 with an Amine Resin: A Molecular Modeling Study of the CO_2 Capturing Process. *Ind. Eng. Chem. Res.* **2017**, *56*, 12297–12304. [CrossRef]
23. Hefti, M.; Mazzotti, M. Modeling water vapor adsorption/desorption cycles. *Adsorption* **2014**, *20*, 359–371. [CrossRef]
24. Marx, D.; Joss, L.; Hefti, M.; Mazzotti, M. Temperature Swing Adsorption for Postcombustion CO_2 Capture: Single-and Multicolumn Experiments and Simulations. *Ind. Eng. Chem. Res.* **2016**, *55*, 1401–1412. [CrossRef]
25. Marx, D.; Joss, L.; Hefti, M.; Pini, R.; Mazzotti, M. The Role of Water in Adsorption-based CO_2 Capture Systems. *Energy Procedia* **2013**, *37*, 107–114. [CrossRef]
26. Gebald, C.; Wurzbacher, J.A.; Borgschulte, A.; Zimmermann, T.; Steinfeld, A. Single-Component and Binary CO_2 and H_2O Adsorption of Amine-Functionalized Cellulose. *Environ. Sci. Technol.* **2014**, *48*, 2497–2504. [CrossRef] [PubMed]
27. Young, J.; Ez, E.G.A.-D.; Garcia, S.; Van Der Spek, M. The impact of binary water-CO_2 isotherm models on the optimal performance of sorbent-based direct air capture processes. *Energy Environ. Sci.* **2021**, *14*, 5377. [CrossRef]
28. Stampi-Bombelli, V.; van der Spek, M.; Mazzotti, M. Analysis of direct capture of CO_2 from ambient air via steam-assisted temperature–vacuum swing adsorption. *Adsorption* **2020**, *26*, 1183–1197. [CrossRef]
29. Wurzbacher, J.A.; Gebald, C.; Brunner, S.; Steinfeld, A. Heat and mass transfer of temperature–vacuum swing desorption for CO_2 capture from air. *Chem. Eng. J.* **2016**, *283*, 1329–1338. [CrossRef]
30. Serna-Guerrero, R.; Da'na, E.; Sayari, A. New insights into the interactions of CO_2 with amine-functionalized silica. *Ind. Eng. Chem. Res.* **2008**, *47*, 9406–9412. [CrossRef]
31. Stuckert, N.R.; Yang, R.T. CO_2 Capture from the Atmosphere and Simultaneous Concentration Using Zeolites and Amine-Grafted SBA-15. *Environ. Sci. Technol.* **2011**, *45*, 10257–10264. [CrossRef] [PubMed]
32. Yu, J.; Chuang, S.S.C. The Structure of Adsorbed Species on Immobilized Amines in CO_2 Capture: An in Situ IR Study Scheme 1. Reaction Mechanism of CO_2 with Aqueous Amines Scheme 2. Reaction Mechanism of CO_2 with Immobilized Amines. *Energy Fuels* **2016**, *30*, 58. [CrossRef]
33. Anderson, R.B.; Hall, W.K. Modifications of the Brunauer, Emmett and Teller equation. *J. Am Chem Soc.* **1948**, *70*, 1727–1734. [CrossRef] [PubMed]
34. Franchi, R.S.; Harlick, P.J.E.; Sayari, A. Applications of Pore-Expanded Mesoporous Silica. 2. Development of a High-Capacity, Water-Tolerant Adsorbent for CO_2. *Ind. Eng. Chem. Res.* **2005**, *44*, 8007–8013. [CrossRef]

35. Hicks, J.C.; Drese, J.H.; Fauth, D.J.; Gray, M.L.; Qi, G.; Jones, C.W. Designing Adsorbents for CO_2 Capture from Flue Gas-Hyperbranched Aminosilicas Capable of Capturing CO_2 Reversibly. *J. Am. Chem. Soc.* **2008**, *130*, 2902–2903. [CrossRef] [PubMed]
36. Su, F.; Lu, C.; Kuo, S.-C.; Zeng, W. Adsorption of CO_2 on Amine-Functionalized Y-Type Zeolites. *Energy Fuels* **2010**, *24*, 1441–1448. [CrossRef]
37. Fauth, D.J.; Gray, M.L.; Pennline, H.W.; Krutka, H.M.; Sjostrom, S.; Ault, A.M. Investigation of Porous Silica Supported Mixed-Amine Sorbents for Post-Combustion CO_2 Capture. *Energy Fuels* **2012**, *26*, 2483–2496. [CrossRef]
38. Xu, X.; Zhao, X.; Sun, L.; Liu, X. Adsorption separation of carbon dioxide, methane and nitrogen on monoethanol amine modified β-zeolite. *J. Nat. Gas Chem.* **2009**, *18*, 167–172. [CrossRef]
39. Lourenço, M.A.O.; Fontana, M.; Jagdale, P.; Pirri, C.F.; Bocchini, S. Improved CO_2 adsorption properties through amine functionalization of multi-walled carbon nanotubes. *Chem. Eng. J.* **2021**, *414*, 128763. [CrossRef]
40. Elfving, J.; Sainio, T. Kinetic approach to modelling CO_2 adsorption from humid air using amine-functionalized resin: Equilibrium isotherms and column dynamics. *Chem. Eng. Sci.* **2021**, *246*, 116885. [CrossRef]

Disclaimer/Publisher's Note: The statements, opinions and data contained in all publications are solely those of the individual author(s) and contributor(s) and not of MDPI and/or the editor(s). MDPI and/or the editor(s) disclaim responsibility for any injury to people or property resulting from any ideas, methods, instructions or products referred to in the content.

Article

Investigation on YSZ- and SiO$_2$-Doped Mn-Fe Oxide Granules Based on Drop Technique for Thermochemical Energy Storage

Yan Ma, Kai Wang, Sikai Liang, Zhongqing Li, Zhiyuan Wang * and Jun Shen

School of Energy and Power Engineering, University of Shanghai for Science and Technology, Shanghai 200093, China; 18395588645@163.com (Y.M.); wang213410@163.com (K.W.); liang18339426475@163.com (S.L.); llq10162457@163.com (Z.L.); kennyshen@vip.163.com (J.S.)
* Correspondence: wangzhiyuan@usst.edu.cn

Abstract: The Mn-Fe oxide material possesses the advantages of abundant availability, low cost, and non-toxicity as an energy storage material, particularly addressing the limitation of sluggish reoxidation kinetics observed in pure manganese oxide. However, scaling up the thermal energy storage (TCES) system poses challenges to the stability of the reactivities and mechanical strength of materials over long-term cycles, necessitating their resolution. In this study, Mn-Fe granules were fabricated with a diameter of approximately 2 mm using the feasible and scalable drop technique, and the effects of Y$_2$O$_3$-stabilized ZrO$_2$ (YSZ) and SiO$_2$ doping, at various doping ratios ranging from 1–20 wt%, were investigated on both the anti-sintering behavior and mechanical strength. In a thermal gravimetric analyzer, the redox reaction tests showed that both the dopants led to an enhancement in the reoxidation rates when the doping ratios were in an appropriate range, while they also brought about a decrease in the reduction rate and energy storage density. In a packed-bed reactor, the results of five consecutive redox tests showed a similar pattern to that in a thermal gravimetric analyzer. Additionally, the doping led to the stable reduction/oxidation reaction rates during the cyclic tests. In the subsequent 120 cyclic tests, the Si-doped granules exhibited volume expansion with a decreased crushing strength, whereas the YSZ-doped granules experienced drastic shrinkage with an increase in the crushing strength. The 1 wt% Si and 2 wt% Si presented the best synthetic performance, which resulted from the milder sintering effects during the long-term cyclic tests.

Keywords: thermochemical energy storage; Mn-Fe oxide granules; doping; packed-bed reactor; sintering inhibition

1. Introduction

Thermal energy storage (TES) is one of the promising solutions for addressing the intermittency and fluctuation of renewable energy sources, such as solar energy, wind energy, tidal energy, et al. [1], which can be classified into three categories based on the mechanisms, i.e., sensible heat storage (SHS) [2], latent heat storage (LHS) [3], and thermochemical heat/energy storage (TCES) [4]. Thermochemical energy storage (TCES) systems based on metal oxides have emerged as prominent solutions due to their advantages of a high energy storage density and the utilization of ambient air both as the gaseous reagent and heat transfer fluid (HTF), compared with other conventional energy storage systems [5]. Among those metal oxide candidates, for example, Co$_3$O$_4$/CoO [6], CuO/Cu$_2$O [7], BaO$_2$/BaO [8], and Mn$_2$O$_3$/Mn$_3$O$_4$ [9], manganese oxides possess the advantages of cost-effectiveness, abundance, non-toxicity, and environmental benignity. In particular, the Mn$_2$O$_3$/Mn$_3$O$_4$ system is a kind of relatively more researched TCES system when compared with the Co$_3$O$_4$/CoO system with the highest energy storage density, which is always controversial due to its potential carcinogenic properties [10]. The primary limitations of Mn$_2$O$_3$/Mn$_3$O$_4$ systems are the sluggish reoxidation rate and gradual degradation resulting from sintering during multiple cycles, as well as the thermal hysteresis characteristics of the redox reactions, the onset temperature gap of which is about approximately 200 °C. All this leads to a decline in the cyclability and exergy efficiency [11–13]. The chemical modification strategy, especially doping Fe cations, can effectively enhance the re-oxidation kinetics and lower the hysteresis when compared with the morphology modification [14,15]. The doping of the Fe cation has been extensively demonstrated in numerous studies as a highly effective approach to expedite the rate of the oxidation reaction and enhance the cycle stability [11,13,16–20]. Examples of Mn/Fe oxides with varying molar ratios, arranged in descending order of the energy storage density, include 3:1 (271 J/g) [19], 2:1 (233 J/g) [11], 4:1 (219 J/g) [17], and 1:2 (~200 J/g) [21].

The TCES system, which is based on solid–gas reactions, has utilized various reactor types including fluidized-bed reactors, moving-bed reactors, and fixed-bed reactors [22–24]. The fixed-bed reactors stand out among these reactor configurations due to the cost-effectiveness, the scalability, and the ability of effectively mitigating particle attrition and erosion when compared to the fluidized-bed reactors. However, the main drawbacks of a fixed-bed reactor involve the pressure drop caused by large quantities of finely powdered material and the sintering effect that occurs during a prolonged exposure to high temperatures [15]. These factors would result in gas channeling and a decrease in the reaction rate in the practical operation. Additionally, the inherent poor heat transfer within the fixed-bed can be addressed by employing air as both a reactant and a heat transfer fluid (HTF). Therefore, the selection of the appropriate thermal storage materials in the fixed-bed reactor is crucial and all factors such as metal oxide compositions, the shape, and the size should be synthetically considered. Lots of investigations have been conducted to enhance the energy storage density of oxide materials by pelletization or granulation [20,21], extruded modules [25,26], and coated structures [27,28]. In the case of a fixed-bed reactor, utilizing mm-sized spherical particles offers advantages such as a reduced void fraction and enhanced homogeneity.

During operation at a high temperature, the micro- and macro-structure of the material may be deteriorated with the cycled reductions and oxidations due to the multitude of stresses (chemical, mechanical, and thermal) experienced in a reactor. Wokon et al. demonstrated that the morphology of Mn-Fe oxide particles, prepared by a build-up granulation technique without any support or binder, underwent significant changes due to severe sintering issues in the 100 high-temperature cyclic tests. Consequently, this resulted in a decrease in the rate of the oxidation reaction and cycle stability. Additionally, a substantial increase in the particle volume was observed after the redox cycles, resulting in the decreased density and heightened fragility of the particles [19]. Xiang et al. observed a 15% decrease in the energy storage capacity of a honeycomb module weighing 110 g, composed of Mn/Fe oxides with a molar ratio of 4:1, after 100 redox cycles [26].

The resolution of these issues has been pursued through various endeavors, including the optimization of particle preparation methods and the addition of support materials as the sintering inhibitor. The addition of 10 wt% alumina was found to significantly alleviate the expansion of the Co_3O_4 honeycomb upon the redox cyclic process, thereby maintaining the superior thermo-mechanical stability of the macro-structural compared to that of the pure Co_3O_4 honeycomb [25]. Preisner et al. studied the impact of incorporating 20 wt% of various supporting materials, namely ZrO_2, CeO_2, and TiO_2, on the mechanical strength of manganese–iron oxide particles prepared by the build-up granulation technique in a moving-bed reactor [29]. Compared to the undoped particles, which exhibited a high tendency for agglomeration and fracture, the incorporation of ZrO_2 and CeO_2 can effectively alleviate this phenomenon and enhance the attrition strength, while TiO_2 doping is not suitable as the formation of a stable phase occurs. Gan et al. investigated the impact of the SiC doping of $(Mn_{0.8}Fe_{0.2})_2O_3$ particles, prepared by the extrusion–spheronization technique, on the attrition resistance and long-term reaction cycle performance of the oxides [23]. They reported that incorporating 0.5–5 wt% SiC into Mn-Fe particles enhanced the attrition resistance and long-term cyclability of the Mn-Fe particles, although the reaction performance was slightly reduced. Gigantino et al. investigated the anti-sintering effect of yttria-stabilized zirconia (YSZ) on the porous CuO-based granules, which were fabricated by the drop technique and tested in a packed reactor for two separate sets of 30 cycles under isobaric and isothermal conditions [30]. The results demonstrated that the doped granules presented the significant mitigation of agglomeration when the doping ratio exceeded 50 wt%. Bielsa et al. further identified the critical parameters of the granule synthesis process based on the previous drop technique, as modified by Gigantino et al., and investigated the TCES performance of Si-doped manganese oxide particles in a lab-scale packed-bed reactor [31]. The author claimed that increasing the temperature to 1100 °C can enhance the mechanical strength by 30%. In addition, Si-doped manganese oxide particles exhibited complete re-oxidation behavior during 100 cyclic tests, indicating that Si helps mitigate the sintering effects, and the results were consistent with their prior findings [32].

In summary, the spherical manganese–iron oxide particles exhibit significant potential as a promising candidate for high-temperature thermal energy storage (TCES) systems in fixed-bed reactors. This also imposes corresponding requirements on the strength and reactivity of the particles under the high-temperature conditions, which can be enhanced through appropriate preparation and doping strategies. In this work, we fabricated Mn-Fe oxide granules with a molar ratio of 3:1 via the drop technique which is a novel, feasible, and scalable granule synthesis method. The effect of doping different mass ratios of SiO_2 (at 1 wt%, 2 wt%, 5 wt%, and 10 wt%) and YSZ (at 5 wt%, 10 wt%, and 20 wt%) on the characteristics of Mn-Fe granules was investigated in this study. Firstly, we examined the redox performance of two kinds of dopants with varying doping mass ratios using a STA and a lab-scale packed-bed reactor, respectively. Secondly, the doped granules were subjected to 120 redox cyclic tests and were sampled every 30 cycles to assess the TECS performance evolution in a simultaneous thermal analyzer (STA). Corresponding analytical techniques were employed before and after the cycles to identify micro- and macro-structural changes. The results obtained from this experiment on granules modification have, to some extent, contributed to the advancement of upscaling TCES systems, which is also applicable to other metal oxide pairs.

2. Results and Discussion
2.1. Effect of Dopants on Redox Reaction Characteristics

The characterization of the phase compositions and the chemical states of the oxygen of the doped Mn-Fe granules are shown in Figure 1. As depicted in Figure 1a, only two distinct phases, $(Mn, Fe)_2O_3$ (ICDD; PDF-2; #41-1442) and $Zr_{0.92}Y_{0.08}O_{1.96}$ (ICDD; PDF-2; #48-0224), are present in the synthesized granules, indicating that no new phase is formed during the calcination process. Furthermore, the intensity of the YSZ diffraction peak gradually increases with the doping ratio. Although some shifts of the characteristic peak of $(Mn, Fe)_2O_3$ are observed with the YSZ doping, the peak shifts are irregular. For example, the peak gradually shifts to the low diffraction angle region with the increase in the doping ratio from 0 to 10 wt%, whereas it shifts to the high diffraction angle region with a further increase in the ratio to 20 wt%. In the XRD tests, many factors could result in the peak shift, such as the structure strain from the doping of a heterogeneous atom, the inhomogeneity and impurity in the sample, the morphological characteristics of the sample (powder, coating, and film), the conditions of the test instrument, etc. Generally, the peak shift would show a regular trend with heterogeneous atoms doping as the change in the cell volume because of the replacement of the host atom by the doped atom [33–35]. Therefore, the shifts of the characteristic peak in Figure 1a may not only be ascribed to the doping effects. In order to clarify the reason for irregular shifts when YSZ is doped in the XRD pattern, a lot of work is thus needed. As the TCES performance is the main concern in this study, a detailed analysis of the XRD peak shift has been out of the scope of this work. In the next work, a more precise characterization should be conducted to investigate the doping effect of YSZ on the characteristic peak of $(Mn, Fe)_2O_3$. In the case of Si-doped granules, a new complex phase, braunite Mn_7SiO_{12} (ICDD; PDF-2; #41-1367), is observed with increasing the SiO_2 content, which also exhibits a characteristic peak next to Mn_2O_3 in the XRD patterns, as depicted in Figure 1b. The magnification of the region of 32–34°

shows the characteristic peak overlap of (Mn, Fe)$_2$O$_3$ and Mn$_7$SiO$_{12}$. Previous studies have reported that the presence of this braunite phase is undesirable because of its exceptional stability within the investigated temperature range [32,36].

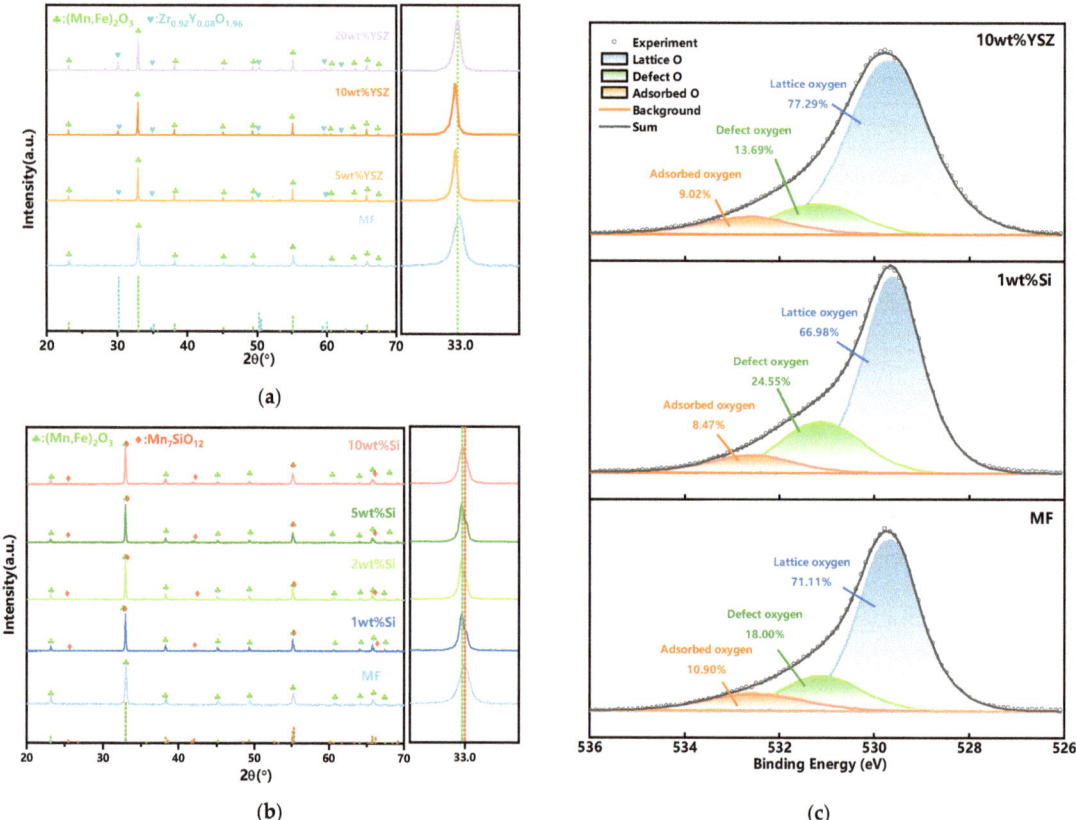

Figure 1. Analyses of the phase composition and chemical state of O of the doped granules. (a) YSZ-doped granules (left) and a magnification of the region of 32–34° (right) (green dotted line: (Mn,Fe)$_2$O$_3$); (b) Si-doped granules (left) and a magnification of the region of 32–34° (right) (green dotted line: (Mn,Fe)$_2$O$_3$, red dotted line: Mn$_7$SiO$_{12}$). (c) O 1s XPS spectra of MF, 1 wt% Si, and 10 wt% YSZ.

The O species in MF, 1 wt% Si, and 10 wt% YSZ is further analyzed using XPS, and the corresponding O 1s spectra are depicted in Figure 1c. Three distinct peaks are observed in the deconvoluted O 1s spectra. The predominant peak at ~530 eV is consistent with lattice oxygen. The peak at ~531 eV can be attributed to defect oxygen or surface oxygen ions, and a peak at ~533 eV is indicative of adsorbed oxygen species such as hydroxyl (OH$^-$) and carbonate (CO$_3^{2-}$). Furthermore, the content of lattice oxygen, defect oxygen, and adsorbed oxygen is calculated based on the area of those oxygen species in the XPS spectra. The lattice oxygen content in MF is 71.11%, while it is increased to 77.29% for 10 wt% YSZ due to the high oxygen ion conductivity of YSZ. The incorporation of YSZ has been demonstrated to enhance the concentration of lattice oxygen and facilitate the conversion of a portion of adsorbed oxygen into surface lattice oxygen, which serves as the rate-determining step in the redox reaction [37]. The concentration of defect oxygen is increased to 24.55% when Si is doping. This means that an approximate amount of Si may lead to the enhanced adsorption of oxygen at the surface as more defect oxygen species would facilitate the adoption of oxygen in the gas phase. Therefore, the reoxidation process could be improved with the more facilitated transport of oxygen ions within the crystal lattice [38,39].

The thermogravimetric (TG) curves of synthesized granules, incorporating various dopants and ratios, were experimentally obtained within a temperature range of 650 °C to 1050 °C using a ramp rate of 10 °C min^{-1}, as illustrated in Figure 2. The results obtained from the TG curves are summarized in Table 1. It is found that the discrepancy in the mass change between the experimental and theoretical results for YSZ-doped samples is negligible, with a maximum deviation of only 0.07%, indicating that YSZ serves an inert function in redox reactions. However, the observed mass changes of Si-doped granules exhibit a significant deviation from their corresponding theoretical values, which gradually diminishes with an increase in the Si content. The influence of doping on the redox reaction rate is depicted in Figure 3. It is found that when the doping ratio of Si is 10 wt%, the reduction conversion is only 41.9% and the reoxidation conversion is 33.98%, as depicted in Figure 3a. The reason for this significant reduction in the reaction conversions can be attributed to the formation of a stable manganese silicon phase known as braunite Mn_7SiO_{12}, which consumes a portion of the active oxide. It is also supported by the XRD analyses in Figure 1b.

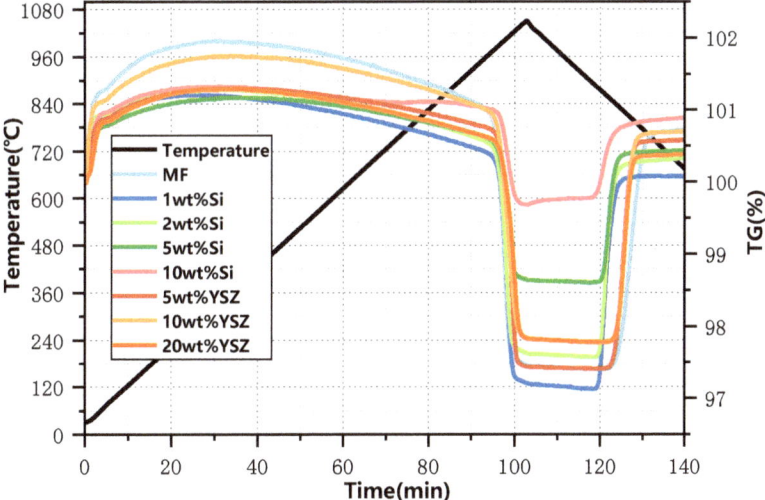

Figure 2. TG profiles of the doped granules at a heating rate of 10 °C min^{-1}.

Table 1. Thermochemical performance of the doped granules based on the TG testing.

Sample	Estimated Mass Change (%)	Experimental Mass Change (%)	T_{red}/T_{ox} (°C)	$\Delta T_{Hysteresis}$ (°C)	$\Delta H_{red}/\Delta H_{ox}$ (J/g)
MF	3.37	3.37	983.3/830.3	153	123.4/−144
1 wt%Si	3.33	3.09	990.9/877.1	113.8	90.12/−91.35
2 wt%Si	3.30	2.80	989.5/875.2	114.3	74.95/−103
5 wt%Si	3.20	1.87	992.3/868.3	124	51.22/−67.11
10 wt%Si	3.03	1.27	991.9/883.4	108.5	37.97/−33.9
5 wt%YSZ	3.19	3.21	997.7/840.3	157.4	85.14/−114.2
10 wt%YSZ	3.03	2.96	988.1/835.6	152.5	77.19/−111.5
20 wt%YSZ	2.69	2.64	1000.6/834.5	166.1	65.39/−101.4

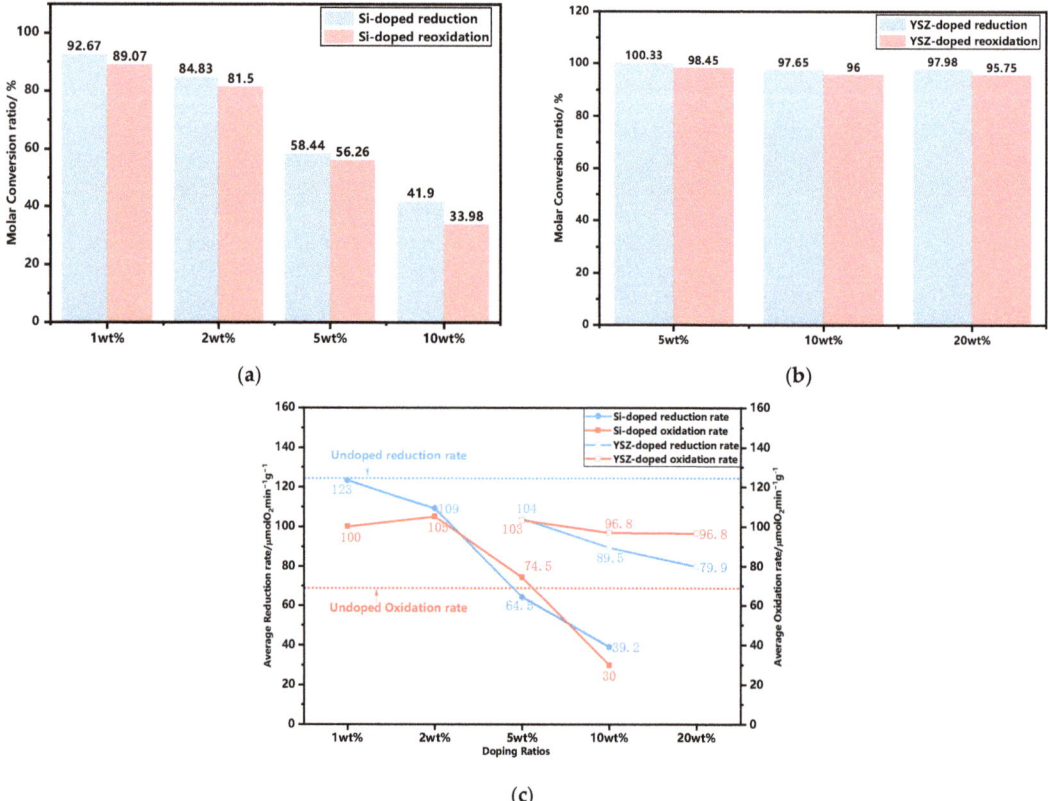

Figure 3. The influence of doping on the redox reaction rate. (**a**) Molar conversion ratio of Si-doped granules; (**b**) molar conversion ratio of YSZ-doped granules; (**c**) the average reduction and oxidation rate at different doping ratios.

As depicted in Figure 3c, the average reduction and oxidation rates of YSZ exhibit a gradual decline as the doping ratio increases from 5 wt% to 20 wt%, resulting in a minimum rate of 79.86 µmol O_2 min^{-1} g^{-1} for reduction, while stabilizing at 96.78 µmol O_2 min^{-1} g^{-1} for oxidation. The overall trend for Si doping exhibits similarities to that of YSZ doping, while within the doping range of 5–10 wt%, Si doping demonstrates a higher susceptibility to adverse effects in terms of both reduction and oxidation rates compared to YSZ doping at the same scale. The average reduction and oxidation rate of 10 wt% Si, in particular, exhibits a minimum value of only 39.21 µmol O_2 min^{-1} g^{-1} and 29.96 µmol O_2 min^{-1} g^{-1}, respectively, while YSZ-doped samples maintain significantly higher rates of 89.54 µmol O_2 min^{-1} g^{-1} and 96.84 µmol O_2 min^{-1} g^{-1}. Doping with a low Si content of 1 wt% results in an increase in the oxidation rate to 100.12 µmol O_2 min^{-1} g^{-1}, while the reduction rate remains high at 123.32 µmol O_2 min^{-1} g^{-1}, which is comparable to that of the undoped sample. When the mass ratio reaches 2 wt%, the reduction rate declines to 109.375 µmol O_2 min^{-1} g^{-1}, while the oxidation rate continues to increase and reaches its maximum level of 105.18 µmol O_2 min^{-1} g^{-1}, which is comparable to that of YSZ at 5 wt%.

The incorporation of dopants also influences the initial temperature of the redox reaction, as demonstrated in Table 1. The onset temperatures of both the reduction and oxidation have an increase with doping YSZ and SiO_2. The onset temperatures for the reduction and oxidation reactions exhibit an average increase of 12 °C and 6.5 °C, respectively, upon YSZ doping, resulting in a slight increase in the thermal hysteresis except for the case of 10 wt% YSZ where it remains constant. The SiO_2 doping also exhibits a similar effect, with an increase in the reduction onset temperature of approximately 7.85 °C compared to pure Mn-Fe granules. Particularly, it is noteworthy that there is a significant elevation in the oxidation onset temperature by about 45.7 °C upon SiO_2 doping, leading to a decrease in hysteresis values and consequently enhancing thermal efficiency. It is also found that the overall reaction enthalpy of the MF group exhibits a lower value than the theoretical value 202 J/g. The reported reaction enthalpy values always vary among the existing research studies, which can be ascribed to the variations in synthesis routes and experimental conditions. Overall, the trend of enthalpy values observed in this study demonstrates a clear decreasing trend with an increasing doping ratio. Therefore, it is crucial to carefully select appropriate doping ratios to maintain an optimal energy storage density.

2.2. Redox Performance in Packed-Bed Reactor

The redox performance of approximately 2 g doped Mn-Fe granules was further assessed in a laboratory-scale packed-bed reactor in the temperature range of 600–1050 °C with a ramp rate of 10 °C min^{-1} and a constant air flow rate of 1 NL·min^{-1}. The profiles of the evolution of the outlet O_2 concentration are continuously monitored throughout the redox reactions, as depicted in Figure 4. Evidently, the oxygen concentration profiles consistently maintain a uniform shape throughout the redox reaction process across all assays, displaying distinct peaks of oxygen release and uptake in each cycle. This observation highlights a remarkable level of repeatability between cycles for the doped Mn-Fe granules, albeit with a slight drift observed during the non-reactive stage.

The trends of oxygen release and uptake in each cycle are illustrated in Figure 5. The thermochemical energy storage performance of the doped granules is also concluded in Table 2. In the case of oxygen release, the pure MF granules undoubtedly exhibit the highest average O_2 release capability, as depicted in Table 2, which is the same to the TG test results (see Figure 3c). The addition of Si dopant fails to contribute to the O_2 release process. The Si-doped granules exhibited lower theoretical release values of 79.86% and 70.66% for 1 wt% Si and 2 wt% Si, respectively. When the doping ratio is increased to 5 wt% and 10 wt%, the average O_2 release values are decreased to 50.57% and 16.14%, respectively. The reason for the decreased amounts of released O_2 is mainly due to the formation of some kind of manganese–silicon compound, and the deceased reaction process is also consistent with the TG test results (see Figure 3c). The YSZ-doped granules, however, even exhibit a higher theoretical released value of 86.5–90% when compared with the pure and Si-doped granules. It is mainly because YSZ doping has brought about the enhanced concentration of lattice oxygen in the granule, which results in the improved O_2 storage capacity [37]. Subsequently, the O_2 release in the reduction reaction process is thus increased. In terms of the average oxygen uptake, all groups with doped granules exhibit a superior performance compared to pure Mn-Fe granules, except for the 5 wt% Si and 10 wt% Si groups, which is also similar to the TG test results. The appropriate levels of doping significantly enhance the oxidation reaction.

Table 2. Thermochemical energy storage performance of the granules in packed-bed reactor.

Sample	Average O_2 Release/Uptake (µmol/g)	Theoretical Release/Uptake (%)	Reactor Average Reduction/Oxidation Duration (min)	TG Reduction/Oxidation Duration (min)
MF	902.67/744.89	85.79/70.79	8.90/7.94	8.45/14.75
1 wt%Si	831.82/838.86	79.86/80.53	8.43/8.22	7.83/9.27
2 wt%Si	728.60/751.16	70.66/72.85	8.59/7.66	8.10/7.99
5 wt%Si	505.46/475.34	50.57/47.55	6.60/9.42	9.06/7.55
10 wt%Si	152.86/137.79	16.14/14.55	9.57/12.11	10.12/10.74
5 wt%YSZ	899.68/845.40	90.01/84.58	10.57/15.79	9.63/9.52
10 wt%YSZ	833.92/793.59	88.06/83.80	8.55/9.26	10.33/9.39
20 wt%YSZ	729.35/702.36	86.65/83.44	6.68/13.25	10.33/8.33

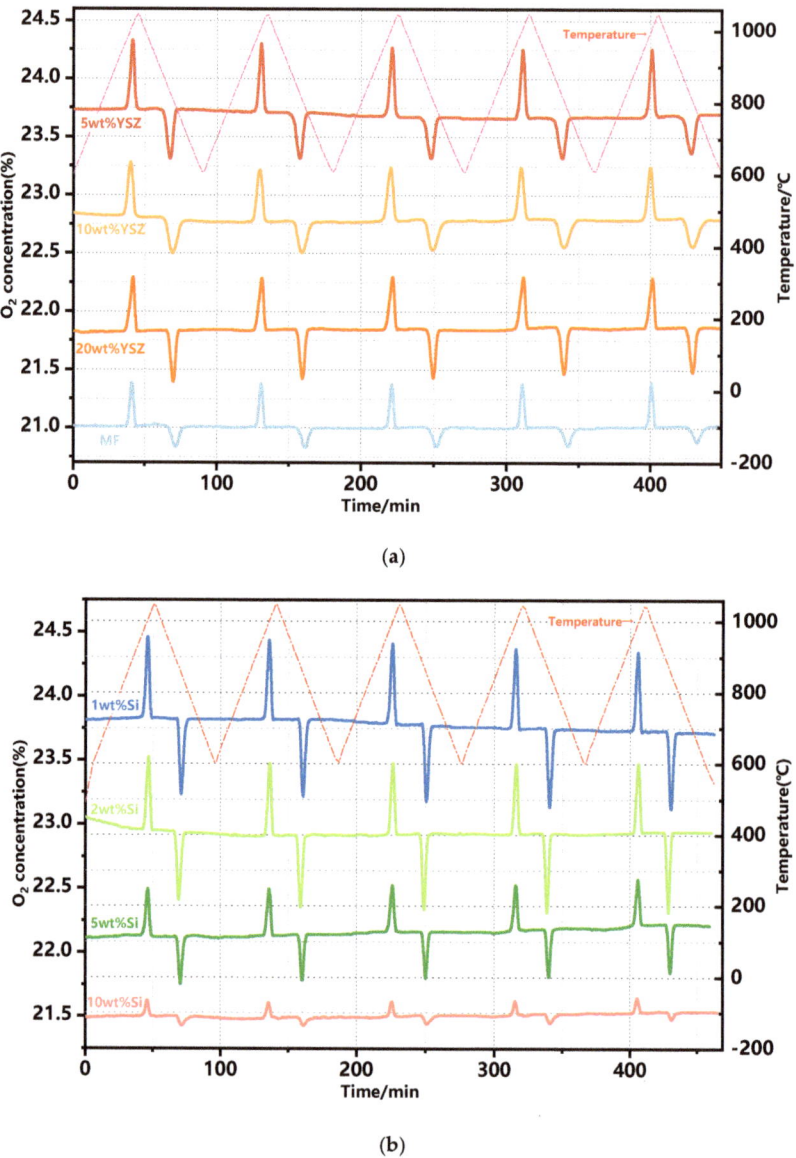

Figure 4. O$_2$ concentration profiles of the doped granules over 5 consecutive redox cycles. (**a**) YSZ−doped and MF granules; (**b**) Si−doped granules.

The average rates of oxygen release and uptake are also illustrated in Figure 6. It is found that the undoped granules exhibit a consistently high average oxygen release rate. Additionally, all doped samples, except for 5 wt% Si and 10 wt% Si, have no significant difference in this rate. When considering the average oxygen uptake rate, all doped granules, with the exception of 10 wt% Si, present a significant enhancement when compared with pure Mn-Fe granules. In particular, 1 wt% Si and 2 wt% Si still possess excellent oxygen uptake rates of 107.1 μmol O$_2$ min^{-1} g^{-1} and 99.1 μmol O$_2$ min^{-1} g^{-1} after five redox cycles, respectively. Meanwhile, the remaining doped groups maintain similar rates ranging from 62 to 70.2 μmol O_2 min^{-1} g^{-1}.

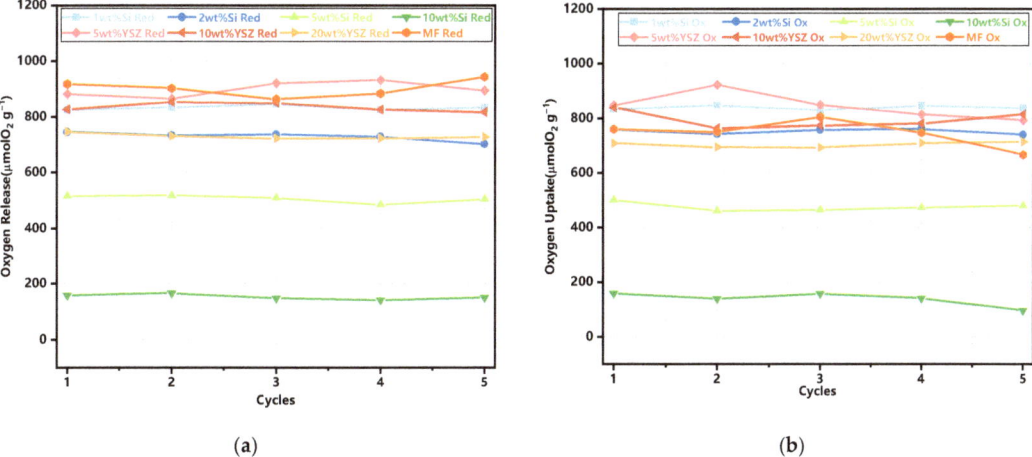

Figure 5. The trends of oxygen release and uptake in the redox reactions. (**a**) The reduction reaction; (**b**) the oxidation reaction.

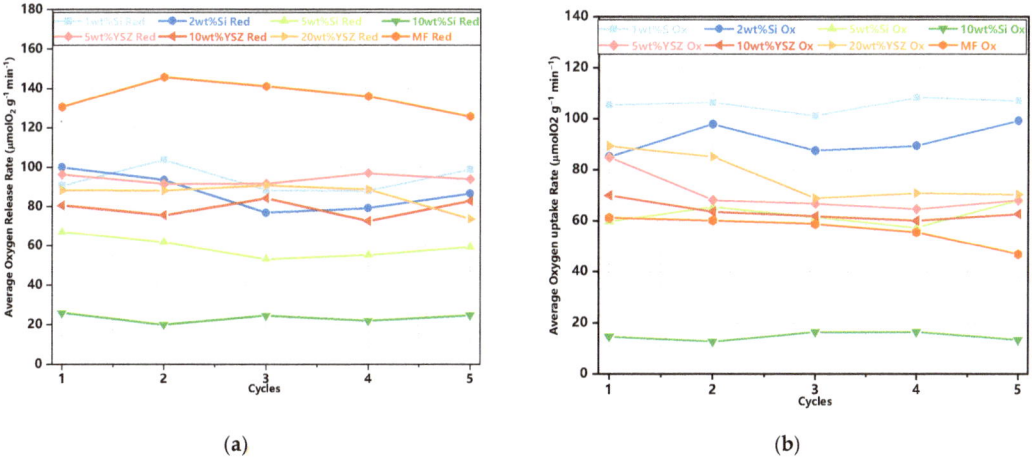

Figure 6. The average oxygen release and uptake rate in the redox cyclic tests. (**a**) The reduction reaction; (**b**) the oxidation reaction.

2.3. Long-Term Redox Cycles' Performance Analysis

In order to study the cyclic stability and evolution of the micro- and macro-structure of granules, the 120 redox reactions were carried out in a muffle furnace under the same temperature procedure used in the reactor tube, and the granules were sampled for further analysis after every 30 cycled tests. Additionally, only group MF, 1 wt% Si, 2 wt% Si, and 10 wt% YSZ were selected for further experimentation based on the consideration of the better redox performance and energy storage density shown in Section 3.2. After the 120th cycle, the XRD, SEM, and MIP characterizations were conducted to investigate the phase transformation and microstructural evolution of the cycled granules. Meanwhile the crushing strength of the granules was tested before and after the cycles.

The XRD patterns of the granules after 120 cyclic tests are presented in Figure 7. All Si-doped granules exhibit the presence of phase Mn_2O_3 (ICDD; PDF-2; #41-1442) as well as a minor phase, braunite Mn_7SiO_{12}, which is consistent with the observations before the cycle process. However, in the MF and 10 wt% YSZ groups, besides Mn_2O_3 (ICDD; PDF-2; #41-1442), there is also evidence of the appearance of a Mn_3O_4 phase (ICCD; PDF-2; #01-089-4837), indicating incomplete conversion during the re-oxidation reaction. The thermogravimetric (TG) curves of the sampled granules after every 30 cyclic tests are depicted in Figure 8. These curves have been adjusted for clarity while preserving their original significance. In Figure 8a,b, a slight mass gain (within the range indicated by the dashed black square) is observed in the initial heating program at approximately 580 °C and 650 °C for MF and 10 wt% YSZ, before the initiation of the reduction reaction, respectively. The emergence of this phenomenon is observed for 10 wt% YSZ after 30 cyclic tests, whereas it only manifests for MF after the 120th cycle. The long-term redox cycles appear to have a minimal impact on the reduction reaction of all groups, as evidenced by the nearly vertical weight loss profile observed over 120 cycles. In terms of re-oxidation evolution, both the MF and 10 wt% YSZ groups display a sluggish trend, while the Si-doped group consistently exhibits a stable vertical change in mass gain throughout the 120 cycles.

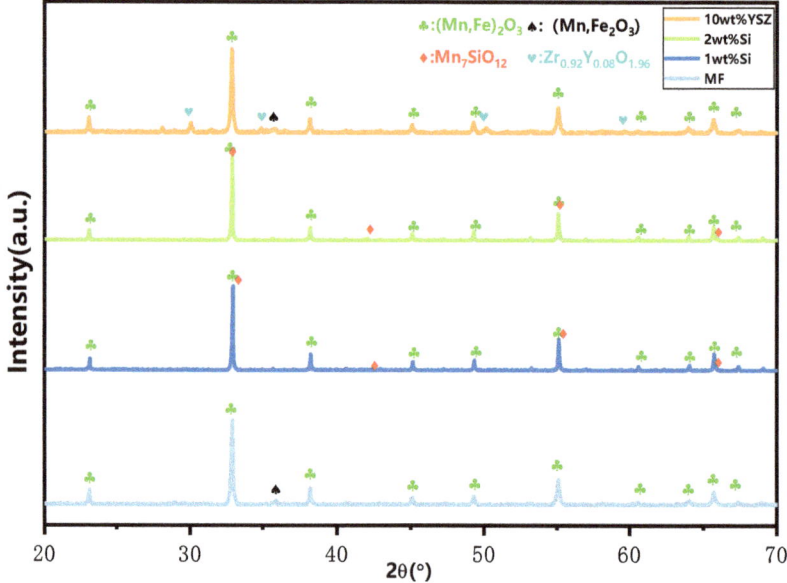

Figure 7. XRD patterns of granules after 120 cyclic tests.

The molar conversions of the sampled granules after every 30 cyclic tests, with the initial cycle serving as a reference, are illustrated in Figure 9. It is found that the reactivities in the reduction reaction are basically consistent for both the doped and undoped groups, with the exception of a slight decrease to 88.2% for the 2 wt% Si after 120 cyclic tests. In terms of the oxidation reaction process, the reactivity of all groups generally decreases over the cycles. However, the Si-doped groups exhibit a more gradual decline trend when compared with both the 10 wt% YSZ and MF groups, which experience rapid declines starting from the 90th and 120th cycle. After 120 cycles, the oxidation reactivities of the Si-doped groups remained at 92.6% and 83.2% for 1 wt% and 2 wt%, respectively, surpassing those of the undoped group (82.2%) and the 10 wt% YSZ group (76.7%).

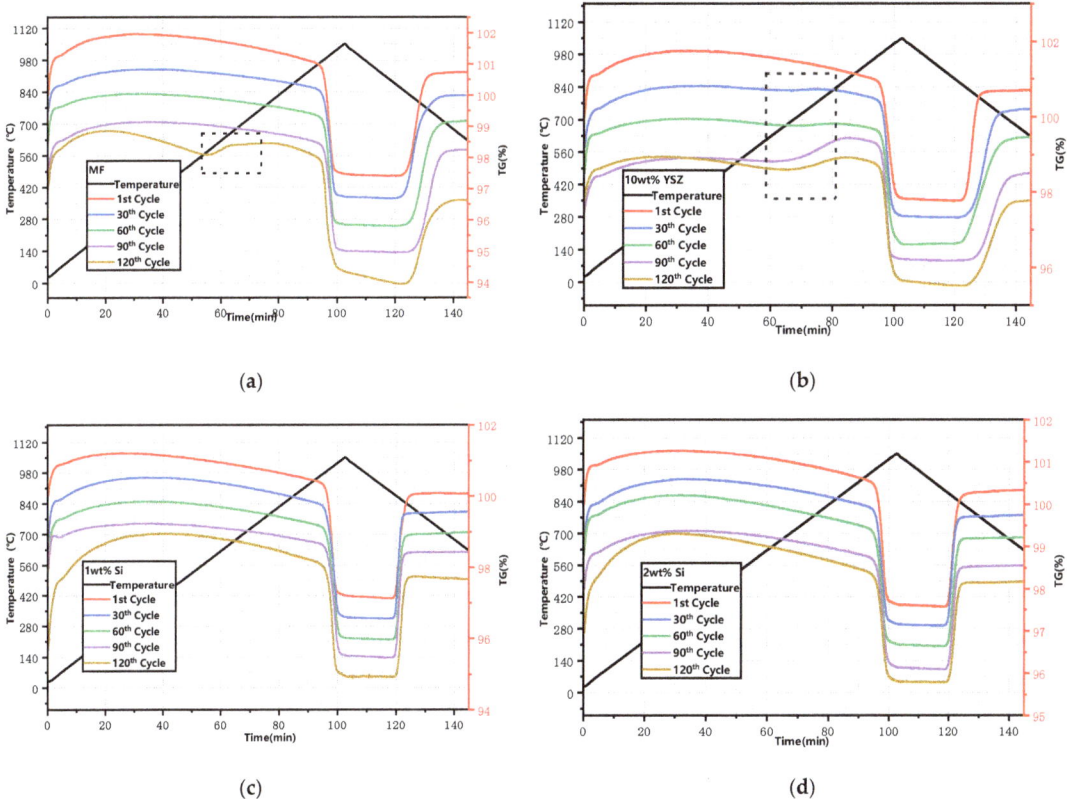

Figure 8. TG curves of the granules after every 30 cyclic tests. (**a**) MF; (**b**) 10 wt% YSZ; (**c**) 1 wt% Si; (**d**) 2 wt% Si (dashed black square: mass gain).

The evolution of the average reaction rate (μmol O_2 min^{-1} g^{-1}) for reduction and oxidation is illustrated in Figure 10. The undoped granules present the highest average reduction rate, reaching 117 μmol O_2 min^{-1} g^{-1} among the other doped groups at the beginning of the 120 cycles. However, there is a noticeable decreasing trend with the cyclic tests. In contrast, the doped granules exhibit a lower but stable reduction reaction rate throughout the cycles. In particular, the 1 wt% Si still exhibits a rate of 89.58 μmol O_2 min^{-1} g^{-1} after 120 cycles, which is comparable to its initial reaction rate of 89.16 μmol O_2 min^{-1} g^{-1}. As for the oxidation process, it is found that doping can enhance the initial average oxidation rate, which is also demonstrated in Section 2.1. But after 30 cycles, both undoped and 10 wt% YSZ groups experience a rapid decline in their reaction rates, which persists until the end of the experiment. In contrast, the Si-doped group maintains its reaction reactivity without any significant decline. Notably, the 1 wt% Si group shows a relatively high and stable rate throughout the 120 cycles, maintaining a value of 134.8 μmol O_2 min^{-1} g^{-1}, which is 3.7 times higher than that of the undoped group. The oxidation reaction rate of the 2 wt% Si group still remains at 93.95 μmolO_2 min^{-1} g^{-1}, second only to that of the 1 wt% Si group and representing a value that is 2.6 times higher than that of the undoped group, despite having the lowest average reduction rate (70.17 μmolO_2 min^{-1} g^{-1}).

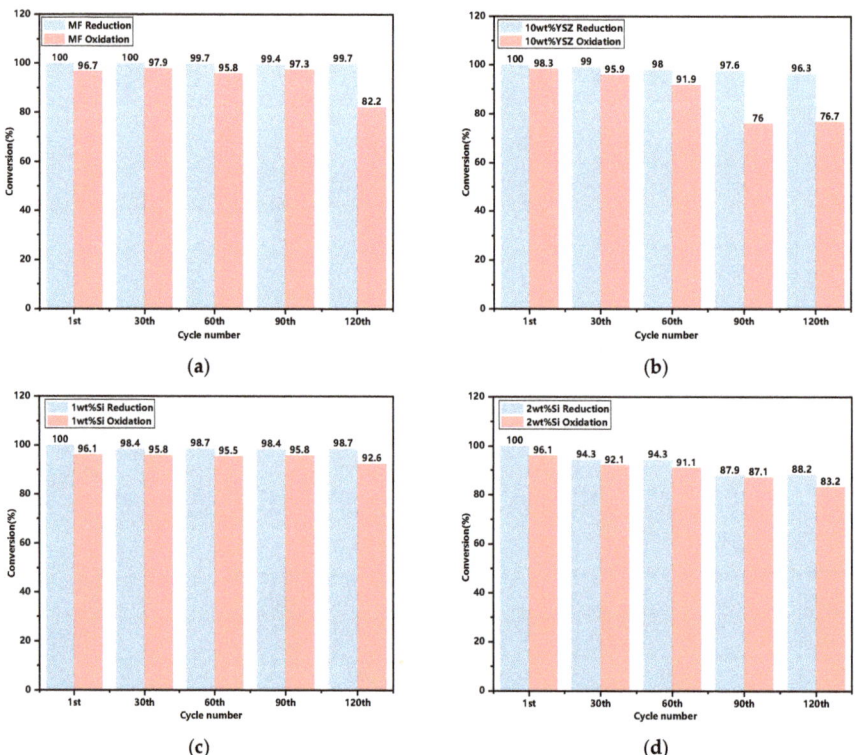

Figure 9. The reduction and oxidation reaction conversions of the granules after every 30 cyclic tests. (**a**) MF; (**b**) 10 wt% YSZ; (**c**) 1 wt% Si; (**d**) 2 wt% Si.

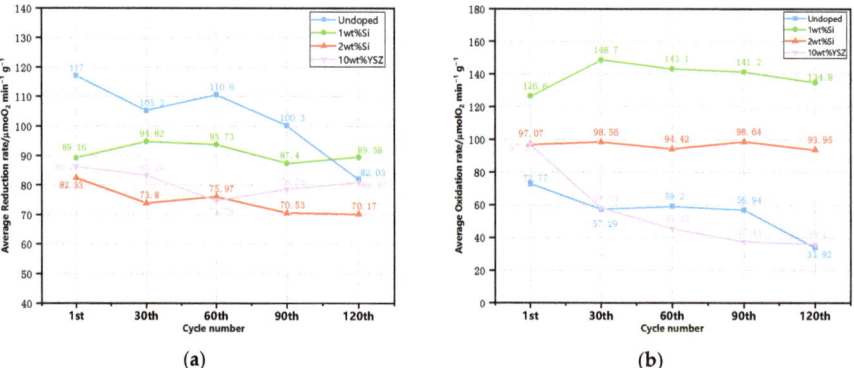

Figure 10. The evolution of average reaction rate of the granules in the cyclic tests. (**a**) The reduction reaction; (**b**) the oxidation reaction.

In addition to assessing the redox performance, the evolution of the granule size for each group was evaluated after every 30 cycles using ImageJ software, employing an analysis of more than 20 granules, as depicted in Figure 11. The initial granule sizes of all groups tend to be similar, ranging from approximately 2.36–2.5 mm prior to the cyclic process. However, noticeable variations in size are observed among the groups after the completion of the cycles. The granule size of the Si-doped groups exhibited a nearly linear increase with the number of cycles, reaching a maximum value of 3.4 mm for 2 wt% Si and the second highest value of 3.28 mm for 1 wt% Si at the end of the cycles, which is nearly 40% and 38.9% larger than its initial state. The increase in size of the Si-doped groups is significantly more pronounced compared to the relatively inconspicuous changes observed in the granule sizes of the undoped group. The 10 wt% YSZ group exhibited an anomalous phenomenon characterized by a contraction in the size of the granules. During the initial 60 cycles, significant and discernible alterations in the granule sizes were observed in both the 10 wt% YSZ and 2 wt% Si groups, with one group exhibiting contraction while the other demonstrated expansion. Subsequently, the contraction rate of 10 wt% YSZ gradually decelerated and eventually stabilized at 2.02 mm, indicating a reduction of nearly 20% compared to its initial state.

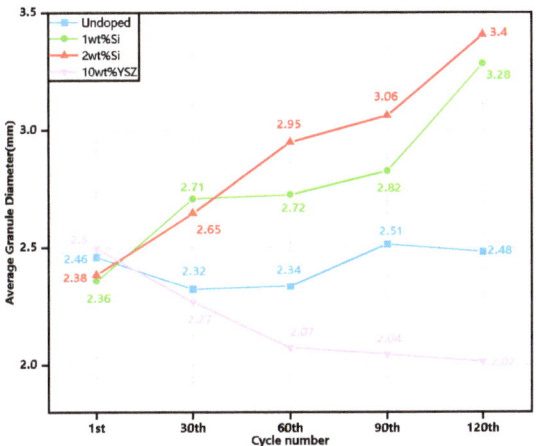

Figure 11. The evolution of granule sizes from the 1st cycle to the 120th cycle.

Furthermore, MIP characterizations were conducted to investigate the porosity changes of the granules both pre- and post-120 cycles. The findings are summarized in Table 3, accompanied by their corresponding differential and normalized cumulative Hg porosimetry curves (in cc/cc, enabling the calculation of porosity), as well as the associated granule images presented in Figure 12. The initial total porosity is approximately 30% for all groups, except for the 1 wt% Si group which exhibits a slightly lower value of 26.85% compared to the other groups. Also, the initial mean pore size of the doped granules measures less than 1 μm (green dotted line), while the undoped granules have a mean pore size of 1.31 μm (yellow dotted line). The porosity of the Si-doped groups exhibited an approximate increase of 16.1% and 40% for 1 wt% Si and 2 wt% Si, respectively, compared to their initial porosity after 120 cycles. In contrast, the undoped group only showed a negligible increase in porosity. The porosity in the 10 wt% YSZ group exhibited a significant decrease, dropping from 32.91% to 18.5%. The overall porosity trend presents a high degree of consistency with the evolution of granule size.

The mechanical strength of the granules was also assessed before and after 120 cycles, as presented in Table 3. The initial average crushing strength of all of the groups is approximately ~0.36 N, surpassing the previously reported values of 0.166 N (Bielsa et al. [32]). Notably, the 2 wt% Si group exhibited the highest value of 0.92 N. As previously discussed, the macro-structure of the granules undergoes significant changes in terms of expansion or contraction after the cycles, resulting in considerable variations in their corresponding crushing strength. The crushing strength of all of the groups exhibited a decrease, except for the 10 wt% YSZ group which demonstrated an increase to 2.44 N after the cycles. The crushing strength of the 2 wt% Si and 1 wt% Si groups decreased to 0.48 N and 0.28 N, respectively, surpassing the value of the undoped group (0.11 N), indicating that the Si-doped groups lead to a significant improvement in mechanical enhancement after long-term cycles.

Table 3. Analyses of MIP and mechanical strength (main parameters of interest) of fresh and used granules.

Sample	Bulk Density Fresh/Used (g/cm^3)	Total Porosity Fresh/Used (%)	d_{50} Fresh/Used (μm)	Average Crushing Strength Fresh/Used (N)
MF	0.99/0.98	30.87/29.43	1.3/30.18	0.36/0.11
1 wt%Si	1.12/0.56	26.85/31.17	0.55/5.17	0.39/0.28
2 wt%Si	1.08/0.41	29.22/40.92	0.43/2.89	0.92/0.47
10 wt%YSZ	0.94/1.61	32.91/18.55	0.67/89.6	0.37/2.44

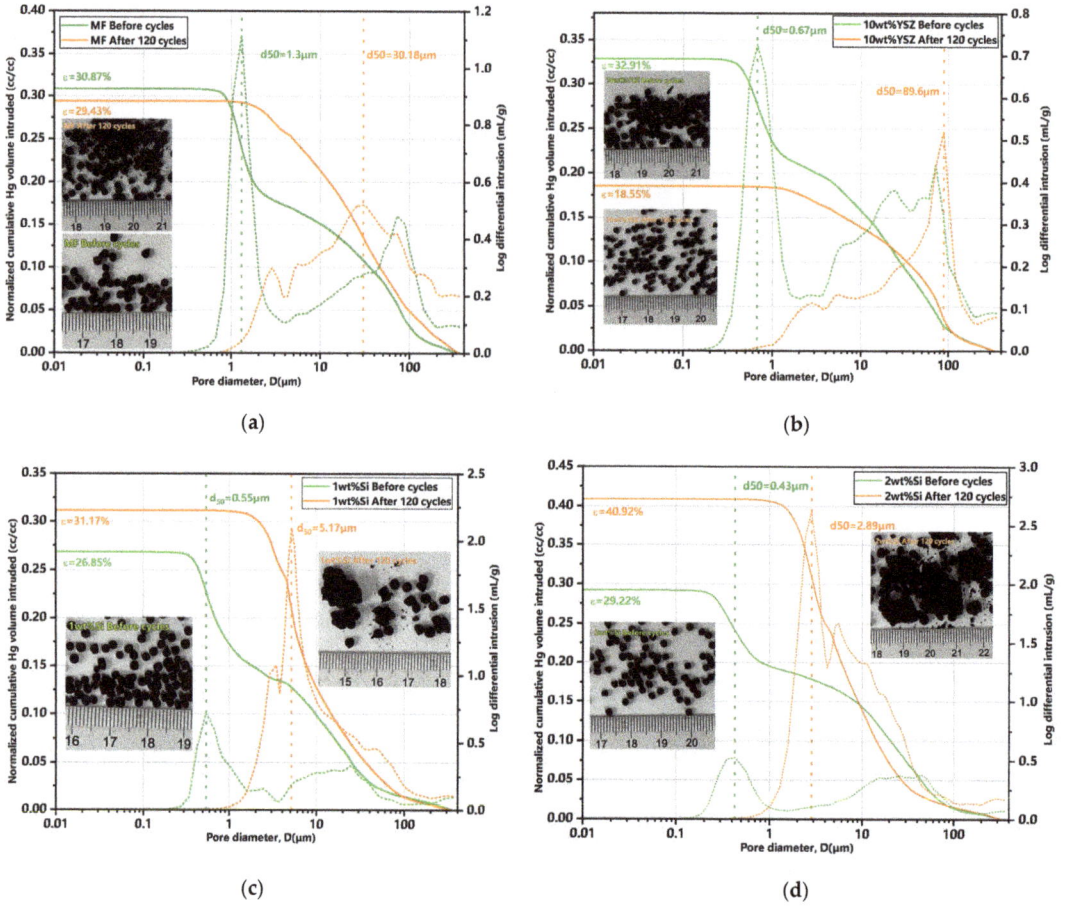

Figure 12. MIP curves of granules for pre- and post-120 cycles and their corresponding granules images. (**a**) MF granules; (**b**) 10 wt% YSZ granules; (**c**) 1 wt% Si granules; (**d**) 2 wt% Si granules (green dotted line: mean pore size before cycles; yellow dotted line: mean pore size after cycles).

In order to investigate the microstructural evolution of the granules before and after 120 cycles, we also conducted SEM characterizations. The SEM images of cross-sectional cuts of granules are depicted in Figure 13 in consecutive magnifications.

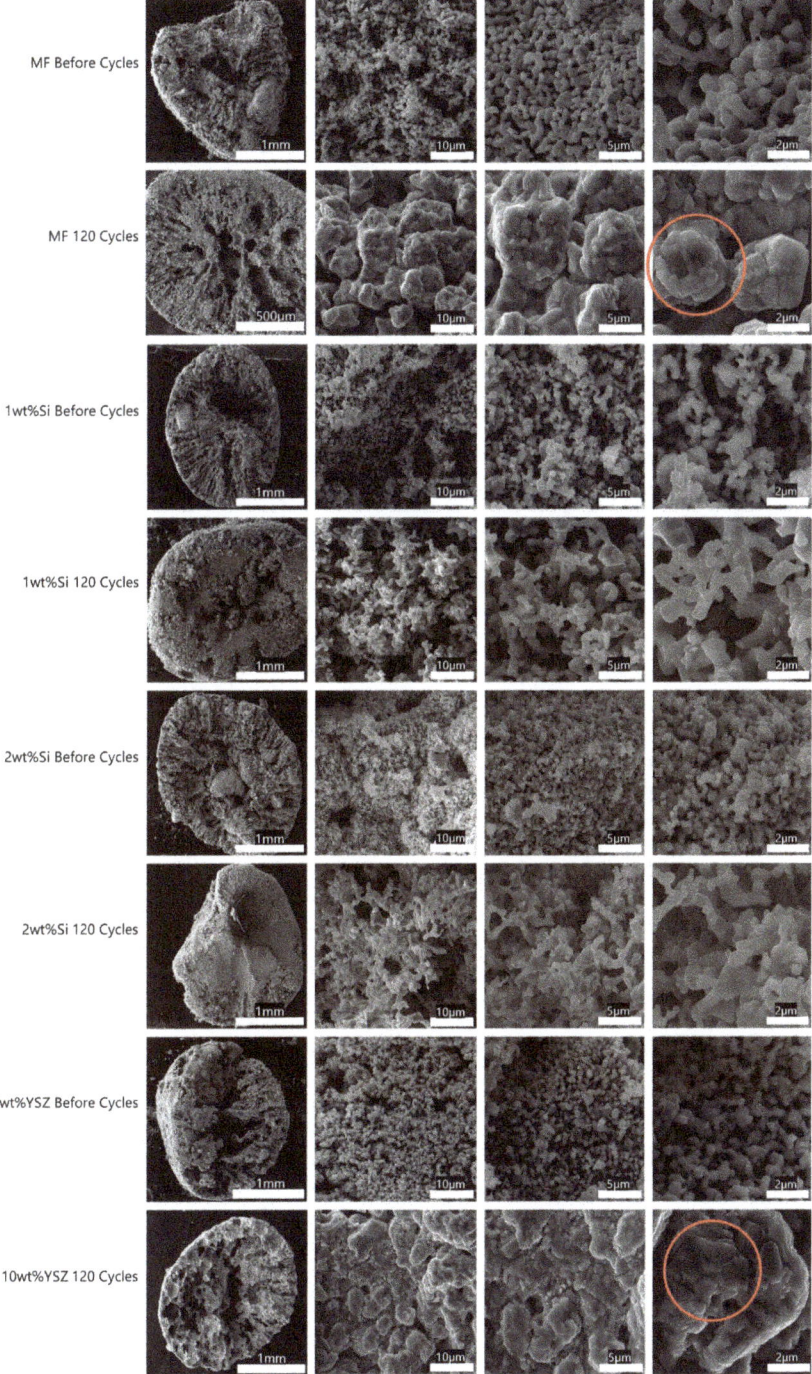

Figure 13. The SEM images of cross-sectional cuts of granules before and after 120 cycles, accompanied by consecutive magnification (red circle: sintering area).

From the SEM images, it is evident that the granulation process results in the dissolution of organic solvents in water, leading to the formation of channels extending from the surface to the interior and exhibiting grooved patterns at cross-sections. Consequently, the presence of these channels significantly facilitates the ingress of oxygen from the air. However, after undergoing consecutive cycles, the internal structure underwent significant changes under various stresses. The channels in Si-doped granules were clearly eliminated due to expansion, while the internally predominant structure of doped YSZ particles exhibited an interconnected porous architecture despite experiencing significant volume shrinkage during the cyclic process. By further magnifying the local area of the cross-section, it is possible to observe that the grains grow larger than its initial size in all granules. In contrast to the MF and 10 wt% YSZ samples where the grain morphology appears coarse, sharp, and angular due to sintering (the area signed by the red circle), the Si-doped granules, particularly those containing 1 wt% Si, exhibit a porous coral-like structure which indicated a pronounced anti-sintering effect. In general, the sintering process can be classified into liquid-phase sintering and solid-state sintering, with a focus on the latter in this study. The predominant phenomenon observed during solid state sintering is the simultaneous occurrence of densification and grain growth (coarsening), both of which rely on mass transport along the grain boundaries [40]. In fact, the chemical modification of grain boundaries through the introduction of doping cations has been proven to be an effective strategy. SiO_2, traditionally used as a dopant, aids in preventing sintering. Previously, Bielsa et al. [32] investigated the lower doping ratio of Si cations on Mn oxides and proposed that the sintering rate in Mn_2O_3 is governed by ion grain boundary diffusivity, which can be hindered by the segregation of Si^{4+} dopants at the grain boundaries. Additionally, due to its smaller size and higher valence, Si^{4+} dopants can also enhance re-oxidation kinetics and mitigate sintering effects. Recently, Huang et al. [41] investigated the impact of incorporating a surface modifier $MnSiO_3$ into $(Mn_{0.8}Fe_{0.2})_2O_3$ on its anti-sintering properties and the nanoscale $MnSiO_3$ particles were effectively immobilized on the surface of $(Mn_{0.8}Fe_{0.2})_2O_3$, thereby impeding crystal growth and ensuring stability over 1000 cycles.

3. Materials and Methods

3.1. Granules Synthesis

The preparation process of Mn-Fe oxide granules is shown in Figure 14. The process of granulation primarily comprises three stages: synthesis of Mn-Fe oxides, preparation of granules, and calcination of the granules.

Synthesis of Mn-Fe oxides. The metal oxides were synthesized using a modified Pechini method as described by Sunde et al. [42] and Jana et al. [43]. The metal precursors, consisting of Mn $(NO_3)_2 \cdot 4H_2O$ (98%, Macklin, Shanghai, China) and Fe $(NO_3)_3 \cdot 9 H_2O$ (>98%, S Macklin), were introduced into an aqueous solution of citric acid (CA, ≥99.5%, Macklin, China) with a molar ratio of Me: CA as 1:5 under continuous stirring for 3 h at a temperature of 70 °C. Subsequently, in order to facilitate the polymerization process, ethylene glycol (EG, ≥99.5%, Macklin, China) was introduced at a molar ratio of CA:EG = 3:2 and the solution was subjected to continuous stirring at 90 °C for 2 h. The gel was subsequently subjected to drying at a temperature of 200 °C for a duration of 3 h, followed by air calcination at 450 °C for a period of 4 h. The calcined gel was finely ground into powder and subsequently subjected to further static air calcination at 700 °C for 4 h with a heating/cooling rate of 2 °C/min, ensuring the formation of the bixbyite crystal structure.

Preparation of granules. The granulation process in this work is implemented using a novel, feasible, and scalable technique known as the "drop technique", which has been modified by Gigantino et al. [30] and Bielsa et al. [31]. The granules are doped with SiO_2 (99.9%, Aladdin, Wuhan, China) and 8 mol% Y_2O_3-stabilized ZrO_2 (YSZ, Aladdin, China), respectively. The Mn-Fe oxides powder, mixed with the dopants of SiO_2 (at 1 wt%, 2 wt%, 5 wt%, and 10 wt%) and YSZ (at 5 wt%, 10 wt%, and 20 wt%), was homogenized in a planetary ball mill operating at a speed of 600 rpm for 30 min, resulting in the powder with an average particle size of 0.05 μm. In this work, MF stands for pure Mn-Fe oxide granules and the abbreviation for doped granules is represented by the doping mass ratio combined with the dopant, such as 1 wt% Si or 5 wt% YSZ, etc.

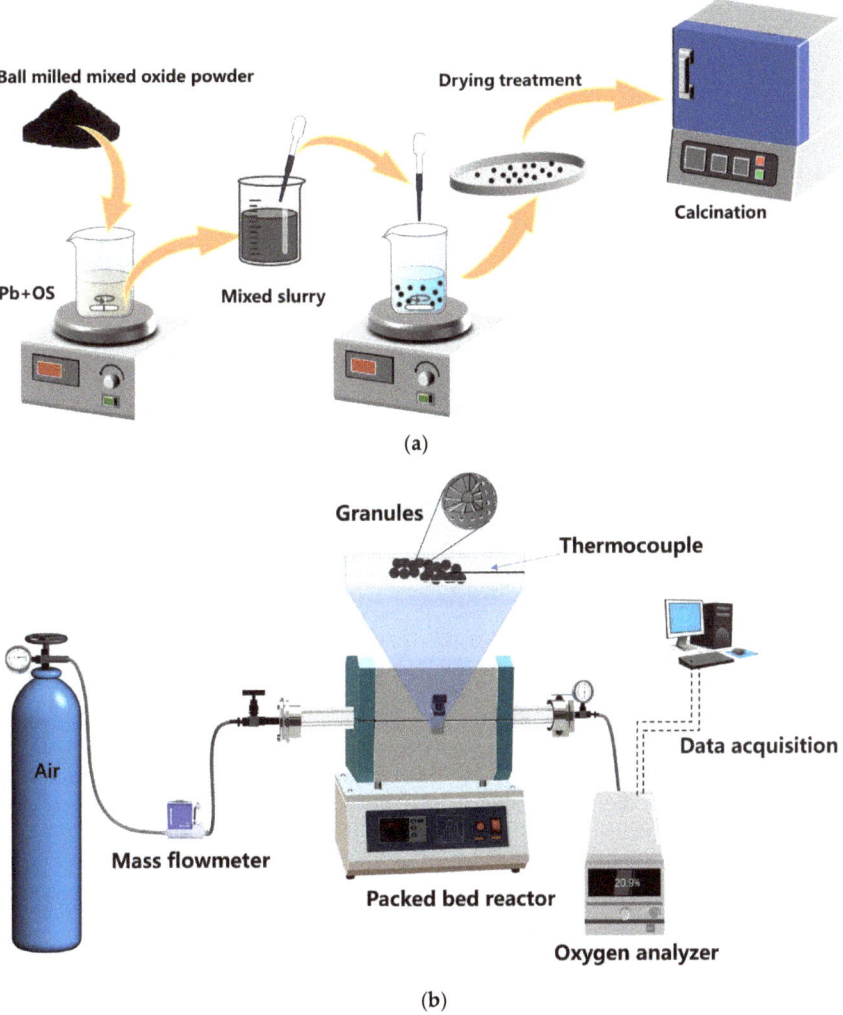

Figure 14. Schematic diagram of the preparation process of Mn-Fe granules and the cyclic tests in a packed reactor. (**a**) The granulation process; (**b**) the cyclic test process.

The polymeric binder (PB) and organic solvent (OS) were selected from ethyl cellulose (viscosity of 18–22 cP, 5% in toluene/ethanol 80:20 vol%, Macklin, China) and 1-methyl-2-pyrrolidinone (>99% purity, Macklin, China), respectively. The mixed metal oxides powder (MO) was added to a solution comprising a polymeric binder (PB) dissolved in an organic solvent (OS) with a mass ratio of MO:PB:OS = 3:1:9. The slurry was preheated to 45 °C in order to reduce its viscosity before being dispensed through a syringe needle with a tip diameter of 2 mm into a precipitating bath composed of deionized water and the surfactant Tween 80 (Macklin, China) (0.2 mL/L), which effectively lowers the surface tension of water. In practical experiments, we observed that the length of the metal needle plays a critical role in slurry flow dynamics. In particular, the utilization of longer needles has been observed to induce a rapid decrease in slurry temperature, leading to an elevation in viscosity and subsequent obstruction that hinders slurry flow. Moreover, achieving precise control over the rate of dropping by manipulating the syringe plunger becomes challenging, particularly when encountering needle blockage. Therefore, we selected a disposable dropper made of polyethylene plastic with a 2 mm tip diameter instead of employing a syringe equipped with a metal needle, as it provides enhanced ease and intuitive control over the drop rate. Moreover, in cases of blockage, it is more convenient to trim the obstructed portion of the dropper rather than replacing the metal needle. The height between the dropper tip and the precipitation bath is maintained at 3 cm to ensure that the granules possess sufficient kinetic energy to penetrate the bath surface and any deformation upon reaching the surface of the water was prevented. The polymer within the droplets underwent solidification while the solvent diffused into the surrounding bath water.

Calcination of granules. The granules, which possessed a regular spherical shape and were filtered using a sieve, were dried overnight in ambient air. Subsequently, the granules were subjected to calcination in air at a temperature of 1050 °C for 4 h, employing a heating ramp of 2 °C/min. This thermal treatment aimed to eliminate the organic matter, enhance the granules' strength, and induce the development of a porous structure within them.

3.2. Characterization of Granules

The powder X-ray diffraction (XRD) analyses were conducted using a Bruker D8 Advance diffractometer with Cu Kα radiation (λ = 1.5406 Å), covering a diffraction angle (2θ) range of 10–80° with a step size of 0.02°. The crystal phases were identified using Jade 6.0 software, with reference to the ICDD PDF-2 database. The microstructure and morphology of the granules were examined before and after redox cycles using scanning electron microscopy (SEM) with a TESCAN MIRA microscope equipped with a tungsten source at an accelerating voltage of 15 kV. X-ray Photoelectron Spectroscopy (XPS) of the granules was recorded using nonmonochromatic Al Kα radiation (1486.8 eV) using a Thermo Fisher Nexsa X-ray Photoelectron Spectrometer (Thermo Fisher Scientific, Waltham, MA, USA). All binding energies were referred to the C 1s peak at 284.6 eV to compensate for the effect of surface charge. The core-level spectra were curve-fitted into their possible components using Gaussian–Lorentzian peaks after subtracting a Shirley background with the Avantage program 5.9. The granules were subjected to pore size characterization using an AutoPore V 9600 mercury intrusion porosimeter from Micromeritics Instrument Corporation (Norcross, GA, USA), with a sample mass of approximately 600 mg, a contact angle of 130°, and a maximum pressure of 61,000 psia. The average diameter of granules was determined by analyzing more than 20 granules using ImageJ software v1.54, a widely used image processing program. The crushing strength of selected granules before and after 120 cycles was also examined using the SENS CMT6000 apparatus.

3.3. Redox Reactivity of Granules

The redox reaction of a selected composition with a Mn/Fe molar ratio of 3:1 can be described as follows [19]:

$$6(Mn_{0.75}Fe_{0.25})_2O_3 \rightleftharpoons 4(Mn_{0.75}Fe_{0.25})_3O_4 + O_2 \tag{1}$$

The reaction kinetics and chemical stability of the granules at different doping ratios were investigated using the simultaneous thermal analyzer STA 449 F3 Jupiter (Netzsch, Bayern, Germany). Each sample, consisting of 3–4 granules weighing approximately 10 mg, was placed into 85 µL open Al$_2$O$_3$ crucibles (Netzsch). These samples underwent charging and discharging cycles under a synthetic air stream flowing at a rate of 100 mL/min. The molar conversion for reduction and oxidation is defined as

$$X_{Red} = \frac{6 \cdot \Delta n_{O_2} M_{(Mn_{0.75}Fe_{0.25})_2O_3}}{m_{(Mn_{0.75}Fe_{0.25})_2O_3 + dopant, tot} \cdot w_{(Mn_{0.75}Fe_{0.25})_2O_3}} \tag{2}$$

$$X_{Ox} = \frac{4 \cdot \Delta n_{O_2} M_{(Mn_{0.75}Fe_{0.25})_3O_4}}{m_{(Mn_{0.75}Fe_{0.25})_3O_4 + \text{dopant, tot}} \cdot w_{(Mn_{0.75}Fe_{0.25})_3O_4}} \quad (3)$$

where the M_i, w_i, and $m_{i+\text{dopant, tot}}$ represent the molar mass, mass fraction, and total mass, respectively, of species i. Δn_{O_2} stands for the amount of reacted moles of oxygen which can be calculated by $\Delta n_{O_2} = \Delta m_{O_2}/M_{O_2}$.

As for the cyclic tests in a lab-scale packed-bed reactor, a tube reactor type was adopted. The quartz tube reactor (Φ 13 mm × 1200 mm) was positioned within an electrical furnace, with both ends of the tube hermetically sealed using T-type flanges to facilitate the insertion of a Type-K thermocouple (Φ = 1.5 mm) and efficient gas efflux towards an oxygen analyzer. Approximately 2 g of granules were placed in the middle of the tube reactor. The temperature of the granules was measured by the thermocouple which was inserted at the center of the granules, and the signal was transmitted to a temperature programmer/controller Yudian AI-888 with an accuracy of ±0.01 °C during the experiment. The oxygen concentration downstream of the reactor during the redox cycles was measured using an oxygen analyzer CI-PC926 (Changai, Shanghai, China). The process of the cyclic tests in the packed-bed reactor is depicted in Figure 14b. The molar conversions of granules during the cyclic tests in reactor can also be determined by Equations (2) and (3), and the corresponding Δn_{O_2}, which is determined by integrating the molar flow rate of O_2 release/uptake during the reduction/oxidation, defined as:

$$\Delta n_{O_2} = \int_{\text{begin}}^{\text{end}} \dot{n}_{O_2,\text{reacted}}(t) dt \quad (4)$$

$$\dot{n}_{O_2,\text{reacted}} = \frac{\dot{V}_{\text{air,in}}}{\hat{V}} \cdot \frac{y_{O_2,\text{ out}} - y_{O_2,\text{ in}}}{1 - y_{O_2,\text{ out}}} \quad (5)$$

where $\dot{V}_{\text{air,in}}$, $\hat{V}$, and y_i represent the inlet's overall air flow rate at the inlet, the molar volume of an ideal gas at T = 273.15 K and total pressure p = 101.325 kPa (Standard Temperature and Pressure = STP), and the oxygen molar fraction, respectively. The average rate of reduction and oxidation (μmolO$_2$ min^{-1} g^{-1}) can be defined as the amount of oxygen released or taken up during the reaction divided by the corresponding reaction time per unit mass of granules.

4. Conclusions

In the TCES application, it is essential for energy storage materials to have adequate mechanical strength and stable reactivity to ensure the long-term operation under the conditions of the high temperature and alternating redox atmosphere. In this study, we synthesized Mn-Fe oxide granules with an average particle size of approximately 2 mm using the drop technique, incorporating SiO_2 and YSZ at the mass ratios ranging from 1 wt% to 20 wt%. In the TG test, the average reduction rate of the doped granules exhibited a deceleration, while the average oxidation rate experienced a significant enhancement when compared to undoped granules, except for 5 wt% Si and 10 wt% Si. Throughout the five consecutive cycles in the packed-bed reactor experiment, the doped oxides presented stable enhanced reoxidation rates when the doping ratios were in an appropriate range, and decreased reduction rates and energy storage densities. After 120 long-term cyclic tests, the Si-doped granules possessed a better anti-sintering property than that of YSZ-doped granules. Meanwhile, the 1 wt% Si granules exhibited the highest re-oxidation ratio of 92.6%, followed by the second highest ratio of 83.2% for 2 wt% Si. In summary, the incorporation of a small amount ($\leq$1 wt%) of SiO_2 demonstrates the most optimal TCES performance when considering all factors. However, in the long-term operation under high temperatures, the volume expanding of the Si-doped granules would lead to an increment of pressure drop in the packed-bed reactor, which should be focused. In addition, further dedicated efforts towards enhancing the mechanical strength in the Mn-Fe-Si system are necessary.

Author Contributions: Y.M.: conceptualization, data curation, formal analysis, validation, investigation, methodology, writing—original draft, writing—review and editing. Z.W.: conceptualization, resources, software, project administration, writing—review and editing, funding acquisition. J.S.: project administration, resources, conceptualization. K.W.: visualization, investigation, validation. S.L.: visualization, validation. Z.L.: visualization, validation. All authors have read and agreed to the published version of the manuscript.

Funding: This research received no external funding.

Institutional Review Board Statement: Not applicable.

Informed Consent Statement: Not applicable.

Data Availability Statement: Data are contained within the article.

Conflicts of Interest: The authors declare no conflicts of interest.

References

1. Yan, J.; Lu, L.; Ma, T.; Zhou, Y.; Zhao, C.Y. Thermal management of the waste energy of a stand-alone hybrid PV-wind-battery power system in Hong Kong. *Energy Convers. Manag.* **2020**, *203*, 112261. [CrossRef]
2. Abdalla, A.N.; Nazir, M.S.; Tao, H.; Cao, S.; Ji, R.; Jiang, M.; Yao, L. Integration of energy storage system and renewable energy sources based on artificial intelligence: An overview. *J. Energy Storage* **2021**, *40*, 102811. [CrossRef]
3. Yan, T.; Wang, R.Z.; Li, T.X.; Wang, L.W.; Fred, I.T. A review of promising candidate reactions for chemical heat storage. *Renew. Sustain. Energy Rev.* **2015**, *43*, 13–31. [CrossRef]
4. Zhao, C.; Yan, J.; Tian, X.; Xue, X.; Zhao, Y. Progress in thermal energy storage technologies for achieving carbon neutrality. *Carbon Neutrality* **2023**, *2*, 10. [CrossRef]
5. Wu, S.; Zhou, C.; Doroodchi, E.; Nellore, R.; Moghtaderi, B. A review on high-temperature thermochemical energy storage based on metal oxides redox cycle. *Energy Convers. Manag.* **2018**, *168*, 421–453. [CrossRef]
6. Neises, M.; Tescari, S.; De Oliveira, L.; Roeb, M.; Sattler, C.; Wong, B. Solar-heated rotary kiln for thermochemical energy storage. *Sol. Energy* **2012**, *86*, 3040–3048. [CrossRef]
7. Jafarian, M.; Arjomandi, M.; Nathan, G.J. Thermodynamic potential of molten copper oxide for high temperature solar energy storage and oxygen production. *Appl. Energy* **2017**, *201*, 69–83. [CrossRef]
8. Fahim, M.A.; Ford, J.D. Energy storage using the BaO_2 BaO reaction cycle. *Chem. Eng. J.* **1983**, *27*, 21–28. [CrossRef]
9. Miguel, S.Á.D.; Gonzalez-Aguilar, J.; Romero, M. 100-Wh Multi-purpose Particle Reactor for Thermochemical Heat Storage in Concentrating Solar Power Plants. *Energy Procedia* **2014**, *49*, 676–683. [CrossRef]
10. Block, T.; Knoblauch, N.; Schmücker, M. The cobalt-oxide/iron-oxide binary system for use as high temperature thermochemical energy storage material. *Thermochim. Acta* **2014**, *577*, 25–32. [CrossRef]
11. Block, T.; Schmücker, M. Metal oxides for thermochemical energy storage: A comparison of several metal oxide systems. *Sol. Energy* **2016**, *126*, 195–207. [CrossRef]
12. Agrafiotis, C.; Roeb, M.; Sattler, C. Exploitation of thermochemical cycles based on solid oxide redox systems for thermochemical storage of solar heat. Part 4: Screening of oxides for use in cascaded thermochemical storage concepts. *Sol. Energy* **2016**, *139*, 695–710. [CrossRef]
13. Agrafiotis, C.; Block, T.; Senholdt, M.; Tescari, S.; Roeb, M.; Sattler, C. Exploitation of thermochemical cycles based on solid oxide redox systems for thermochemical storage of solar heat. Part 6: Testing of Mn-based combined oxides and porous structures. *Sol. Energy* **2017**, *149*, 227–244. [CrossRef]
14. Wong, B. *General Atomics Thermochemical Heat Storage for Concentrated Solar Power, Thermochemical System Reactor Design for Thermal Energy Storage*; DOE/GO18145 TRN: US201209%%526; OSTI.GOV: San Diego, CA, USA, 2011. [CrossRef]
15. Carrillo, A.J.; Serrano, D.P.; Pizarro, P.; Coronado, J.M. Thermochemical heat storage based on the Mn_2O_3/Mn_3O_4 redox couple: Influence of the initial particle size on the morphological evolution and cyclability. *J. Mater. Chem. A* **2014**, *2*, 19435–19443. [CrossRef]
16. André, L.; Abanades, S.; Cassayre, L. High-temperature thermochemical energy storage based on redox reactions using Co-Fe and Mn-Fe mixed metal oxides. *J. Solid State Chem.* **2017**, *253*, 6–14. [CrossRef]
17. Carrillo, A.J.; Serrano, D.P.; Pizarro, P.; Coronado, J.M. Improving the Thermochemical Energy Storage Performance of the Mn_2O_3/Mn_3O_4 Redox Couple by the Incorporation of Iron. *ChemSusChem* **2015**, *8*, 1947–1954. [CrossRef] [PubMed]
18. Carrillo, A.J.; Serrano, D.P.; Pizarro, P.; Coronado, J.M. Understanding Redox Kinetics of Iron-Doped Manganese Oxides for High Temperature Thermochemical Energy Storage. *J. Phys. Chem. C* **2016**, *120*, 27800–27812. [CrossRef]
19. Wokon, M.; Block, T.; Nicolai, S.; Linder, M.; Schmücker, M. Thermodynamic and kinetic investigation of a technical grade manganese-iron binary oxide for thermochemical energy storage. *Sol. Energy* **2017**, *153*, 471–485. [CrossRef]

20. Wokon, M.; Kohzer, A.; Linder, M. Investigations on thermochemical energy storage based on technical grade manganese-iron oxide in a lab-scale packed bed reactor. *Sol. Energy* **2017**, *153*, 200–214. [CrossRef]
21. Hamidi, M.; Bayon, A.; Wheeler, V.M.; Kreider, P.; Wallace, M.A.; Tsuzuki, T.; Catchpole, K.; Weimer, A.W. Reduction kinetics for large spherical 2:1 iron-manganese oxide redox materials for thermochemical energy storage. *Chem. Eng. Sci.* **2019**, *201*, 74–81. [CrossRef]
22. Carrillo, A.J.; Gonzalez-Aguilar, J.; Romero, M.; Coronado, J.M. Solar Energy on Demand: A Review on High Temperature Thermochemical Heat Storage Systems and Materials. *Chem. Rev.* **2019**, *119*, 4777–4816. [CrossRef] [PubMed]
23. Gan, D.; Sheng, H.; Zhu, P.; Xu, H.; Xiao, G. Long-term replenishment strategy of SiC-doped Mn-Fe particles for high-temperature thermochemical energy storage. *Sol. Energy* **2023**, *262*, 111842. [CrossRef]
24. Zsembinszki, G.; Solé, A.; Barreneche, C.; Prieto, C.; Fernández, A.; Cabeza, L. Review of Reactors with Potential Use in Thermochemical Energy Storage in Concentrated Solar Power Plants. *Energies* **2018**, *11*, 2358. [CrossRef]
25. Pagkoura, C.; Karagiannakis, G.; Zygogianni, A.; Lorentzou, S.; Konstandopoulos, A.G. Cobalt Oxide Based Honeycombs as Reactors/Heat Exchangers for Redox Thermochemical Heat Storage in Future CSP Plants. *Energy Procedia* **2015**, *69*, 978–987. [CrossRef]
26. Xiang, D.; Gu, C.; Xu, H.; Xiao, G. Self-Assembled Structure Evolution of Mn-Fe Oxides for High Temperature Thermochemical Energy Storage. *Small* **2021**, *17*, 2101524. [CrossRef] [PubMed]
27. Agrafiotis, C.; Roeb, M.; Schmücker, M.; Sattler, C. Exploitation of thermochemical cycles based on solid oxide redox systems for thermochemical storage of solar heat. Part 2: Redox oxide-coated porous ceramic structures as integrated thermochemical reactors/heat exchangers. *Sol. Energy* **2015**, *114*, 440–458. [CrossRef]
28. Agrafiotis, C.; Tescari, S.; Roeb, M.; Schmucker, M.; Sattler, C. Exploitation of thermochemical cycles based on solid oxide redox systems for thermochemical storage of solar heat. Part 3: Cobalt oxide monolithic porous structures as integrated thermochemical reactors/heat exchangers. *Sol. Energy* **2015**, *114*, 459–475. [CrossRef]
29. Preisner, N.C.; Block, T.; Linder, M.; Leion, H. Stabilizing Particles of Manganese-Iron Oxide with Additives for Thermochemical Energy Storage. *Energy Technol.* **2018**, *6*, 2154–2165. [CrossRef]
30. Gigantino, M.; Sas Brunser, S.; Steinfeld, A. High-Temperature Thermochemical Heat Storage via the CuO/Cu_2O Redox Cycle: From Material Synthesis to Packed-Bed Reactor Engineering and Cyclic Operation. *Energy Fuels* **2020**, *34*, 16772–16782. [CrossRef]
31. Bielsa, D.; Oregui, M.; Arias, P.L. New insights into Mn_2O_3 based metal oxide granulation technique with enhanced chemical and mechanical stability for thermochemical energy storage in packed bed reactors. *Sol. Energy* **2022**, *241*, 248–261. [CrossRef]
32. Bielsa, D.; Zaki, A.; Arias, P.L.; Faik, A. Improving the redox performance of Mn_2O_3/Mn_3O_4 pair by Si doping to be used as thermochemical energy storage for concentrated solar power plants. *Sol. Energy* **2020**, *204*, 144–154. [CrossRef]
33. Yu, X.; Wang, L.; Chen, M.; Fan, X.; Zhao, Z.; Cheng, K.; Chen, Y.; Sojka, J.; Wei, Y.; Liu, J. Enhanced activity and sulfur resistance for soot combustion on three-dimensionally ordered macroporous-mesoporous $Mn_xCe_{1-x}O_\delta/SiO_2$ catalysts. *Appl. Catal. B Environ.* **2019**, *254*, 246–259. [CrossRef]
34. Han, Y.; Zhao, J.; Quan, Y.; Yin, S.; Wu, S.; Ren, J. Highly Efficient $La_xCe_{1-x}O_{2-x/2}$ Nanorod-Supported Nickel Catalysts for CO Methanation: Effect of La Addition. *Energy Fuels* **2021**, *35*, 3307–3314. [CrossRef]
35. Onrubia-Calvo, J.A.; Pereda-Ayo, B.; De-La-Torre, U.; González-Velasco, J.R. Key factors in Sr-doped $LaBO_3$ (B=Co or Mn) perovskites for NO oxidation in efficient diesel exhaust purification. *Appl. Catal. B Environ.* **2017**, *213*, 198–210. [CrossRef]
36. Yilmaz, D.; Darwish, E.; Leion, H. Investigation of the combined Mn-Si oxide system for thermochemical energy storage applications. *J. Energy Storage* **2020**, *28*, 101180. [CrossRef]
37. Zhang, Z.; Yu, J.; Zhang, J.; Ge, Q.; Xu, H.; Dallmann, F.; Dittmeyer, R.; Sun, J. Tailored metastable Ce–Zr oxides with highly distorted lattice oxygen for accelerating redox cycles. *Chem. Sci.* **2018**, *9*, 3386–3394. [CrossRef] [PubMed]
38. Zou, D.; Yi, Y.; Song, Y.; Guan, D.; Xu, M.; Ran, R.; Wang, W.; Zhou, W.; Shao, Z. The $BaCe_{0.16}Y_{0.04}Fe_{0.8}O_{3-\delta}$ nanocomposite: A new high-performance cobalt-free triple-conducting cathode for protonic ceramic fuel cells operating at reduced temperatures. *J. Mater. Chem. A* **2022**, *10*, 5381–5390. [CrossRef]
39. Zhou, J.; Xiang, D.; Zhu, P.; Deng, J.; Gu, C.; Xu, H.; Zhou, J.; Xiao, G. ZrO_2-Doped Copper Oxide Long-Life Redox Material for Thermochemical Energy Storage. *ACS Sustain. Chem. Eng.* **2023**, *11*, 47–57. [CrossRef]
40. German, R.M. Coarsening in Sintering: Grain Shape Distribution, Grain Size Distribution, and Grain Growth Kinetics in Solid-Pore Systems. *Crit. Rev. Solid State Mater. Sci.* **2010**, *35*, 263–305. [CrossRef]
41. Huang, Y.; Zhu, P.; Xu, H.; Gu, C.; Zhou, J.; Xiao, G. Mn-based oxides modified with $MnSiO_3$ for thermochemical energy storage. *Chem. Eng. J.* **2024**, *483*, 149437. [CrossRef]
42. Sunde, T.O.L.; Grande, T.; Einarsrud, M.-A. Modified Pechini Synthesis of Oxide Powders and Thin Films. In *Handbook of Sol-Gel Science and Technology*; Springer: Berlin/Heidelberg, Germany, 2018; pp. 1089–1118.
43. Jana, P.; de la Peña O'Shea, V.A.; Coronado, J.M.; Serrano, D.P. Cobalt based catalysts prepared by Pechini method for CO_2-free hydrogen production by methane decomposition. *Int. J. Hydrogen Energy* **2010**, *35*, 10285–10294. [CrossRef]

Disclaimer/Publisher's Note: The statements, opinions and data contained in all publications are solely those of the individual author(s) and contributor(s) and not of MDPI and/or the editor(s). MDPI and/or the editor(s) disclaim responsibility for any injury to people or property resulting from any ideas, methods, instructions or products referred to in the content.

Article

Air-Purge Regenerative Direct Air Capture Using an Externally Heated and Cooled Temperature-Swing Adsorber Packed with Solid Amine

Heak Vannak [1], Yugo Osaka [2], Takuya Tsujiguchi [2] and Akio Kodama [3,*]

[1] Graduate School of Natural Science and Technology, Kanazawa University, Kanazawa 920-1192, Japan; heak.vannak999@stu.kanazawa-u.ac.jp
[2] Faculty of Mechanical Engineering, Institute of Science and Engineering, Kanazawa University, Kanazawa 920-1192, Japan
[3] Institute for Frontier Science Initiative, Kanazawa University, Kanazawa 920-1192, Japan
* Correspondence: akodama@se.kanazawa-u.ac.jp

Abstract: CO_2 capture from air is crucial in achieving negative emissions. Based on conventional or newly developed high-enriching processes, we investigated the rough enrichment of CO_2 from air via an externally heated or cooled adsorber (temperature-swing adsorption, TSA), along with air purge using double-pipe heat exchangers packed with low-volatility polyamine-loaded silica. A simple adsorption–desorption cycle was attempted in a TSA experiment, by varying the temperature from 20 °C to 60 °C using moist air, yielding an average CO_2 concentration of product gas that was ~17 times higher than the feed air, but the CO_2 recovery rate was poor. A double-step adsorption process was applied to increase CO_2 adsorption and recovery simultaneously. In this process, substantial-CO_2-concentration gas was used as the product gas, and the remaining gas was used as the reflux feed gas for adsorber. This method can provide a product gas with ~100 times higher CO_2 concentration than raw gas, with a recovery ratio ~60% under the shortest adsorption/desorption time and the longest refluxing time of cycle operation. Therefore, the refluxing step significantly helped to enhance CO_2 capture via adsorption from elevated-CO_2-concentration recirculating gas. With this CO_2 concentration, the product gas can serve as the CO_2 supplement for the growing plant processes.

Keywords: direct air capture; temperature swing adsorption; solid amine adsorbent; carbon dioxide; waste heat

1. Introduction

The use of fossil fuels in various applications has increased the concentration of CO_2 in atmospheric air from 325 to 410 parts per million (ppm) over the last 50 years. Consequently, in the coming decades, global temperature is expected to be 1.5 °C higher than pre-industrial levels [1,2]. To combat this global concern, governments are actively implementing policies that are aimed at limiting CO_2 outflow and striving for net-zero CO_2 emissions. Moreover, approximately 800 GtCO_2 emissions need to be avoided between now and 2050, and 120–160 GtCO_2 will have to be sequestered to achieve the said emission reductions during this period [3]. Until recently, only emissions from industries that release high CO_2 concentrations were considered for CO_2 capture. However, their emissions account for only 50% of the total amount of greenhouse gases emitted into the atmosphere. The remaining proportion consists of distributed emissions, including those from vehicle exhaust, agriculture, and habitation [4].

Capturing CO_2 directly from the ambient air (direct air capture, DAC) is a widely used process for solving the issue of distributed CO_2 emissions and managing CO_2 buildup from past outgassing. The main challenge for capturing CO_2 at ultralow concentrations is energy

Citation: Vannak, H.; Osaka, Y.; Tsujiguchi, T.; Kodama, A. Air-Purge Regenerative Direct Air Capture Using an Externally Heated and Cooled Temperature-Swing Adsorber Packed with Solid Amine. *Separations* 2023, 10, 415. https://doi.org/10.3390/separations10070415

Academic Editors: Zilong Liu, Meixia Shan, Yakang Jin and Zhiqian Jia

Received: 10 June 2023
Revised: 13 July 2023
Accepted: 19 July 2023
Published: 21 July 2023

Copyright: © 2023 by the authors. Licensee MDPI, Basel, Switzerland. This article is an open access article distributed under the terms and conditions of the Creative Commons Attribution (CC BY) license (https://creativecommons.org/licenses/by/4.0/).

consumption. Temperature-swing adsorption (TSA) using solid adsorbents is a promising technique to overcome this challenge, as it can operate with a low-temperature heat source using either solar or waste heat energy [5]. Furthermore, this process has substantially low environmental impacts.

To reduce the energy required for CO_2 capture from air, there is a need to enhance natural-CO_2 removal techniques. Afforestation and reforestation are simple methods for removing CO_2 from air, but they cannot handle the rising anthropogenic emissions. Enhanced weathering and ocean alkalinity enhancement can also be used to reduce CO_2 concentrations in atmospheric air. However, these methods are rarely implemented, due to their poor commercial viability and potential risks [6]. Bioenergy with carbon capture and storage (BECCS) has been introduced for CO_2 emission reduction. However, enhancing BECCS plants requires large amounts of land and water and, thus, may affect food security [7]. In this context, improving the CO_2 separation performance of the TSA process, operated with a low-grade heat source, can contribute to energy savings in CO_2 emission treatment.

In the TSA process, the trade-off between the regeneration temperature of the adsorbent and separation performance is a major challenge. To overcome this challenge, numerous studies have focused on improving adsorbent capacity. Among various adsorbents, solid amine adsorbents are viable for DAC, due to their comparatively high specific CO_2 capacities and adsorption rates at ultradilute CO_2 concentrations [8]. Moreover, a specific system of adsorption processes and operational conditions are required for the optimum performance of the adsorbents. Nevertheless, only a few studies have considered the design of adsorption process and operational conditions of TSA in the DAC field.

Generally, air or inert-gas purge is applied for regenerating the adsorbent process in conventional or direct-heating TSA. In such processes, a large amount of purge gas is supplied to the adsorber to increase the temperature of adsorbent within a short time, thereby decreasing the CO_2 concentration at the regeneration outlet. To reduce the flow rate of the purge and cycle time, electric-swing adsorption (ESA), temperature-vacuum-swing adsorption (TVSA), and indirect heating using hollow-fiber adsorbents or heat exchangers have been investigated [9–12]. ESA and TVSA cannot be used with low-grade heat sources, as they consume large amounts of electricity during the heating and depressurizing processes. Furthermore, using hollow-fiber adsorbents in the TSA process lowers the cycle time to less than 4 min [13]. However, the pressure drop of a hollow-fiber adsorbent bed is higher than that of a compact heat exchanger; hence, preparing the adsorbent bed requires expertise [14]. A simple way to prepare adsorbers is to use heat exchangers packed with an adsorbent for the indirect heating and cooling of TSA; this can reduce the adsorber heating time and the amount of purge gas.

In addition to improving the heat transfer in the adsorber, multiple cycle steps and operating multiple adsorbers are employed to improve the TSA performance for regeneration at low temperatures. Ntiamoah et al. applied a three-step cycle, including adsorption, hot-gas purge, and cooling steps, to a single adsorber to separate CO_2 from flue gas [15]. They achieved more than a 91% CO_2 concentration of product gas and a maximum CO_2 recovery of 55.5%, using a regeneration temperature of 150 °C. Masuda et al. used two adsorbers for a two-stage CO_2 separation process from post-combustion [16]. Air purge under a regeneration temperature of 80 °C yielded a maximum CO_2 concentration of 95% in the product gas and an overall recovery ratio of 60%. Multiple steps of adsorption processes under swing temperatures have been reported in the DAC [17,18]. However, these studies employed a vacuum step to improve the separation performance. To date, applying air purge for the regeneration process in the multiple-stage process using multiple adsorbers for CO_2 capture from the air has not been studied.

To develop a system for adsorption processes using low-grade heat sources for DAC, this study aimed to investigate the design of cycle operations required for CO_2 separation from wet ambient air, using an externally heated and cooled TSA-packed adsorbent in a heat exchanger. Here, a double-tube heat exchanger filled with a functionalized polyamine adsorbent, which has not yet been commercialized, was used as the adsorber. This process improved the CO_2 separation from wet air via a regeneration process at a temperature of 60 °C with air purge. Regardless of the vacuum processes, multistage processes and multiple adsorbers were combined. This study investigated (i) the difference in separation performance between a simple adsorption–desorption process and multistage processes and (ii) the effect of different durations of cycle steps, feed gas, and purge gas on CO_2 purity and the recovery ratio. The design of the TSA cycle in this study would produce the product gas, which could be used for microbial and algae cultivation for food and fuel and in greenhouses.

2. Experimental Methods

2.1. Adsorbent Packed Column

The studied material was a functionalized polyamine adsorbent derived from a previous study [19]. The adsorbent particle had a diameter of approximately 0.6 mm and could be regenerated at a moderate temperature of 60 °C, which is the maximum regenerated temperature that the adsorbent can withstand.

Figure 1 shows a schematic of the reactor. The adsorbent was packed in the copper tube serving as the inner pipe of the double-pipe heat exchanger for applying both sorption steps. Glass wool was inserted into both ends of the adsorbent section to provide a flexible space for changing the adsorbent volume. A filter was placed outside each piece of glass wool to stabilize the position of the adsorbent. Indirect heat transfer was applied during adsorption and desorption by circulating cooling and hot water, respectively. The feed and purge gases had opposite directions, and the desorption outlet was placed at the bottom of the reactor. If the regeneration outlet gas contains a large amount of water vapor, it tends to condense. Setting the desorption outlet at the bottom of the adsorber prevents condensate from flowing back into the adsorber, because the condensate is discharged by gravity. Moreover, an electric heating cable was applied to avoid this condensation at the regeneration outlet.

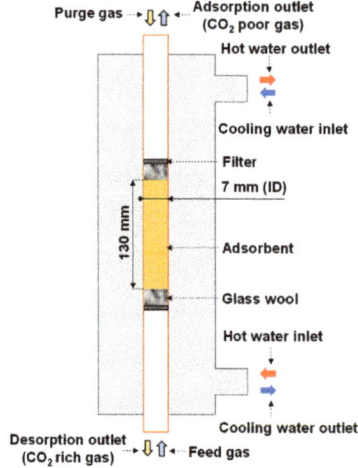

Figure 1. Schematic of double-pipe heat exchanger (reactor).

2.2. Breakthrough Curves of the Adsorbent

Breakthrough curve experiments were conducted to determine the effectiveness of the adsorption and desorption processes. In the experiments, 1.36 g (2.1 mL) of the adsorbent was packed in a copper tube (outer diameter: 6.35 mm; inner diameter: 4.57 mm), which served as the inner pipe of the double-pipe heat exchanger. A standard gas with a CO_2 concentration of 400 ppm was used as the feed gas at a flow rate of 3.9 L STP/min for adsorption at 20 °C. Desorption was performed at 60 °C with a purge-air flow rate of

0.04 L STP/min. The dew-point temperature (Td) of both inlet gases was controlled at −50 °C (dry condition), −5 °C, 0 °C, 5 °C, and 8 °C.

Figure 2 shows the CO_2 adsorption rate at different levels of humidity applied to the feed gas. Additionally, the packed adsorbent was tested, starting with a dry condition and followed by a wet condition with Td = 5 °C, 0 °C, 8 °C, and −5 °C. The adsorption breakthrough started immediately after the completion of the preregeneration process. Since the adsorbent temperature was controlled by cooling and heating water, it was stabilized for only 1 min. The result indicated that low-humidity condition significantly increased the CO_2 adsorption rate during the first 100 min, as observed at Td of −5 °C and 0 °C. However, high humidity (the case of Td = 8 °C) was responsible for the lowest adsorption rate in the first 60 min and longest adsorption cycle time. This indicated that CO_2/H_2O adsorption selectivity decreases when moisture is oversupplied to the feed gas. However, because amine adsorbents increased CO_2 adsorption capacity under wet conditions, although there was no competition between CO_2 and water adsorption under dry conditions, the dry adsorption was saturated first.

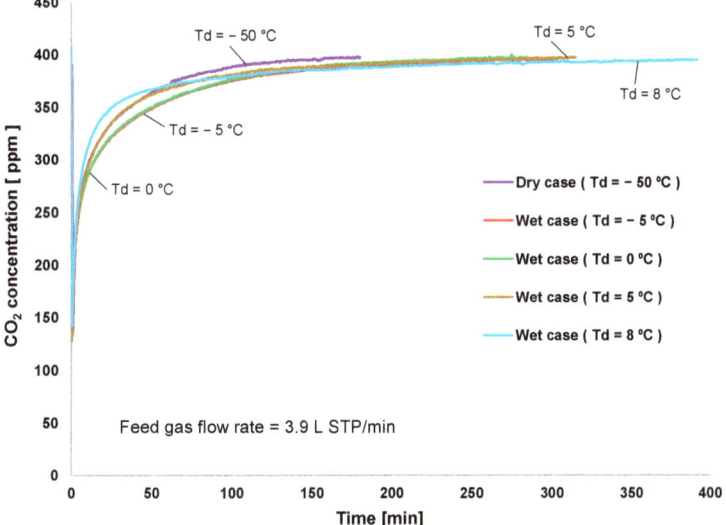

Figure 2. Breakthrough curve of CO_2 adsorption.

Figure 3 shows the performance of CO_2 desorption breakthroughs for the various adsorption cases. Desorption occurred immediately after the saturation of CO_2 adsorption. The highest CO_2 desorption rate was obtained under dry conditions, although the lowest amount of CO_2 uptake was also obtained under this condition, as shown in Figure 4. Even though high humidity in the feed gas reduced the rate of CO_2 adsorption, the overall amount of adsorbed CO_2 remained significantly higher than that under dry conditions. This indicated that amine reacts with CO_2 to form carbamates, and these carbamates continue to react with CO_2 to form bicarbonate species in the presence of moisture. However, high humidity causes the oversaturation of the materials with water, which could block CO_2 access to the amine group [20]. Hence, the case of Td = 0 °C showed the highest adsorption capacity. The error between the amounts of CO_2 adsorbed and desorbed for each case was smaller than 5%. Furthermore, this error was caused by the slight variations in the CO_2 concentration in the feed gas and by error in converting data between the data logger and CO_2 analyzer. The total amount of CO_2 adsorbed did not decrease according to the sequence of the experiment. This revealed that the variation of the amount of CO_2 adsorbed is completely influenced by the humidity conditions, rather than by amine oxidation or degradation of the adsorbent.

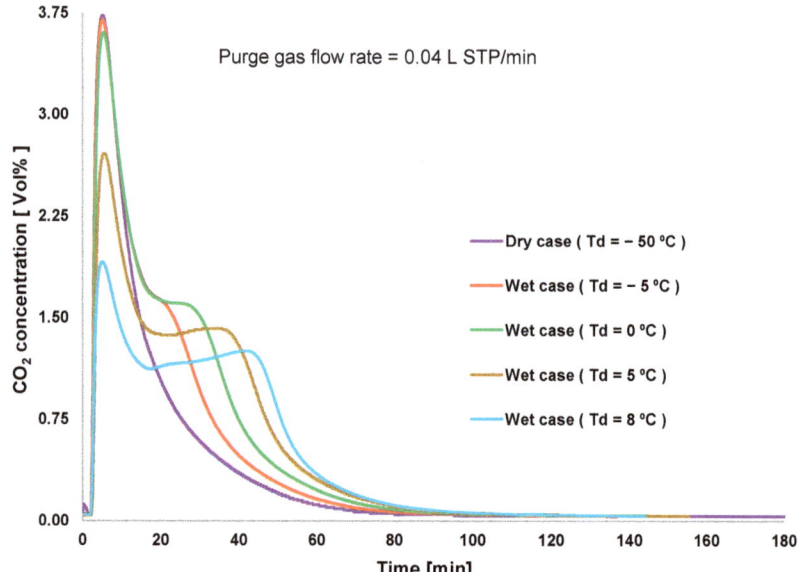

Figure 3. Breakthrough curves of CO_2 desorption.

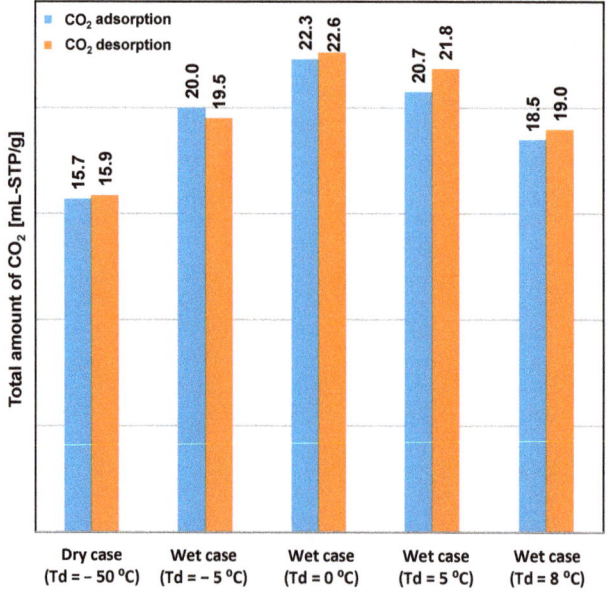

Figure 4. Total amounts of CO_2 adsorbed and desorbed, as obtained from the breakthrough curves.

Figure 3 shows that the absence of moisture allows adsorbents to desorb CO_2 rapidly, because the supplied heat is consumed only for the CO_2 desorption process. The second peak in the desorption curve was observed in all wet conditions. An increase in the humidity of the feed gas reduced the first peak and increased the second peak of the desorption curve. The heat supplied for the desorption process was partially consumed during water desorption, resulting in the reduced first peak of the CO_2 desorption rate. An increase in the second peak of CO_2 desorption was attributed to the completion of the decomposition of bicarbonate species.

2.3. Experimental Apparatus and Conditions for Simple Adsorption–Desorption Cycle

To investigate the cycle operation, the copper tube (outer diameter: 8 mm, inner diameter: 7 mm, as indicated in Figure 1) used as the inner pipe of the heat exchanger was filled with 3.24 g (5 mL) of the adsorbent. The flow rate of both inlet gases increased proportionally with an increase in the adsorbent amount. As shown in Figure 5, the apparatus supported two reactors: one for adsorption and the other for desorption. Thus, the simultaneous use of both reactors for the same case study accelerated the confirmation of experimental results. The entire system combined two water circulations (hot water and cooling water) and two gas flows (process gas and regeneration gas). Eight three-way valves alternatingly supplied the cooling water and process gas to the adsorption reactor and the hot water and regeneration gas to the desorption reactor. A membrane humidifier (HFB-02-100/BNP, AGC Engineering Co., Ltd., Chiba, Japan) was used to control the amount of moisture in the gas, and the Td of the gas was checked at the humidifier outlet using a humidity sensor (Vaisala Co., Ltd., HMP, Vantaa, Finland) at a pressure slightly higher than atmospheric pressure. The flow rates of the feed and purge gases were controlled using mass flow controllers (Azbil Co., Ltd., MQV0500, Tokyo, Japan; Horiba STEC Co., Ltd., SEC-E40, Kyoto, Japan, respectively). Pressure indicators (Nidec Co., Ltd., PA-750, Kyoto, Japan) were placed at the outlets of both mass flow controllers to determine the inlet pressures during adsorption and desorption. Moreover, four humidity sensors were installed at the gas inlets and outlets of both reactors to monitor the varying moisture levels of the gases after they passed through the adsorbent column during both steps. Thermocouples were inserted into the adsorbent column to confirm the sorption temperatures. The outlet flow rates during sorption and the CO_2 concentrations were checked using dry-type gas flowmeters (Shinagawa Co., Ltd., DC1, Tokyo, Japan) and CO_2 concentration analyzers (Shimadzu Co., Ltd., CGT-7100, Kyoto, Japan), respectively. Before measurement, the CO_2-concentration-measurement device was calibrated using CO_2 and N_2 as the standard gases. A data logger (Graphtec, GL820, Yokohama, Japan) was used to record the measured data. The reactors and the tubes and fittings of the gas and water systems were covered with a thermal isolation material.

The simple adsorption–desorption cycle was conducted under the conditions in Table 1. For each reactor, desorption immediately started after adsorption stopped. Moreover, the simple adsorption–desorption cycle was operated in two scenarios. In the first scenario, the adsorption and desorption periods were equal. In the second scenario, desorption was longer than adsorption, the duration of which was 5 min. After completing one scenario, the tested adsorbent was replaced with a new one. Furthermore, wet adsorption and desorption were chosen to investigate the cycle time of operation, because dry air rarely occurs in the actual atmosphere. To avoid condensation at the regeneration outlet when supplying highly humid feed gas to the adsorption step, humid air with a Td of 5 °C (approximate relative humidity at the adsorption temperature = 35%, 8644 ppmv) was used as the feed and purge gases. The CO_2 concentration of the applied air was slightly inconsistent (420–500 ppm). After the quasi-steady state, the desorption outlet gas was collected in an aluminum bag to check its CO_2 concentration, and the required data were recorded.

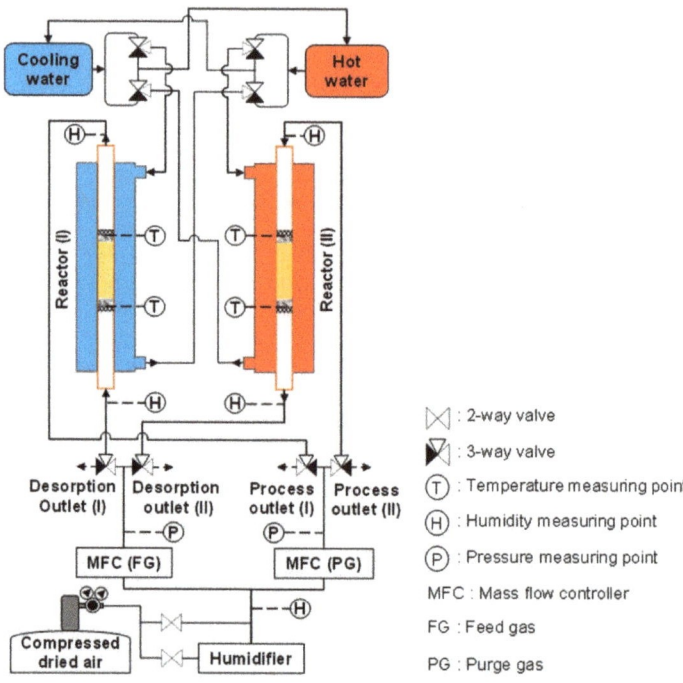

Figure 5. Schematic of experimental apparatus for simple adsorption–desorption cycle.

Table 1. Experimental conditions for simple adsorption–desorption cycle.

RUN#	Step	Temperature (°C)	Air Flow Rate (L STP/min)	Air Humidity Dew Point (°C)	Time (min)
1	Adsorption Desorption	20 60	9.3 0.09	5	3 3
2	Adsorption Desorption	20 60	9.3 0.09	5	5 5
3	Adsorption Desorption	20 60	9.3 0.09	5	10 10
4	Adsorption Desorption	20 60	9.3 0.09	5	30 30
5	Adsorption Desorption	20 60	9.3 0.09	5	60 60
6	Adsorption Desorption	20 60	9.3 0.09	5	5 10
7	Adsorption Desorption	20 60	9.3 0.09	5	5 15
8	Adsorption Desorption	20 60	9.3 0.09	5	5 20
9	Adsorption Desorption	20 60	9.3 0.09	5	5 25
10	Adsorption Desorption	20 60	9.3 0.09	5	5 30

2.4. Experimental Apparatus and Conditions for Double-Step Adsorption

The feed and purge gas pipeline systems of the simple adsorption–desorption cycle were modified to support the refluxing step in the double-step adsorption process. During refluxing, the regeneration gas (whose CO_2 concentration was greatly higher than that of the room air) from the desorption column flowed into the adsorber. In this modification, three two-way valves and one three-way valve were added to the system, as shown in Figure 6. A two-way valve was placed along the pipe connecting the desorption outlets of both adsorbers. Two other two-way valves were installed before the two sampling points of the desorbed gas. A three-way valve was placed at the outlet of the mass flow controller of the feed gas to release the feed gas during refluxing.

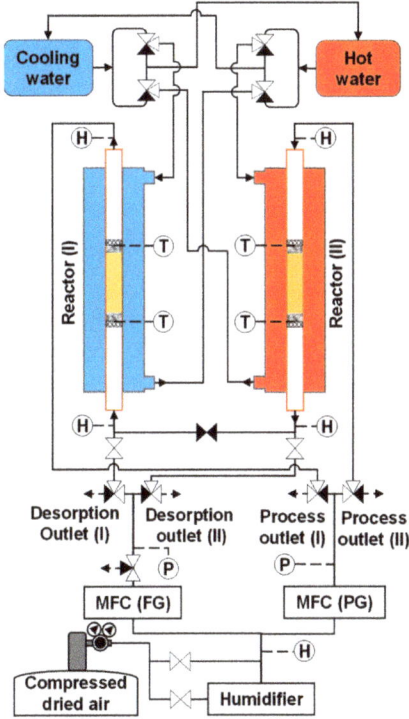

Figure 6. Schematic of experimental apparatus for double-step adsorption.

The loop of the double-step adsorption procedure is illustrated in Figure 7. As seen in Figure 7a, hot water and purge gas were supplied to reactor I for desorption, while cooling water and feed gas were transferred into reactor II for adsorption. In this step, the substantial-CO_2-concentration gas at the desorption outlet of reactor I was collected as the product gas. As shown in Figure 7b, refluxing began immediately after the ambient air (feed gas) was discharged into reactor II. Simultaneously, the desorption outlets of both reactors were connected, and the regeneration gas of reactor I was refluxed to reactor II for secondary adsorption. Then, the regeneration outlets of both reactors were disconnected. In the next step, as shown in Figure 7c, the experimental system worked in reverse to the procedure in Figure 7a. During this step, the substantial-CO_2-concentration gas regenerated from reactor II was collected as the product gas. Figure 7d presents the last step of the loop. The feed gas was released from the system to terminate the step in Figure 7c, and both reactors were joined again at the desorption outlet. Then, the working process in Figure 7b was reversed. The experimental conditions are described in Table 2. The double-step adsorption comprised three scenarios: scenario 1 (from RUN11-1 to 13-4), scenario 2 (from RUN14-1 to 16-4), and scenario 3 (from RUN17-1 to 19-4). Because the tested adsorbent was replaced with a new adsorbent, the decrease in the adsorbent's performance was not investigated in this study. Therefore, the duration for which the adsorbent can maintain the operation is not indicated by the experiment.

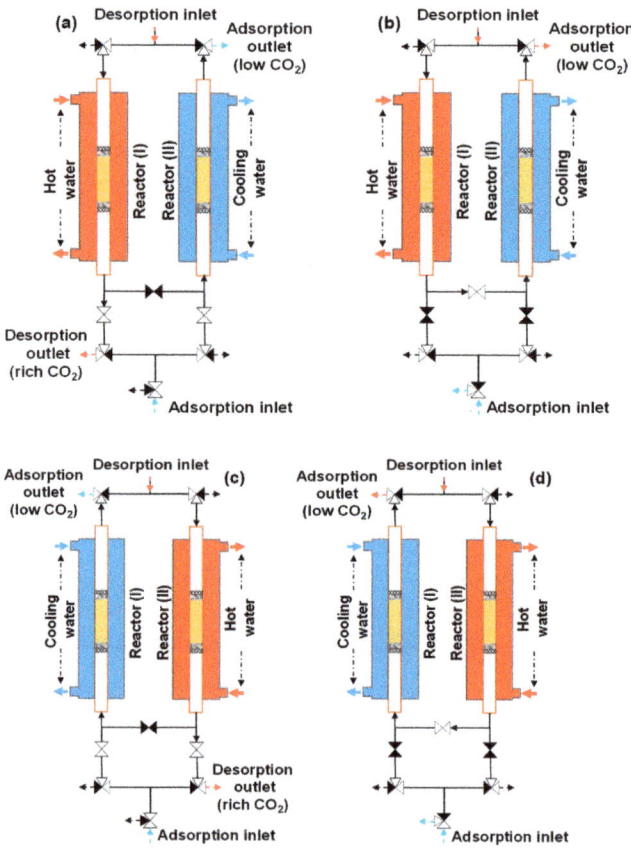

Figure 7. Experimental procedure for double-step adsorption: (**a**) reactor I (desorption) and reactor II (adsorption), (**b**) refluxing desorption gas from reactor I to II, (**c**) reactor I (adsorption) and reactor II (desorption), and (**d**) refluxing desorption gas from reactor II to I.

Table 2. Experimental conditions for double-step adsorption.

RUN#	Step	Temperature (°C)	Air Flow Rate (L STP/min)	Air Humidity Dew Point (°C)	Time (min)
11-1	Adsorption	20	9.3		3
	Desorption	60	0.09	5	3
	Refluxing	60 & 20	0.09		3
11-2	Adsorption	20	9.3		3
	Desorption	60	0.09	5	3
	Refluxing	60 & 20	0.09		30
11-3	Adsorption	20	9.3		3
	Desorption	60	0.09	5	3
	Refluxing	60 & 20	0.09		55

Table 2. Cont.

RUN#	Step	Temperature (°C)	Air Flow Rate (L STP/min)	Air Humidity Dew Point (°C)	Time (min)
11-4	Adsorption	20	9.3		3
	Desorption	60	0.09	5	3
	Refluxing	60 & 20	0.09		110
12-1	Adsorption	20	9.3		5
	Desorption	60	0.09	5	5
	Refluxing	60 & 20	0.09		5
12-2	Adsorption	20	9.3		5
	Desorption	60	0.09	5	5
	Refluxing	60 & 20	0.09		30
12-3	Adsorption	20	9.3		5
	Desorption	60	0.09	5	5
	Refluxing	60 & 20	0.09		55
12-4	Adsorption	20	9.3		5
	Desorption	60	0.09	5	5
	Refluxing	60 & 20	0.09		110
13-1	Adsorption	20	9.3		10
	Desorption	60	0.09	5	10
	Refluxing	60 & 20	0.09		10
13-2	Adsorption	20	9.3		10
	Desorption	60	0.09	5	10
	Refluxing	60 & 20	0.09		30
13-3	Adsorption	20	9.3		10
	Desorption	60	0.09	5	10
	Refluxing	60 & 20	0.09		55
13-4	Adsorption	20	9.3		10
	Desorption	60	0.09	5	10
	Refluxing	60 & 20	0.09		110
14-1	Adsorption	20	9.3		3
	Desorption	60	0.045	5	3
	Refluxing	60 & 20	0.045		3
14-2	Adsorption	20	9.3		3
	Desorption	60	0.045	5	3
	Refluxing	60 & 20	0.045		30
14-3	Adsorption	20	9.3		3
	Desorption	60	0.045	5	3
	Refluxing	60 & 20	0.045		55
14-4	Adsorption	20	9.3		3
	Desorption	60	0.045	5	3
	Refluxing	60 & 20	0.045		110
15-1	Adsorption	20	9.3		5
	Desorption	60	0.045	5	5
	Refluxing	60 & 20	0.045		5
15-2	Adsorption	20	9.3		5
	Desorption	60	0.045	5	5
	Refluxing	60 & 20	0.045		30
15-3	Adsorption	20	9.3		5
	Desorption	60	0.045	5	5
	Refluxing	60 & 20	0.045		55

Table 2. Cont.

RUN#	Step	Temperature (°C)	Air Flow Rate (L STP/min)	Air Humidity Dew Point (°C)	Time (min)
15-4	Adsorption Desorption Refluxing	20 60 60 & 20	9.3 0.045 0.045	5	5 5 110
16-1	Adsorption Desorption Refluxing	20 60 60 & 20	9.3 0.045 0.045	5	10 10 10
16-2	Adsorption Desorption Refluxing	20 60 60 & 20	9.3 0.045 0.045	5	10 10 30
16-3	Adsorption Desorption Refluxing	20 60 60 & 20	9.3 0.045 0.045	5	10 10 55
16-4	Adsorption Desorption Refluxing	20 60 60 & 20	9.3 0.045 0.045	5	10 10 110
17-1	Adsorption Desorption Refluxing	20 60 60 & 20	7.45 0.045 0.045	5	3 3 3
17-2	Adsorption Desorption Refluxing	20 60 60 & 20	7.45 0.045 0.045	5	3 3 30
17-3	Adsorption Desorption Refluxing	20 60 60 & 20	7.45 0.045 0.045	5	3 3 55
17-4	Adsorption Desorption Refluxing	20 60 60 & 20	7.45 0.045 0.045	5	3 3 110
18-1	Adsorption Desorption Refluxing	20 60 60 & 20	7.45 0.045 0.045	5	5 5 5
18-2	Adsorption Desorption Refluxing	20 60 60 & 20	7.45 0.045 0.045	5	5 5 30
18-3	Adsorption Desorption Refluxing	20 60 60 & 20	7.45 0.045 0.045	5	5 5 55
18-4	Adsorption Desorption Refluxing	20 60 60 & 20	7.45 0.045 0.045	5	5 5 110
19-1	Adsorption Desorption Refluxing	20 60 60 & 20	7.45 0.045 0.045	5	10 10 10
19-2	Adsorption Desorption Refluxing	20 60 60 & 20	7.45 0.045 0.045	5	10 10 30
19-3	Adsorption Desorption Refluxing	20 60 60 & 20	7.45 0.045 0.045	5	10 10 55
19-4	Adsorption Desorption Refluxing	20 60 60 & 20	7.45 0.045 0.045	5	10 10 110

2.5. Performance Indices

The performance of the cycle operation time was mainly indexed by the CO_2 concentration and the CO_2 recovery ratio of the desorption outlet gas, R_{CO_2}. This ratio is determined by the following equation:

$$R_{CO_2}[\%] = \frac{V_{CO_2}^{DOG}}{V_{CO_2}^{FG}} \times 100 \quad (1)$$

where $V_{CO_2}^{FG}$ and $V_{CO_2}^{DOG}$ are the volumes of CO_2 in the feed gas and the desorption outlet gas, respectively.

3. Results and Discussion

3.1. Influence of Half-Cycle Time on Performance of Simple Adsorption–Desorption Cycle

The small size of the adsorbent and the high flow rate of the supplied feed gas led to a high adsorption inlet pressure of approximately 120 kPa, but the desorption inlet pressure was low (about 1 kPa), due to the small amount of the supplied purge gas. As shown in Figure 8, a short cycle time considerably increased the CO_2 concentration at the desorption outlet. However, the CO_2 recovery ratio in a short desorption period was poor, despite an increase in the CO_2 concentration of the product gas in this case. When the cycle time was under 10 min, the increasing CO_2 concentration at the desorption outlet could not overcome the reducing volume of the regeneration gas to stabilize the CO_2 recovery ratio. Under these experimental conditions, the maximum CO_2 concentration in the adsorbent material was almost 17 times higher than that in the ambient air, but the CO_2 recovery ratio was low.

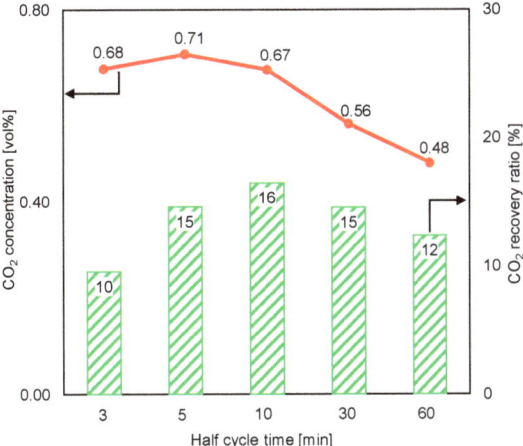

Figure 8. Influence of cycle time on time-averaged CO_2 concentrations at desorption outlet and CO_2 recovery ratios (RUN#:1-5).

The rapid adsorption in this experiment was attributed to the high flow rate of the feed gas, as indicated in Figure 9, yielding the optimal CO_2 concentration in a short cycle time. Thus, the flow rate of the feed gas may have affected the cycle time. However, such an extremely short cycle time was insufficient for the adsorbent to regenerate CO_2. Despite the fast adsorption, the CO_2 recovery ratio was low, due to the high flow rate of the feed gas (approximately 100 times that of the purge gas). This also compromised the CO_2 capture over a prolonged cycle time. Moreover, highly prolonged adsorption time decreased the adsorption ratio before the end of adsorption; hence, the CO_2 recovery ratio reduced considerably. In addition, the amount of packed adsorbent could not completely adsorb the CO_2 contained in the feed gas during the extended adsorption period.

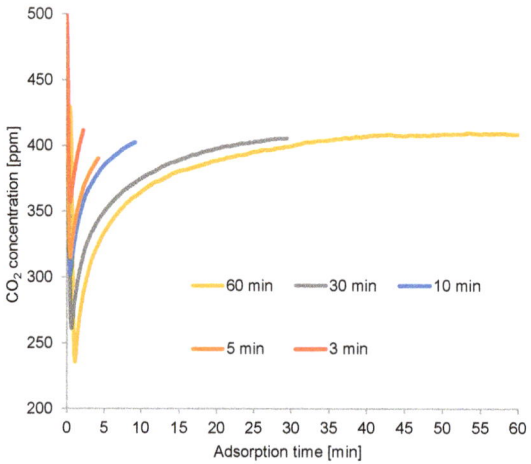

Figure 9. Time profiles of adsorption outlet CO_2 concentrations (RUN#:1-5).

A longer cycle time extended the regeneration process and provided a higher CO_2-concentration peak, as illustrated in Figure 10. However, the CO_2 concentration rapidly declined from its peak, diluting the CO_2 in the desorption gas. Moreover, the excessively high purge gas flow rate contributed to the CO_2 dilution under long cycle times. Under an exceedingly short cycle time, the regeneration process terminated around the time when the CO_2 concentration at desorption outlet was at its peak, resulting in high CO_2 concentration at the desorption outlet under a short cycle time. The second peak of CO_2 concentration continued to appear under a long cycle time, although the diameter of the inner tube of the double-pipe heat exchanger increased proportionally with the amount of adsorbent used in this case, which was higher than that used for the breakthrough curve experiment. Thus, the second peak was attributed to the adsorbent's performance, rather than to the effect of a specific reactor design.

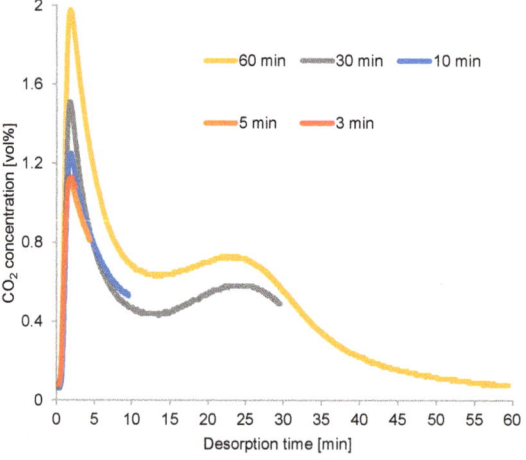

Figure 10. Time profiles of CO_2 concentrations at desorption outlet (RUN#:1-5).

Furthermore, the CO_2 recovery ratio result at a short cycle time was unfavorable, as shown in Figure 8. This result was attributed to the reading error of the mass flowmeter (the measurement of total regeneration outlet gas volume). Nevertheless, because the CO_2 concentration in the desorption outlet was lower than 0.80% (the volume of CO_2 in the product gas was very small compared to the total volume), the peak and variation trend of the recovery ratio remained similar to those shown in Figure 8, even though the estimated total volume of regeneration outlets (the estimation is based on the amount of supplied inlet air for regeneration per each cycle time) was used in the CO_2 recovery ratio calculation. In this regard, the total gas volume was almost the same between the regeneration inlet and the regeneration outlet.

3.2. Effect of Extending Regeneration Time beyond Adsorption Time

The desorption time was extended beyond the adsorption duration to identify the effect of complete regeneration on the desorption outlet CO_2 concentration and the CO_2 recovery ratio. Figure 11 depicts the variations in the desorption outlet CO_2 concentration and the CO_2 recovery ratio under a 5 min adsorption time and a longer desorption period. The CO_2 concentration and CO_2 recovery ratio varied in opposite directions. When the desorption time was 30 min, the CO_2 concentration decreased three times, while the recovery ratio increased two times. Because the purge gas significantly decreased the CO_2 concentration following the extended desorption time, reducing the purge-gas flow rate may have minimized the decrease in CO_2 concentration. Additionally, reducing the purge-gas flow rate could have resulted in a slight decrease in the CO_2 recovery ratio.

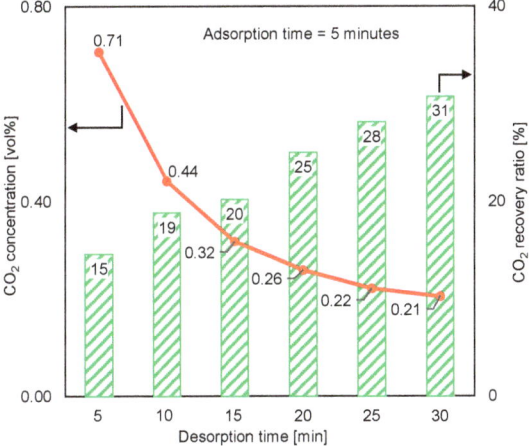

Figure 11. Effect of extending regeneration time beyond adsorption time on time-averaged CO_2 concentrations at desorption outlet and CO_2 recovery ratios (RUN#:2, 6-10).

3.3. Double-Step Adsorption

The result of the simple adsorption–desorption cycle suggested that regeneration was a control step of the cycle operation. As mentioned in Section 3.2, the CO_2 concentration at the desorption outlet could be significantly concentrated by shortening the cycle, but this greatly decreased the CO_2 recovery ratio. The simple adsorption–desorption cycle could not overcome the fluctuation in the time profile of the desorption outlet CO_2 concentration, as illustrated in Figure 10, to increase CO_2 concentration and recovery, simultaneously. The first peak of the desorption outlet CO_2 concentration could potentially be used as the product gas. The second peak of the CO_2 concentration at the desorption outlet could be used to enhance the first peak, resulting in an enrichment of the CO_2 concentration of the product gas. Therefore, we suggest a method that divides the adsorption gas over time, with the substantial-CO_2-concentration gas serving as the product gas and the remaining portion serving as the reflux feed gas to the adsorber. This method can improve CO_2 adsorption and recovery, simultaneously.

3.4. Influence of Cycle Operation on the Performance of Double-Step Adsorption

The CO_2 concentrations at the desorption outlet of both reactors during double-step adsorption were checked. The result was deemed acceptable when the difference between the two concentrations was lower than or equal to 5%. Figure 12 shows the effect of adsorption/desorption (ads/des) and refluxing time variations on the time-averaged regeneration outlet CO_2 concentrations and the CO_2 recovery ratios of both reactors. The best performance belonged to the shortest adsorption/desorption time, which matched the longest refluxing time. Exceedingly long adsorption/desorption times allowed the adsorbent to adsorb more CO_2, but more purge air was added into the product gas, thereby diluting the CO_2 concentration of the product gas. Moreover, since the amount of feed gas was 100 times that of the purge gas, the greatly prolonged adsorption period decreased the CO_2 recovery ratio. Because increasing the CO_2 concentration in the feed gas theoretically increases the adsorption capacity, the adsorption from the elevated-CO_2-concentration refluxing gas was a control step of the cycle. Hence, extending the refluxing period crucially increased the CO_2 concentration at the desorption outlet and the CO_2 recovery ratio. When the adsorption/desorption time equaled the refluxing time, the CO_2 concentrations of the product gases were almost identical, but the highest CO_2 recovery ratio occurred under 5 min adsorption/desorption and 5 min refluxing. Thus, setting equal periods of adsorption/desorption and refluxing had less influence on the CO_2 concentration of the product gas, whereas significant variation in the CO_2 concentration was observed when reflux time was extended beyond the adsorption/desorption time, and maximizing the CO_2 recovery ratio required an appropriate cycle-operation time for a particular amount of supplied feed and purge gas.

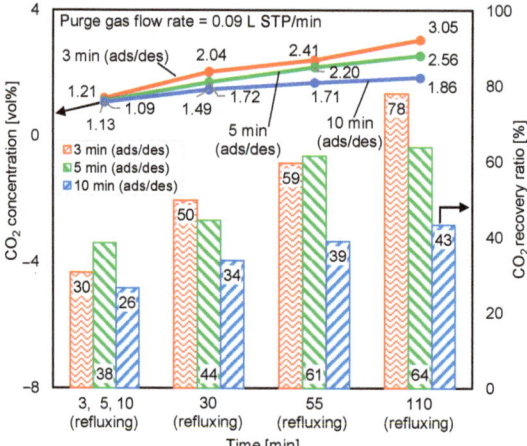

Figure 12. Influence of adsorption/desorption and refluxing times on time-averaged CO_2 concentrations at desorption outlet and CO_2 recovery ratios.

3.5. Influence of Regeneration Air Flow Rate

Figure 13 depicts the variations in the desorption outlet CO_2 concentration and CO_2 recovery ratio at a purge gas flow rate of 0.045 L STP/min. Compared with Figure 12, in Figure 13, when half of the regeneration air that was supplied for the experiment in Figure 12 was reduced, the desorption outlet CO_2 concentration in each case, as shown in Figure 13, increased by more than 40%. However, the low purge gas flow rate was insufficient for the regeneration of the adsorbent; hence, the CO_2 recovery ratio decreased. When the purge gas flow rate was reduced, the longest desorption time yielded a higher increase rate in CO_2 concentration and a lower decrease rate in the CO_2 recovery ratio than the shortest desorption time. Thus, regeneration was slowed by the reduction in the regeneration air flow rate.

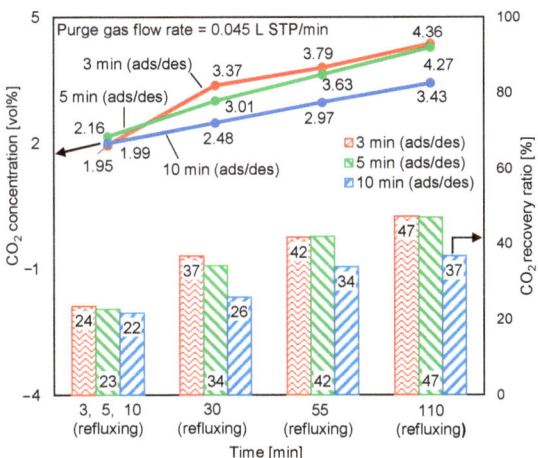

Figure 13. Influence of regeneration air flow rate on time-averaged CO_2 concentrations at desorption outlet and CO_2 recovery ratios.

3.6. Influence of Feed Gas Flow Rate

The supplied feed gas flow rate was reduced by 20% to clarify the excess supplied feed gas, as discussed in Section 3.1. The CO_2 concentrations at desorption outlet and CO_2 recovery ratios at a feed gas flow rate of 7.45 L STP/min and a purge gas flow rate of 0.045 L STP/min are illustrated in Figure 14. According to Figures 13 and 14, this reduction in the feed gas flow rate decreased the CO_2 concentration at the regeneration outlet by less than 5%. Nevertheless, the CO_2 recovery ratio increased by approximately 15%. Additionally, the adsorption pressure inlet reduced from approximately 120 kPa to 85 kPa. In double-step adsorption, the second step (adsorption from elevated-CO_2-concentration refluxing gas) played a more important role in enhancing adsorption capacity than the first step (adsorption from CO_2 in air). Therefore, despite the 20% reduction in the feed gas, the CO_2 concentration of the product gas remained almost stable.

The time profiles of the CO_2 concentrations at the adsorption and desorption outlets are shown in Figures 15 and 16, respectively, to gain more insight into the above results. According to these figures, the outlet gas needed 2 min to reach the detector; hence, the CO_2 concentrations of the gases remaining at the pipe inlet of the CO_2 analyzer were detected. Although the time profiles contained measurement delays, the averaged concentrations were accurate, as the regeneration outlet gases were collected into sample bags near the desorption outlet of the column to check their average CO_2 concentrations. Thus, the dilution of the regeneration outlet CO_2 concentration caused by the remaining gas in the pipeline was negligible. Figure 15 shows a zoomed-in view of the starting state of the CO_2 concentration profile, revealing the variation of the CO_2 concentration at the adsorption outlet during the first 15 min. At the adsorption/desorption time, the adsorbent adsorbed the CO_2 from ambient air at a high flow rate; as a result, the CO_2 concentration at adsorption outlet increased rapidly. Afterward, this adsorbent captured the CO_2 from the low-flow-rate elevated-CO_2-concentration regeneration gas refluxed from the other column; consequently, almost all the CO_2 contained in the refluxing gas was adsorbed.

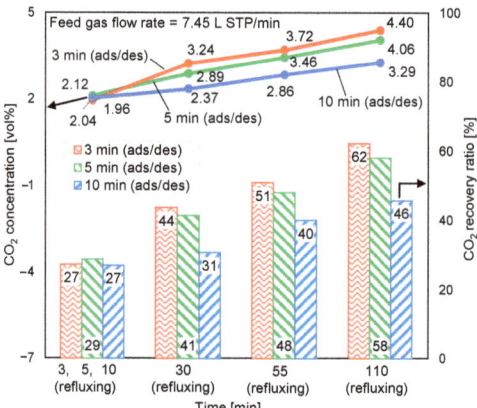

Figure 14. Influence of feed gas flow rate on time-averaged CO_2 concentrations at desorption outlet and CO_2 recovery ratios.

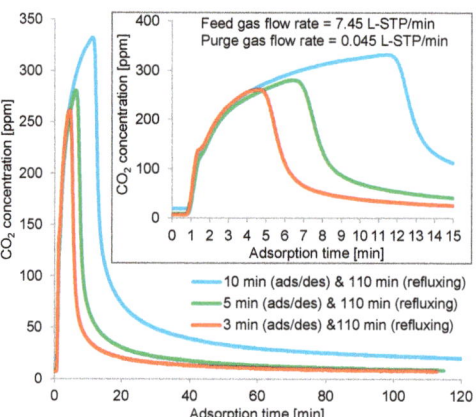

Figure 15. Time profiles of CO_2 concentrations at adsorption outlet in double-step adsorption.

As shown in Figure 16, at the beginning of desorption, the CO_2 flow rate profile narrowed and rose, indicating rapid desorption. The regeneration outlet CO_2-concentration profile showed that the optimum average CO_2 concentration was obtained from the shorter adsorption/desorption time, as discussed in Section 3.1.

During the refluxing step, the CO_2 concentration in the refluxing gas varied, following the fluctuation behavior of the time profiles of CO_2 desorption shown in Figure 10. Moreover, the adsorption capacity of the adsorbent greatly depended on the CO_2 concentration in the adsorption inlet gas. Despite the constant refluxing duration, the CO_2 concentration sent to the adsorber during this step differed, as it depended on the variations in the adsorption/desorption time. As shown in Figure 16, a short adsorption/desorption time allowed the adsorber to capture less CO_2 from the room air, but the adsorber could adsorb CO_2 of the highest concentration from the refluxing gas. Given a longer adsorption/desorption time, the adsorber captured more CO_2 from the room air, but the adsorber adsorbed a low concentration of CO_2 from the refluxing gas when the refluxing step began. Furthermore, an extended refluxing time enabled the adsorber to capture more CO_2 from the second peak of CO_2 regeneration, as indicated by the time profile in Figure 10. Therefore, the peak in the time profile of the desorption outlet CO_2 concentration shown in Figure 16 varied with the refluxing time. Nevertheless, the influence of adsorbed moisture remaining in the adsorbent at the peak of the regeneration outlet CO_2 concentration required investigation.

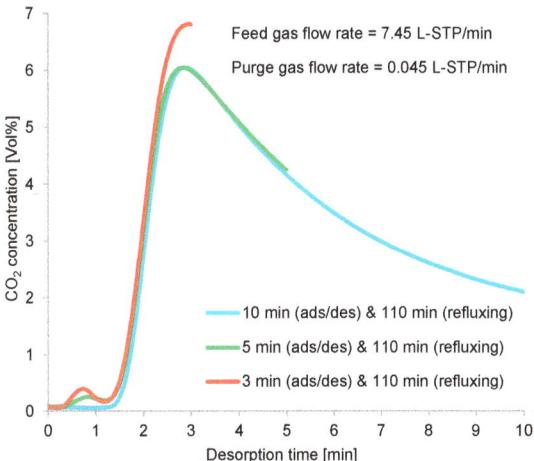

Figure 16. Time profiles of CO_2 concentrations at desorption outlet in double-step adsorption.

4. Conclusions

The capture of CO_2 from ambient air using a functionalized polyamine-impregnated solid adsorbent was investigated via TSA equipped with indirect heating and cooling to evaluate the possibility of integrating this capture method into waste heat or solar energy operation. An adsorption process was developed, and its separation performance was assessed under different cycle-operation times, regeneration air-flow rates, and feed air-flow rates. The focus was on achieving a high regeneration outlet CO_2 concentration and CO_2 recovery ratio. The conclusions are summarized as follows:

1. The adsorption step consumed a much shorter time than the desorption step, which indicated that the cycle operation was controlled by the latter. Shortening the cycle time greatly concentrated the CO_2 at the desorption outlet, due to the rapid CO_2 desorption rate, but the CO_2 recovery ratio decreased significantly because the total amount of regeneration air per cycle time was much smaller than that of feed air. Hence, a trade-off existed between the regeneration outlet CO_2 concentration and the CO_2 recovery ratio.
2. The proposed double-step adsorption process simultaneously improved CO_2 adsorption and recovery. This method divided desorption gas over time, with high-CO_2-concentration gas serving as the product gas and the remaining portion serving as the reflux feed gas to the adsorber. In this case, minimizing the adsorption/desorption time and prolonging the refluxing time significantly improved CO_2 separation. Refluxing played a crucial role in enhancing CO_2 capture because of the adsorption from high-CO_2-concentration recirculation gas. Furthermore, a lower regeneration air flow rate during desorption increased the CO_2 concentration.

In future studies, the influence of adsorbed moisture on CO_2 regeneration behavior must be clarified. Additionally, the development of a continuous adsorption process will be considered.

Author Contributions: Investigation, Methodology, Data curation, Writing—original draft, H.V.; Writing—review and editing, Y.O. and T.T.; Conceptualization, Supervision, A.K. All authors have read and agreed to the published version of the manuscript.

Funding: This research was funded by [New Energy and Industrial Technology Development Organization] grant number [JPNP18016].

Data Availability Statement: Research data can be provided as needed.

Acknowledgments: This work is based on results obtained from a project, JPNP18016, commissioned by the New Energy and Industrial Technology Development Organization (NEDO). The authors would like to thank Enago (www.enago.jp) for the English language review.

Conflicts of Interest: The authors declare no conflict of interest.

References

1. Houghton, J.T. Intergovernmental Panel on Climate Change. In *Climate Change 2001: The Scientific Basis*; Cambridge University Press: Cambridge, UK, 2001.
2. Masson-Delmotte, V.; Zhai, P.; Pirani, A.; Connors, S.L.; Péan, C.; Berger, S.; Caud, N.; Chen, Y.; Goldfarb, L.; Gomis, M.I.; et al. Working Group I Contribution to the Sixth Assessment Report of the Intergovernmental Panel on Climate Change Edited by. 2021. Available online: www.ipcc.ch (accessed on 2 March 2022).
3. Mac Dowell, N.; Fennell, P.; Shah, N.; Maitland, G.C. The role of CO_2 capture and utilization in mitigating climate change. *Nat. Clim. Chang.* **2017**, *7*, 243–249. [CrossRef]
4. Seipp, C.A.; Williams, N.J.; Kidder, M.K.; Custelcean, R. CO_2 Capture from Ambient Air by Crystallization with a Guanidine Sorbent. *Angew. Chem.* **2016**, *129*, 1062–1065. [CrossRef]
5. Joss, L.; Gazzani, M.; Mazzotti, M. Rational design of temperature swing adsorption cycles for post-combustion CO_2 capture. *Chem. Eng. Sci.* **2016**, *158*, 381–394. [CrossRef]
6. Bach, L.T.; Gill, S.J.; Rickaby, R.E.M.; Gore, S.; Renforth, P. CO_2 Removal with Enhanced Weathering and Ocean Alkalinity Enhancement: Potential Risks and Co-benefits for Marine Pelagic Ecosystems. *Front. Clim.* **2019**, *1*, 7. [CrossRef]
7. Smith, P.; Davis, S.J.; Creutzig, F.; Fuss, S.; Minx, J.; Gabrielle, B.; Kato, E.; Jackson, R.B.; Cowie, A.; Kriegler, E.; et al. Biophysical and economic limits to negative CO_2 emissions. *Nat. Clim. Chang.* **2015**, *6*, 42–50. [CrossRef]
8. Varghese, A.M.; Karanikolos, G.N. CO_2 capture adsorbents functionalized by amine—Bearing polymers: A review. *Int. J. Greenh. Gas Control.* **2020**, *96*, 103005. [CrossRef]
9. Zhao, Q.; Wu, F.; Men, Y.; Fang, X.; Zhao, J.; Xiao, P.; Webley, P.A.; Grande, C.A. CO_2 capture using a novel hybrid monolith (H-ZSM5/activated carbon) as adsorbent by combined vacuum and electric swing adsorption (VESA). *Chem. Eng. J.* **2019**, *358*, 707–717. [CrossRef]
10. Jiang, N.; Shen, Y.; Liu, B.; Zhang, D.; Tang, Z.; Li, G.; Fu, B. CO_2 capture from dry flue gas by means of VPSA, TSA and TVSA. *J. CO_2 Util.* **2019**, *35*, 153–168. [CrossRef]
11. Hoysall, D.C.; Determan, M.D.; Garimella, S.; Lenz, R.D.; Leta, D.P. Optimization of Carbon Dioxide Capture Using Sorbent-Loaded Hollow-Fiber Modules. *Int. J. Greenh. Gas Control.* **2018**, *76*, 225–235. [CrossRef]
12. Zainol, N.I.; Osaka, Y.; Tsujiguchi, T.; Kumita, M.; Kodama, A. Separation and enrichment of CH_4 and CO_2 from a dry biogas using a thermally regenerative adsorbent-packed heat exchanger. *Adsorption* **2019**, *25*, 1159–1167. [CrossRef]
13. Rezaei, F.; Subramanian, S.; Kalyanaraman, J.; Lively, R.P.; Kawajiri, Y.; Realff, M.J. Modeling of rapid temperature swing adsorption using hollow fiber sorbents. *Chem. Eng. Sci.* **2014**, *113*, 62–76. [CrossRef]
14. Masuda, S.; Osaka, Y.; Tsujiguchi, T.; Kodama, A. Carbon dioxide recovery from a simulated dry exhaust gas by an internally heated and cooled temperature swing adsorption packed with a typical hydrophobic adsorbent. *Sep. Purif. Technol.* **2021**, *284*, 120249. [CrossRef]
15. Ntiamoah, A.; Ling, J.; Xiao, P.; Webley, P.A.; Zhai, Y. CO_2 capture by temperature swing adsorption: Use of hot CO_2-rich gas for regeneration. *Ind. Eng. Chem. Res.* **2016**, *55*, 703–713. [CrossRef]
16. Masuda, S.; Osaka, Y.; Tsujiguchi, T.; Kodama, A. High-purity CO_2 recovery following two-stage temperature swing adsorption using an internally heated and cooled adsorber. *Sep. Purif. Technol.* **2023**, *309*, 123062. [CrossRef]
17. Sinha, A.; Darunte, L.A.; Jones, C.W.; Realff, M.J.; Kawajiri, Y. Systems Design and Economic Analysis of Direct Air Capture of CO_2 through Temperature Vacuum Swing Adsorption Using MIL-101(Cr)-PEI-800 and mmen-Mg_2(dobpdc) MOF Adsorbents. *Ind. Eng. Chem. Res.* **2017**, *56*, 750–764. [CrossRef]
18. Stampi-Bombelli, V.; van der Spek, M.; Mazzotti, M. Analysis of direct capture of CO_2 from ambient air via steam-assisted temperature–vacuum swing adsorption. *Adsorption* **2020**, *26*, 1183–1197. [CrossRef]
19. Fujiki, J.; Chowdhury, F.A.; Yamada, H.; Yogo, K. Highly efficient post-combustion CO_2 capture by low-temperature steam-aided vacuum swing adsorption using a novel polyamine-based solid sorbent. *Chem. Eng. J.* **2017**, *307*, 273–282. [CrossRef]
20. Goeppert, A.; Czaun, M.; May, R.B.; Prakash, G.K.S.; Olah, G.A.; Narayanan, S.R. Carbon Dioxide Capture from the Air Using a Polyamine Based Regenerable Solid Adsorbent. *J. Am. Chem. Soc.* **2011**, *133*, 20164–20167. [CrossRef] [PubMed]

Disclaimer/Publisher's Note: The statements, opinions and data contained in all publications are solely those of the individual author(s) and contributor(s) and not of MDPI and/or the editor(s). MDPI and/or the editor(s) disclaim responsibility for any injury to people or property resulting from any ideas, methods, instructions or products referred to in the content.

Article

Modeling of Nitrogen Removal from Natural Gas in Rotating Packed Bed Using Artificial Neural Networks

Amiza Surmi [1,2], Azmi Mohd Shariff [3,*] and Serene Sow Mun Lock [3]

[1] Chemical Engineering Department, Universiti Teknologi PETRONAS, Bandar Seri Iskandar 32610, Perak, Malaysia; amiza_surmi@petronas.com
[2] Group Research & Technology, Petroliam Nasional Berhad (PETRONAS), Lot 3288 & 3289, off Jalan Ayer Itam, Kawasan Institusi Bangi, Kajang 43000, Selangor Darul Ehsan, Malaysia
[3] Institute of Contaminant Management, CO2 Research Centre (CO2RES), Universiti Teknologi PETRONAS, Bandar Seri Iskandar 32610, Perak, Malaysia; sowmun.lock@utp.edu.my
* Correspondence: azmish@utp.edu.my; Tel.: +60-5-3687530

Abstract: Novel or unconventional technologies are critical to providing cost-competitive natural gas supplies to meet rising demands and provide more opportunities to develop low-quality gas fields with high contaminants, including high carbon dioxide (CO_2) fields. High nitrogen concentrations that reduce the heating value of gaseous products are typically associated with high CO_2 fields. Consequently, removing nitrogen is essential for meeting customers' requirements. The intensification approach with a rotating packed bed (RPB) demonstrated considerable potential to remove nitrogen from natural gas under cryogenic conditions. Moreover, the process significantly reduces the equipment size compared to the conventional distillation column, thus making it more economical. The prediction model developed in this study employed artificial neural networks (ANN) based on data from in-house experiments due to a lack of available data. The ANN model is preferred as it offers easy processing of large amounts of data, even for more complex processes, compared to developing the first principal mathematical model, which requires numerous assumptions and might be associated with lumped components in the kinetic model. Backpropagation algorithms for ANN Lavenberg–Marquardt (LM), scaled conjugate gradient (SCG), and Bayesian regularisation (BR) were also utilised. Resultantly, the LM produced the best model for predicting nitrogen removal from natural gas compared to other ANN models with a layer size of nine, with a 99.56% regression (R^2) and 0.0128 mean standard error (MSE).

Keywords: artificial neural networks; carbon dioxide; nitrogen; liquefied natural gas

1. Introduction

In Malaysia, over 13 trillion standard cubic feet per day (Tscfd) of undeveloped gas fields have been identified [1]. Commonly, high carbon dioxide (CO_2) fields are associated with high nitrogen concentrations, which makes the treatment of natural gas more challenging. Nevertheless, removing CO_2 to meet the 6.5-mole percentage (mol%) or less pipeline specification might result in greater nitrogen concentrations than what customers require. Since nitrogen is inert, removing the gas maximises the calorific value of the product gas while minimising any safety issues, particularly during liquefied natural gas (LNG) shipping due to stratification and rollover of the product [2]. Consequently, most LNG production plants limit nitrogen to a maximum of 1.0 mol% in their LNG outputs. Nonetheless, removing or venting nitrogen into the atmosphere necessitates a small amount of methane (typically under 1%), which could present a safety hazard as methane is a combustible gas. Methane is also a greenhouse gas that requires control during venting or flaring [3].

In a study [2], it was found that only cryogenic technologies could remove nitrogen to under 1.0 mol% and produce liquefied methane at large plant capacities. Meanwhile, non-cryogenic technologies, such as absorption and adsorption, reportedly remove nitrogen at higher product specifications for feed rates under 15 million standard cubic feet per day (MMSCFD). Consequently, gas separation utilising membranes has been extensively studied and applied. Nonetheless, unlike CO_2 removal from natural gas, nitrogen removal with membranes is reportedly minimal due to the inefficient methane and nitrogen separation to meet the under 1 mol% due to their similar kinetic sizes, 0.36 nm (nitrogen) and 0.38 nm (methane) [2,4].

A general nitrogen removal and liquefaction process employs a flash vessel to liquefy natural gas with methane to meet the LNG nitrogen requirements. Nonetheless, utilising the off-gas from the flash vessel as fuel gas results in hydrocarbon and product losses. Consequently, the cryogenic distillation concept maximises the hydrocarbon recovery from the off-gas. The tall distillation column and larger equipment required for the technique render it unattractive to offshore and onshore plants with limited plant size. Thus, process intensification (PI) is more appealing given its potential to overcome the challenges of maintaining dominance as the primary energy supplier while simultaneously attaining environmentally friendly and sustainable evolutions in the industry [5].

The PI utilises novel concepts and principles that improve chemical industry processes for sustainable product manufacturing. This method reportedly dramatically increased mass and transfer, reduced volume, equipment size and footprint, and operational and capital costs, allowed more sustainable material applications, simplified processes, and offered safer operations [5–16]. In the late 1970s, Professor Colin Ramshaw introduced PI to Imperial Chemical Industries (ICI). He employed a high gravitational force, Higee, to improve the mass transfer in the separation stage, intensify the process, and reduce the equipment size [7,8,16]. The technology utilised a novel rotating device to improve the gravitational force by over 100 times. The advantages of the Higee technology include intensified mass transfer with a very thin liquid film that attracts polymerisation, absorption, synthesis, conversion, and distillation [17].

High-gravity or centrifugal technologies include spinning disk reactor [11,18–20], static mixer [21–23], agitated slurry reactor [24], rotating zig-zag bed [25–28], and rotating packed bed (RPB). Guo et al. (2019) suggested over five RPB commercial applications in the industry, including desulphurisation, denitrification, particulate removal, and emission control. The approach offered equipment size, performance, and cost advantages over conventional technologies [29]. Furthermore, Boodhoo and Harle (2013) discovered that the RPB provided the highest mass and heat transfer compared to other PI technologies with significant equipment size and footprint reductions that benefit pilots or commercial applications [30].

Over the last few years, numerous experiments [31–34], simulations and modelling [35–38], and computational fluid dynamics (CFD)-based hydrodynamic studies [39–42] were conducted to procure a better understanding of the mass transfer within the RPB in high gravity environments and to improve the technology. Nonetheless, a majority of the reports were based on the absorption process and were operated under five bars and above zero temperatures. Limited studies also focused on RPB distillation, which included vacuum distillation and alcohol and heptane–hexane separations. No investigations have been conducted on nitrogen removal from natural gas using RPB distillation under cryogenic conditions and high pressures.

Reports focusing on RPB optimisation via process simulations and modelling, such as CFD and mathematical modelling, are available. Nonetheless, CFD and mathematical modelling are complex and require significant input conditions to obtain accurate prediction models. Consequently, the artificial neural network (ANN) model is preferred due to its simplicity and ability to process massive amounts of data, even for more complex processes, compared to developing first principal mathematical models, which require numerous assumptions and might be associated with lumped components in kinetic models [43]. Furthermore, ANN does not require a more complex mathematical understanding to interpret data procured for analysis; hence, it is more suitable for process optimisation in several engineering fields [44–47].

The ANN is an example of machine learning that employs a nonlinear regression algorithm approach. The model is designed based on the behaviours of the human brain, thus requiring experience and training for more accurate prediction via interconnected neurons. The ANN consists of an input component that receives external signals and data, an output unit that outputs the system processing results, and a hidden segment that is not observable outside the system, situated between the input and output pieces [48].

The ANN is widely employed in chemical engineering for thermodynamic application studies, process design, control, optimisation, safety, experimental data fitting, and machine learning [46,47,49,50]. Moreover, research interest in machine learning in PI for chemical and process engineering has increased exponentially by 26% from 2015 to 2020 [51]. For example, Popoola and Susu (2014) utilised ANN with multiple inputs and outputs to determine the temperature cut-off points of kerosene, diesel, and naphtha products [52]. The input in the study was employed to design a crude distillation column control.

Controlling the temperature of distillation columns is especially crucial in the industry. The process is very dynamic due to incoming process feed condition uncertainties, hence requiring a more robust control to maintain the pressure and temperature and achieve a good mass and energy balance to meet product specifications [49,53]. Wang and co-workers applied a neural network in the temperature proportional–integral–derivative (PID) controller design of a distillation column, considering that neural networks possess better adaptive and fault tolerance abilities [54]. Moreover, [47,49,55–57] reported a successful ANN implementation in distillation controls. Nevertheless, only a few applications are reported on RPB related to ANN model research, while none are available on studies or applications of ANN for cryogenic nitrogen removal utilising RPB.

Saha (2009) developed a prediction model with ANN radial basis function to predict the volumetric gas side mass transfer coefficient based on experimental data from previous studies [58]. The report discovered an improvement of up to 15% with better accuracy than the empirical equations employing experimental data [58]. In another study, Lashkarbolooki et al. (2012) investigated pressure reductions in RPB equipment. The pressure gradient in an RPB is critical to allow counter-current contacts between the gas and liquid, particularly within the rotor or packing, to achieve a better mass transfer. The study employed an ANN model comprising 14 hidden layers and observed an excellent agreement between the model and experiments, with a 5.27% AARD, 3.0×10^{-5} mean square error (MSE), and 0.9985 regression (R^2) [59]. Some reports focused on investigating contaminant removal efficiency from process gas using RPB technology for CO_2 capture [60], adsorption [61], dust removal [62], and ozonation [63].

The present study aimed to develop ANN prediction models for nitrogen removal from natural gas based on in-house experimental data with various process parameters as inputs to meet the nitrogen removal efficiency output product. This study employed three artificial neural network training algorithms, with ANN models utilised for a given multilayer perceptron (MLP) feedforward neural network: the Lavenberg–Marquardt (LM), Bayesian regularisation (BR), and scaled conjugate gradient (SCG). The LM and BR training algorithms were based on the backpropagation algorithm. The swift convergence of LM is its primary benefit [64], whereas BR offers less probability of it being overfitted [65].

2. Results and Discussion

2.1. The Influences of High Gravity Factor on Removal Efficiency

A high gravity factor is essential in an RPB, which generates a high centrifugal acceleration to facilitate an effective mass transfer. The high gravity factor is determined based on the RPB radius and its rotational speed (see Equation (1)). The centrifugal acceleration resulting from high gravity is a key factor distinguishing RPB from conventional packed columns. Only eight studies have been conducted on ANN for RPB to date. All of the reports utilised a high gravity factor or rotational speed as their ANN modelling input for various applications and achieved over 94% R^2 and under 1% MSE (see Table 1).

$$\beta = \frac{2\omega^2 (r_1^2 + r_1 r_2 + r_2^2)}{3(r_1 + r_2)g} \quad (1)$$

where r_1 and r_2 are the inner and outer radii of the rotor (m), respectively, and ω denotes the angular velocity in the RPB (rad/s).

The high gravity factor in the current study was altered between 10 and 90, while the parameters were maintained throughout the experiment to investigate the effects of high gravity on nitrogen removal efficiency. As shown in Figure 1, an improved nitrogen removal efficiency was observed due to the increased high gravity factor. These findings could result from better contact between the gas and liquid at higher centrifugal accelerations, thereby contributing to better contaminant removal [66]. Similar effects were reported for other contaminant removal applications, such as hydrogen sulphide (H_2S) [67], oxygen [38,68], and volatile organic compounds (VOC) [69,70].

Figure 2 illustrates the effects of the rotational speed and operating pressure on the nitrogen removal efficiency. Increasing the rotational speed enhanced the nitrogen removal efficiency, which decreased when the rotational speed reached 800 rpm. The diminished performance might be due to the reduced contact time between the gas and liquid. Conversely, at 200 rpm, the nitrogen removal performance began to decrease with increasing operating pressure. Nonetheless, the removal efficiency improved when the rotational speed was increased to 500 rpm and demonstrated a more stable condition, even at operating pressure variations. The results of this study are comparable to the findings on the prediction of vacuum distillation performance in the RPB reported by Li et al. (2017) [35].

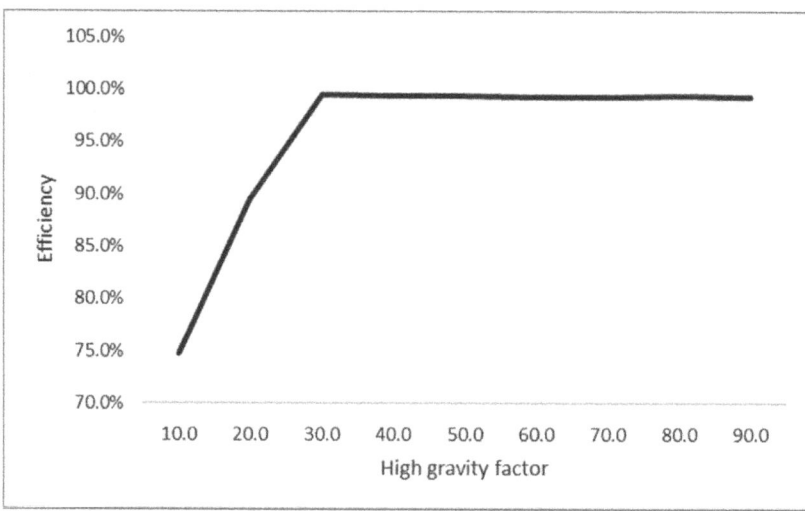

Figure 1. The effects of high gravity on nitrogen removal.

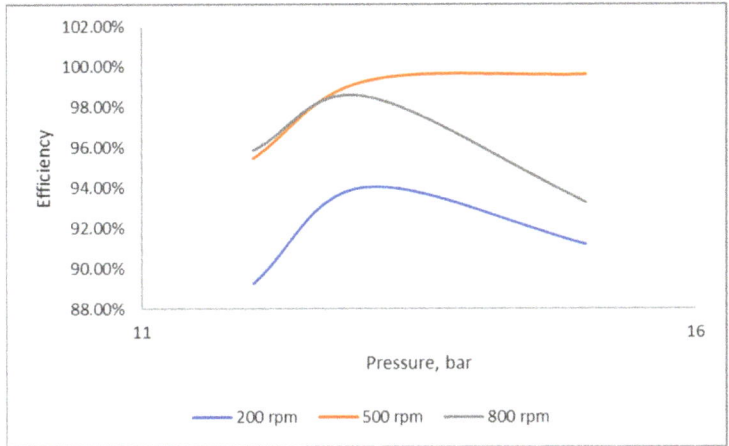

Figure 2. Effects of rotating speed on nitrogen removal at different pressures.

Table 1. The RPB modelling with ANN.

Author	Applications	ANN Modelling		
		Input Parameters	Output Parameter	Results
Wang et al., 2022 [63]	Degradation of bisphenol A (BPA) ozonation	- Concentration - pH - Flowrate - Gravity factor	BPA degradation efficiency	$R^2 = 0.9827$ MSE = 0.0003305
Li, 2021 [70]	Volatile organic compound removal	- Gravity factor - Reynold - Concentration - Henry's constant	VOC removal efficiency	$R^2 = 0.9697$ MSE = 0.0364
Wei et al., 2018 [71]	Biosorption process absorption using agricultural waste	- Gravity factor - Reynold - Contact time - Particle size - Concentration	Biosorption time	$R^2 = 0.996$ MSE = 0.0000904
Li et al., 2017 [62]	Dust removal via absorption process	- Reynold (gas, liquid, rotational) - Particle size	Separation efficiency	$R^2 = 0.9952$ MSE = 0.00013
Li et al., 2016 [72]	Wastewater treatment using adsorption process	- Gravity factor - Reynold number - Contact time - Concentration	Adsorption efficiency	$R^2 = 0.9965$ MSE = 0.00016
Zhao et al., 2014 [60]	CO_2 capture in RPB using absorption process	- Reynold - Schmidt - Grashof - Diffusion and mass transfer	CO_2 capture efficiency	$R^2 = 0.9457$ MSE = 0.0012
Lashkarbolooki et al., 2012 [59]	Prediction of pressure drop	- Reynold (gas, liquid, and rotational)	Pressure drops	$R^2 = 0.9985$ MSE = 0.00003
Saha, 2009 [58]	Mass transfer coefficient prediction	- Liquid velocity - Gas velocity - rotational	MTC	Not reported

Singh et al. (1992) proposed the area transfer unit–number of transfer unit (ATU-NTU) concept to explain the changes in fluid loading along the annular packing radius of an RPB [73]. The NTU is one of the most critical parameters to consider when assessing separation targets for distillation. Generally, determining the number of transfer units (NTU) involves evaluating the integrals in the equation describing the rectification and stripping sections, as shown in Equations (5)–(7). An increased NTU is anticipated if the performance target is set too low or if a thorough contaminant removal is attempted, which could result in an extremely tall column for conventional distillation. On the other hand, high centrifugal acceleration during separation leads to shorter residence times and process intensification, thereby requiring smaller equipment.

Figure 3 depicts ATU variations at different rotating speeds. According to the results, lower ATU values were documented at high operating speeds; hence, a shorter column was required to meet the product specifications. These observations are consistent with the findings reported by Qammar et al. (2018) on ethanol–water separation via total reflux distillation [34].

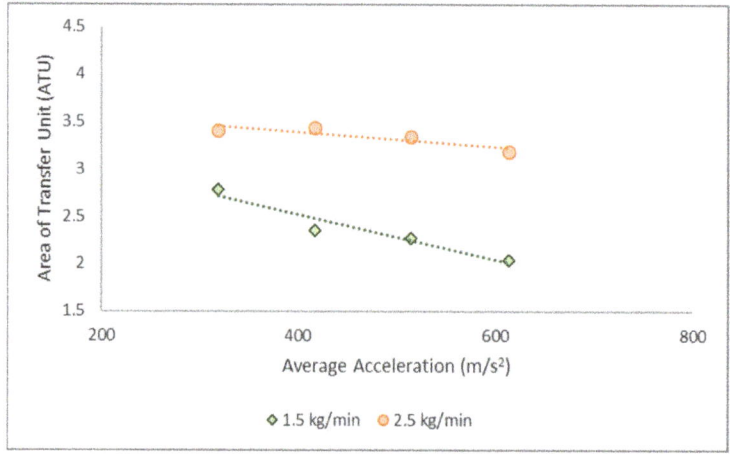

Figure 3. The ATU at varying average accelerations and flow rates.

$$\pi\left(r_o^2 - r_i^2\right) = (ATU_G)(NTU_G) \quad (2)$$

$$ATU_G = \frac{G}{\rho_G h K_G a} \quad (3)$$

$$NTU_G = \int_{y2}^{y1} \frac{dy}{y* - y} \quad (4)$$

2.2. Comparisons between the ANN Models

The neural network architecture correlates with the inputs, hidden layer numbers, and neuron transfer functions [74]. The hidden layers in the current study were varied between 5 and 15 to determine the LM, SCG, and BR prediction models with the highest accuracy. At layer nine, the LM model recorded higher R^2 and lower MSE values of 99.56% and 0.0128, respectively (see Figure 4). Nevertheless, increasing the hidden layer to more than nine led to more prominent MSE figures and a slightly reduced R^2.

The SCG model employed in the present study produced MSE values under 0.2%. Nonetheless, the 5–10 hidden layers in the model documented a good R^2 trend and a slightly diminished R^2 during validation and assessment with an increasing number of hidden layers (see Figure 5). Figure 6 demonstrates the relatively consistent R^2 and MSE for training and validation of the BR model but documented some reduction during evaluations. At nine layers, the R^2 and MSE of the BR model were 98.89 and 0.0493%, respectively, which were slightly better than those of the SCG by 0.0072%.

The modelling results in this study revealed that increasing the number of layers resulted in higher MSE and lower R^2 values, especially for the LM model. These findings could be due to overfitting. The R^2 results were better when fewer hidden layers (6–10) were utilised. The LM model architecture with nine hidden layers was the optimal structure, which also produced a superior prediction than the SCG and BR models. Nonetheless, the accuracy of the model might be compromised if the number of hidden layers is too low, thus reducing the probability of meeting the target figures. Consequently, training, evaluating, and validating the models to ascertain the optimal number of hidden layers are necessary. Figure 7 illustrates the outcomes of the LM model based on the predicted model and experimental data. Based on the results, R^2 yielded 96.66% of the 15% test data; therefore, it can be utilised to predict nitrogen removal from natural gas in the future.

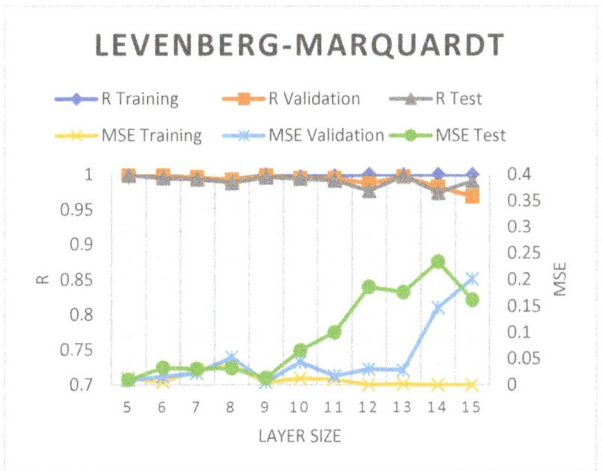

Figure 4. The R^2 and MSE of the LM-ANN model.

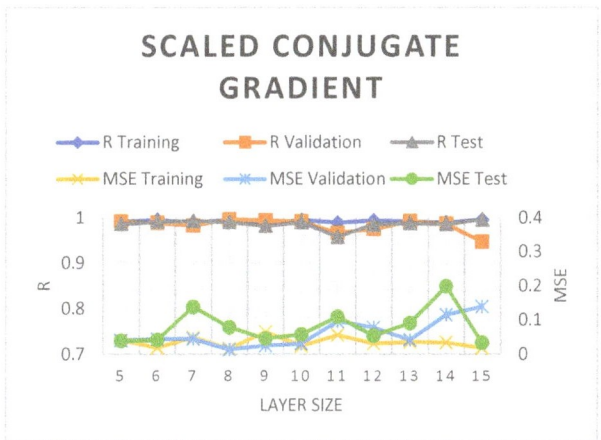

Figure 5. The R^2 and MSE of the SCG-ANN model.

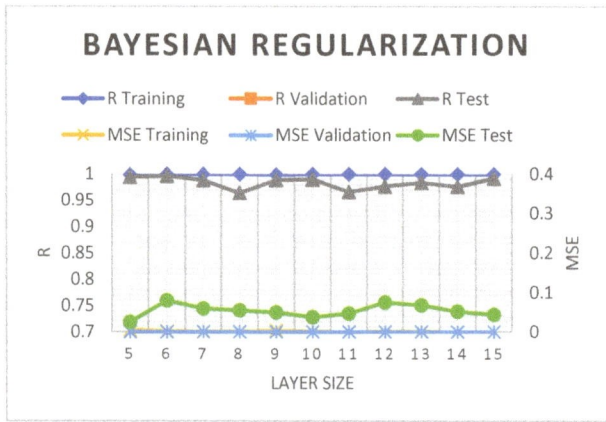

Figure 6. The R^2 and MSE of the BR-ANN model.

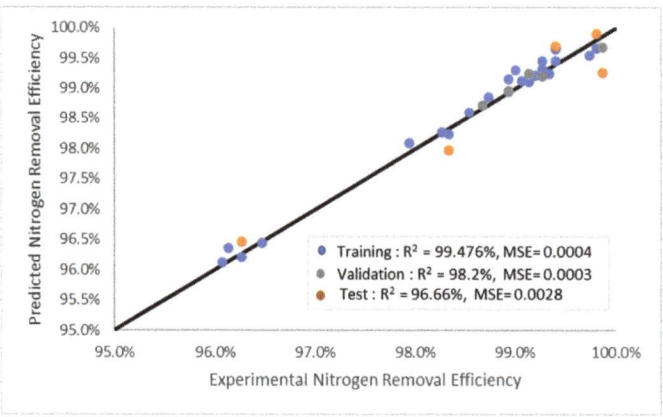

Figure 7. Nitrogen removal efficiencies of the experimental and prediction models.

3. Methodology

3.1. Experimental Setup

The primary components of the RPB utilised in the experimental setup for the cryogenic distillation-RPB (CD-RPB) process to remove nitrogen from natural gas in the current study (see Figure 8). Natural gas with a nitrogen concentration of up to 20 mol% was introduced to the CD-RPB via a liquid inlet post-chilled to ~−120 °C. A liquid distributor was designed for the inlet liquid to procure the spray effect, forming tiny droplets and a thin liquid film on the surface of the packing inside the CD-RPB.

The inlet and reflux liquids from the reflux vessel and vapour flow from the reboiler utilised in this study were contacted in a counter-current manner, thereby providing improved mass transfer and nitrogen removal efficiency. The rich nitrogen was discharged as a gas phase, whereas the main product, LNG, was obtained as the bottom product at −161 °C. In the present study, the CD-RPB motor was installed at the bottom of the pressure vessel. The rotational speed was varied from 100 to 1000 rpm during the experiments.

The RPB was equipped with stationary housing, liquid distributors, inner and outer diameters, packing or rotor, and a motor. The low-temperature and high-pressure evaluations (12–15 bars) employed in this study resulted in a quite complex cryogenic experimental setup. Consequently, other auxiliary equipment was crucial to support the CD-RPB setup and assessment. Equation (5) was employed to determine the nitrogen removal efficiency (n).

$$n = \frac{C_i - C_o}{C_i} \times 100\% \qquad (5)$$

where n is the nitrogen removal efficiency, C_i denotes the inlet concentration, and C_o represents the outlet concentration.

The setup used in this study was intended to process 840 kg of feed gas per day. A methane and nitrogen gas mixture was employed as the feed gas. The gas mixture was fed into the mol-sieve vessel to remove up to 1.0 part per million (ppm) moisture. A Shaw metre inline analyser was installed to monitor the quality of the gas, ensuring that it met the water specifications. The data were critical for low-temperature processes to avoid downstream processes of hydrate formation before sending the gas to the chiller and CD-RPB. Subsequently, distillation separated the nitrogen from methane based on its relative volatility. The rich nitrogen was then removed as the top product, while the bottom yield was LNG. The products were analysed using gas chromatography (GC) with a direct online sampling line connection. Furthermore, the evaluation setup included base layer process control, such as chiller temperature control and CD-RPB pressure control, to maintain the stability of the process.

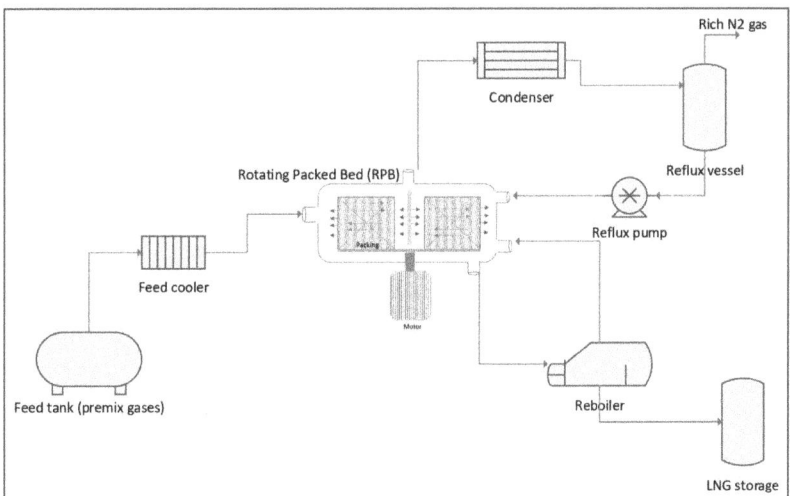

Figure 8. Simplified block diagram of the cryogenic nitrogen removal system experimental setup.

3.2. Model Development with Neural Networks

The ANN model in this study possessed a minimum of three layers: input, hidden, and output. Figure 9 demonstrates the complex interconnection between the layers to produce the outputs [46,66]. The most complex architecture in the ANN was the 'black box', which is the hidden layer connecting the input and output variables. The structure is known as the 'black box' due to the still unknown technical explanation within the hidden layer.

The number of hidden layers depends on the architecture of the ANN model to achieve an acceptable r-square (R^2) and lower means square error (MSE), as described in Equations (6) and (7). The input variables employed in the current study were the process parameters that were adjusted and controlled during the experiments to obtain the targeted values. The parameters and specifications of the model utilised in this study are summarised in Table 2.

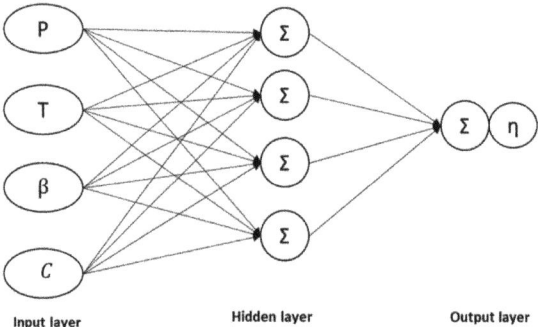

Figure 9. The typical neural network architecture.

Table 2. The parameters and specifications of the models.

Parameters	Model Specifications
Model	• Levenberg–Marquardt (LM) • Bayesian Regularization (BR) • Scaled Conjugate Gradient (SCG)
Samples' distribution [63]	• Training: 75% • Test: 15% • Validation: 15%
Number of inputs	4
Number of outputs	1
Hidden layer transfer function	Sigmoid
Output layer transfer function	Linear
Number of data sets	45

The obtained MSE and regression values, R, reflected the quality of the results in the present study. The average squared variations between the outputs generated by the MATLAB function and the targets (true, measured data corresponding to the inputs provided to the MATLAB functions) are denoted by the MSE. Consequently, since smaller figures were preferred, the algorithm with minimal MSE was considered the most suitable. The MSE was calculated using Equation (6). R correlated the obtained ANN outputs and the targets. An R value of 1 corresponded to a close association, while a value of 0 represented a random link. The R value was obtained by utilising Equation (7). The t value represented the arithmetic mean of the target values.

$$\text{MSE} = \frac{1}{N}\sum_{i=1}^{N}(e_i)^2 = \frac{1}{N}\sum_{i=1}^{N}(t_i - a_i)^2 \quad (6)$$

$$R^2 = 1 - \frac{\sum_{i=1}^{N}(t_i - a_i)^2}{\sum_{i=1}^{N}(\bar{t_i} - t_i)^2} \quad (7)$$

where N represents the sample numbers (input–output pairs) utilised for training the network, and t denotes the target value.

Figure 10 depicts the conceptual structure and the ANN modelling approach utilised in this study. The data collection process was initiated with experimental procedures and continued with data screening and input categorisation. The input data were then fed into the ANN models for training, validations, and testing. Data from the experiment were employed for all consistent ANN models. Inconsistencies were observed when variables, such as R^2 and MSE, were recorded on the hidden layers.

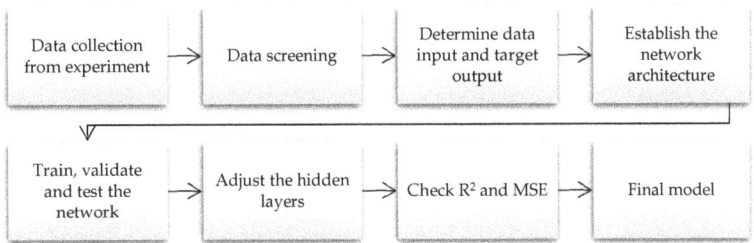

Figure 10. The ANN modelling flowchart.

All the ANN models in the current study followed an identical set of instructions. After data collection and examination, several models were considered to determine which provided the best and most accurate predictions. A MATLAB (R2022a) equipped with a version 8.3 toolbox was employed to perform the evaluations.

4. Conclusions

This study established three types of ANN models for nitrogen removal from natural gas utilising an RPB based on experimental data. The high gravity factor is a critical condition in RPB applications as it promotes vapour–liquid contact and enhances fluid mass transfers during contaminant removal. Consequently, the high gravity factor in the current study was varied from 30 to 90 to maintain a nitrogen removal efficiency above 97%.

Other process inputs, such as pressure, temperature, and concentration, are also essential for producing a good prediction model to remove nitrogen from natural gas under cryogenic operating conditions. The LM model performed ideally with nine hidden layers, with 99.56% R^2 and 0.0128 MSE compared to the SCG and BY models. Nevertheless, all models produced over 90% R^2 at different hidden layers.

The increased number of layers almost consistently reduced the R^2 and enhanced the MSE across all three ANN models. These findings might be due to data overfitting to meet the output. Nonetheless, the correlation inside the hidden layers remained unknown, hence considered a 'black box.' The results of this study suggested that the LM model was the best model for predicting nitrogen removal efficiency when employing an RPB under cryogenic conditions. Furthermore, reliance on the experimental setup and requirements was minimised. Consequently, the predicted model was reliable and applicable to process optimisations, techno-economic studies, and technology upscaling for commercialisation.

Author Contributions: Conceptualisation, A.S. and S.S.M.L.; methodology, A.S.; software, A.S.; validation, A.S. and S.S.M.L.; formal analysis, A.S.; investigation, A.S. and S.S.M.L.; resources, A.S. and S.S.M.L.; data curation, A.S.; writing—original draft preparation, A.S.; writing—review and editing, A.S., A.M.S. and S.S.M.L.; supervision, A.M.S.; funding acquisition, A.M.S. All authors have read and agreed to the published version of the manuscript.

Funding: This research was funded by PETRONAS Research Sdn Bhd, grant number 015MD0-055, 015MD0-126, and 015MD0-089, and Yayasan Universiti Teknologi PETRONAS, grant number 015LC0-390.

Institutional Review Board Statement: Not applicable.

Informed Consent Statement: Not applicable.

Data Availability Statement: Not applicable.

Acknowledgments: The authors would like to acknowledge the cost centres 015MD0-055, 015MD0-126, and 015LC0-390 for supporting the research project, and 015MD0-089 for financial assistance with the postgraduate study.

Conflicts of Interest: The authors declare no conflict of interest.

Sample Availability: Samples of the compounds are not available from the authors.

References

1. Lal, B.; Shariff, A.M.; Mukhtar, H.; Nasir, Q.; Qasim, A. An overview of cryogenic separation techniques for natural gas with high CO2content. *J. Eng. Appl. Sci.* **2018**, *13*, 2152–2155. [CrossRef]
2. Rufford, T.; Smart, S.; Watson, G.; Graham, B.; Boxall, J.; da Costa, J.D.; May, E. The removal of CO_2 and N_2 from natural gas: A review of conventional and emerging process technologies. *J. Pet. Sci. Eng.* **2012**, *94–95*, 123–154. [CrossRef]
3. Ott, C.; Roberts, M.; Trautmann, S.; Krishnamurthy, G. State-of-the-Art Nitrogen Removal Methods from Air Products for Liquefaction Plants. *LNG J.* **2015**. Available online: http://www.airproducts.com/~/media/Files/PDF/industries/lng/en-LNG-journal-paper.pdf (accessed on 10 November 2022).
4. Meyer, H.S.; Henson, M.S. Methane Selective Membranes for Nitrogen Removal from Low Quality Natural Gas—High Permeation Is Not Enough. In Proceedings of the Natural Gas Technologies II: Ingenuity and Innovation, Phoenix, AZ, USA, 8–11 February 2004.
5. Abdulrahman, I.; Máša, V.; Teng, S.Y. Process intensification in the oil and gas industry: A technological framework. *Chem. Eng. Process.-Process. Intensif.* **2021**, *159*, 108208. [CrossRef]
6. Bucci, V.P. Process Intensification Transforming Chemical Engineering. *Macromolecules* **2000**, 22–34. Available online: https://www.aiche.org/sites/default/files/docs/news/010022_cep_stankiewicz.pdf (accessed on 15 December 2022).
7. Ramshaw, C. The opportunities for exploiting centrifugal fields. *Heat Recover. Syst. CHP* **1993**, *13*, 493–513. [CrossRef]
8. Reay, D.; Ramshaw, C.; Harvey, A. *Process Intensification: Engineering for Efficiency, Sustainability and Flexibility*; Butterworth-Heinemann: Oxford, UK, 2013; Volume 53. [CrossRef]
9. Harmsen, J. Process intensification in the petrochemicals industry: Drivers and hurdles for commercial implementation. *Chem. Eng. Process.-Process. Intensif.* **2010**, *49*, 70–73. [CrossRef]
10. Kim, Y.-H.; Park, L.K.; Yiacoumi, S.; Tsouris, C. Modular Chemical Process Intensification: A Review. *Annu. Rev. Chem. Biomol. Eng.* **2017**, *8*, 359–380. [CrossRef] [PubMed]
11. Bielenberg, J.; El-Halwagi, M.; Ng, K.M. Editorial overview: Engineering and design approaches to process intensification. *Curr. Opin. Chem. Eng.* **2019**, *25*, A4–A5. [CrossRef]
12. Rivas, D.F.; Castro-Hernández, E.; Perales, A.L.V.; van der Meer, W. Evaluation method for process intensification alternatives. *Chem. Eng. Process.-Process. Intensif.* **2018**, *123*, 221–232. [CrossRef]
13. Pask, S.D.; Nuyken, O.; Cai, Z. The spinning disk reactor: An example of a process intensification technology for polymers and particles. *Polym. Chem.* **2012**, *3*, 2698–2707. [CrossRef]
14. Costello, R.C. Process intensification: Think small. *Chem. Eng.* **2004**, *111*, 27–31.
15. Moorthy, R.K.; Baksi, S.; Bisws, S. Process intensification—An insight. *Chem. Eng. World* **2016**, *51*, 41–47. [CrossRef]
16. Ramshaw, C.; Mallinson, R.H. Mass Transfer Process. U.S. Patent 4,283,255, 11 August 1981. Available online: https://patents.google.com/patent/US4283255A/en (accessed on 18 December 2022).
17. Zhao, H.; Shao, L.; Chen, J.-F. High-gravity process intensification technology and application. *Chem. Eng. J.* **2010**, *156*, 588–593. [CrossRef]
18. Boodhoo, K. Spinning Disc Reactor for Green Processing and Synthesis. In *Process Intensification for Green Chemistry*; John Wiley & Sons, Ltd.: Hoboken, NJ, USA, 2013; pp. 59–90. [CrossRef]
19. Meeuwse, M.; van der Schaaf, J.; Schouten, J.C. Mass Transfer in a Rotor−Stator Spinning Disk Reactor with Cofeeding of Gas and Liquid. *Ind. Eng. Chem. Res.* **2009**, *49*, 1605–1610. [CrossRef]
20. Chianese, A.; Picano, A.; Stoller, M. Spinning disc reactor to produce nanoparticles: Applications and best operating variables. *Chem. Eng. Trans.* **2021**, *84*, 121–126. [CrossRef]
21. Al Taweel, A.; Azizi, F.; Sirijeerachai, G. Static mixers: Effective means for intensifying mass transfer limited reactions. *Chem. Eng. Process.-Process. Intensif.* **2013**, *72*, 51–62. [CrossRef]
22. Ghanem, A.; Lemenand, T.; Della Valle, D.; Peerhossaini, H. Static mixers: Mechanisms, applications, and characterization methods—A review. *Chem. Eng. Res. Des.* **2014**, *92*, 205–228. [CrossRef]
23. Thakur, R.; Vial, C.; Nigam, K.; Nauman, E.; Djelveh, G. Static Mixers in the Process Industries—A Review. *Chem. Eng. Res. Des.* **2003**, *81*, 787–826. [CrossRef]
24. Geng, S.; Mao, Z.-S.; Huang, Q.; Yang, C. Process Intensification in Pneumatically Agitated Slurry Reactors. *Engineering* **2021**, *7*, 304–325. [CrossRef]
25. Li, Y.; Lu, Y.; Liu, X.; Wang, G.; Nie, Y.; Ji, J. Mass-transfer characteristics in a rotating zigzag bed as a Higee device. *Sep. Purif. Technol.* **2017**, *186*, 156–165. [CrossRef]
26. Wang, G.; Xu, Z.; Ji, J. Progress on Higee distillation—Introduction to a new device and its industrial applications. *Chem. Eng. Res. Des.* **2011**, *89*, 1434–1442. [CrossRef]
27. Li, Y.; Li, X.; Wang, Y.; Chen, Y.; Ji, J.; Yu, Y.; Xu, Z. Distillation in a Counterflow Concentric-Ring Rotating Bed. *Ind. Eng. Chem. Res.* **2014**, *53*, 4821–4837. [CrossRef]

28. Wang, G.Q.; Zhou, Z.J.; Li, Y.M.; Ji, J.B. Qualitative relationships between structure and performance of rotating zigzag bed in distillation. *Chem. Eng. Process.-Process. Intensif.* **2018**, *135*, 141–147. [CrossRef]
29. Guo, J.; Jiao, W.; Qi, G.; Yuan, Z.; Liu, Y. Applications of high-gravity technologies in gas purifications: A review. *Chin. J. Chem. Eng.* **2019**, *27*, 1361–1373. [CrossRef]
30. Boodhoo, K.; Harvey, A. Process Intensification: An Overview of Principles and Practice. In *Process Intensification for Green Chemistry: Engineering Solutions for Sustainable Chemical Processing*; Newcastle University: Newcastle upon Tyne, UK, 2013; pp. 3–30. [CrossRef]
31. Yue, X.-J.; Luo, Y.; Chen, Q.-Y.; Chu, G.-W.; Luo, J.-Z.; Zhang, L.-L.; Chen, J.-F. Investigation of micromixing and precipitation process in a rotating packed bed reactor with PTFE packing. *Chem. Eng. Process. Process. Intensif.* **2018**, *125*, 227–233. [CrossRef]
32. Sung, W.-D.; Chen, Y.-S. Characteristics of a rotating packed bed equipped with blade packings and baffles. *Sep. Purif. Technol.* **2012**, *93*, 52–58. [CrossRef]
33. Neumann, K.; Hunold, S.; Groß, K.; Górak, A. Experimental investigations on the upper operating limit in rotating packed beds. *Chem. Eng. Process.-Process. Intensif.* **2017**, *121*, 240–247. [CrossRef]
34. Qammar, H.; Hecht, F.; Skiborowski, M.; Górak, A. Experimental investigation and design of rotating packed beds for distillation. *Chem. Eng. Trans.* **2018**, *69*, 655–660. [CrossRef]
35. Li, W.; Song, B.; Li, X.; Liu, Y. Modelling of vacuum distillation in a rotating packed bed by Aspen. *Appl. Therm. Eng.* **2017**, *117*, 322–329. [CrossRef]
36. Lu, X.; Xie, P.; Ingham, D.; Ma, L.; Pourkashanian, M. Modelling of CO_2 absorption in a rotating packed bed using an Eulerian porous media approach. *Chem. Eng. Sci.* **2019**, *199*, 302–318. [CrossRef]
37. Luo, X.; Wang, M.; Lee, J.; Hendry, J. Dynamic modelling based on surface renewal theory, model validation and process analysis of rotating packed bed absorber for carbon capture. *Appl. Energy* **2021**, *301*, 117462. [CrossRef]
38. Yu, C.-H.; Lin, Y.-J.; Wong, D.S.-H.; Chen, C.-C. Process Modeling of CO_2 Absorption with Monoethanolamine Aqueous Solutions Using Rotating Packed Beds. *Ind. Eng. Chem. Res.* **2022**, *61*, 12142–12152. [CrossRef]
39. Chen, W.-C.; Fan, Y.-W.; Zhang, L.-L.; Sun, B.-C.; Luo, Y.; Zou, H.-K.; Chu, G.-W.; Chen, J.-F. Computational fluid dynamic simulation of gas-liquid flow in rotating packed bed: A review. *Chin. J. Chem. Eng.* **2021**, *41*, 85–108. [CrossRef]
40. Yang, Y.; Ouyang, Y.; Zhang, N.; Yu, Q.; Arowo, M. A review on computational fluid dynamic simulation for rotating packed beds. *J. Chem. Technol. Biotechnol.* **2018**, *94*, 1017–1031. [CrossRef]
41. Zhang, W.; Xie, P.; Li, Y.; Teng, L.; Zhu, J. Hydrodynamic characteristics and mass transfer performance of rotating packed bed for CO2 removal by chemical absorption: A review. *J. Nat. Gas Sci. Eng.* **2020**, *79*, 103373. [CrossRef]
42. Yang, Y.; Xiang, Y.; Chu, G.; Zou, H.; Sun, B.; Arowo, M.; Chen, J.-F. CFD modeling of gas–liquid mass transfer process in a rotating packed bed. *Chem. Eng. J.* **2016**, *294*, 111–121. [CrossRef]
43. Al-Shathr, A.; Shakor, Z.M.; Majdi, H.S.; AbdulRazak, A.A.; Albayati, T.M. Comparison between Artificial Neural Network and Rigorous Mathematical Model in Simulation of Industrial Heavy Naphtha Reforming Process. *Catalysts* **2021**, *11*, 1034. [CrossRef]
44. Osuolale, F.N.; Zhang, J. Energy efficiency optimisation for distillation column using artificial neural network models. *Energy* **2016**, *106*, 562–578. [CrossRef]
45. Paul, A.; Prasad, A.; Kumar, A. Review on Artificial Neural Network and its Application in the Field of Engineering. *J. Mech. Eng. Prakash* **2022**, *1*, 53–61. [CrossRef]
46. Cavalcanti, F.M.; Kozonoe, C.E.; Pacheco, K.A.; Alves, R.M.D.B. Application of Artificial Neural Networks to Chemical and Process Engineering. In *Deep Learning Applications*; IntechOpen: London, UK, 2021. [CrossRef]
47. Sun, L.; Liang, F.; Cui, W. Artificial Neural Network and Its Application Research Progress in Chemical Process. *Asian J. Res. Comput. Sci.* **2021**, *11*, 177–185. [CrossRef]
48. Sun, J.; Tang, Q. Review of Artificial Neural Network and Its Application Research in Distillation. *J. Eng. Res. Rep.* **2021**, 44–54. [CrossRef]
49. Kwon, H.; Oh, K.C.; Choi, Y.; Chung, Y.G.; Kim, J. Development and application of machine learning-based prediction model for distillation column. *Int. J. Intell. Syst.* **2021**, *36*, 1970–1997. [CrossRef]
50. Abiodun, O.I.; Jantan, A.; Omolara, A.E.; Dada, K.V.; Mohamed, N.A.; Arshad, H. State-of-the-art in artificial neural network applications: A survey. *Heliyon* **2018**, *4*, e00938. [CrossRef] [PubMed]
51. López-Guajardo, E.A.; Delgado-Licona, F.; Álvarez, A.J.; Nigam, K.D.; Montesinos-Castellanos, A.; Morales-Menendez, R. Process intensification 4.0: A new approach for attaining new, sustainable and circular processes enabled by machine learning. *Chem. Eng. Process.-Process. Intensif.* **2021**, *180*, 108671. [CrossRef]
52. Popoola, L.T.; Susu, A.A. Application of Artificial Neural Networks Based Monte Carlo Simulation in the Expert System Design and Control of Crude Oil Distillation Column of a Nigerian Refinery. *Adv. Chem. Eng. Sci.* **2014**, *4*, 266–283. [CrossRef]
53. Taqvi, S.A.; Tufa, L.D.; Muhadizir, S. Optimization and Dynamics of Distillation Column Using Aspen Plus®. *Procedia Eng.* **2016**, *148*, 978–984. [CrossRef]
54. Li, C.; Wang, C. Application of Artificial Neural Network in the Control and Optimization of Distillation Tower. *arXiv* **2021**, arXiv:2107.13713.
55. Zhao, N.; Lu, J. Review of Neural Network Algorithm and its Application in Temperature Control of Distillation Tower. *J. Eng. Res. Rep.* **2021**, *20*, 50–61. [CrossRef]

56. Shin, Y.; Smith, R.; Hwang, S. Development of model predictive control system using an artificial neural network: A case study with a distillation column. *J. Clean. Prod.* **2020**, *277*, 124124. [CrossRef]
57. Ali, A.; Abdulrahman, A.; Garg, S.; Maqsood, K.; Murshid, G. Application of artificial neural networks (ANN) for vapor-liquid-solid equilibrium prediction for CH_4-CO_2 binary mixture. *Greenh. Gases: Sci. Technol.* **2018**, *9*, 67–78. [CrossRef]
58. Saha, D. Prediction of mass transfer coefficient in rotating bed contactor (Higee) using artificial neural network. *Heat Mass Transf.* **2008**, *45*, 451–457. [CrossRef]
59. Lashkarbolooki, M.; Vaferi, B.; Mowla, D. Using Artificial Neural Network to Predict the Pressure Drop in a Rotating Packed Bed. *Sep. Sci. Technol.* **2012**, *47*, 2450–2459. [CrossRef]
60. Zhao, B.; Su, Y.; Tao, W. Mass transfer performance of CO_2 capture in rotating packed bed: Dimensionless modeling and intelligent prediction. *Appl. Energy* **2014**, *136*, 132–142. [CrossRef]
61. Gupta, K.N.; Kumar, R. Fixed bed utilization for the isolation of xylene vapor: Kinetics and optimization using response surface methodology and artificial neural network. *Environ. Eng. Res.* **2020**, *26*, 200105. [CrossRef]
62. Li, W.; Wu, X.; Jiao, W.; Qi, G.; Liu, Y. Modelling of dust removal in rotating packed bed using artificial neural networks (ANN). *Appl. Therm. Eng.* **2017**, *112*, 208–213. [CrossRef]
63. Wang, L.; Qi, C.; Lu, Y.; Arowo, M.; Shao, L. Degradation of Bisphenol A by ozonation in a rotating packed bed: Modeling by response surface methodology and artificial neural network. *Chemosphere* **2021**, *286*, 131702. [CrossRef]
64. Taherdangkoo, R.; Tatomir, A.; Taherdangkoo, M.; Qiu, P.; Sauter, M. Nonlinear Autoregressive Neural Networks to Predict Hydraulic Fracturing Fluid Leakage into Shallow Groundwater. *Water* **2020**, *12*, 841. [CrossRef]
65. Guzman, S.M.; Paz, J.O.; Tagert, M.L.M. The Use of NARX Neural Networks to Forecast Daily Groundwater Levels. *Water Resour. Manag.* **2017**, *31*, 1591–1603. [CrossRef]
66. Tamhankar, Y. Design of a High Gravity Distillation. Bachelor's Thesis, University of Pune, Pune, India, 2006.
67. Qian, Z.; Xu, L.-B.; Li, Z.-H.; Li, H.; Guo, K. Selective Absorption of H_2S from a Gas Mixture with CO_2 by Aqueous N-Methyldiethanolamine in a Rotating Packed Bed. *Ind. Eng. Chem. Res.* **2010**, *49*, 6196–6203. [CrossRef]
68. Cheng, H.-H.; Tan, C.-S. Removal of CO_2 from indoor air by alkanolamine in a rotating packed bed. *Sep. Purif. Technol.* **2011**, *82*, 156–166. [CrossRef]
69. Lin, C.-C.; Wei, T.-Y.; Liu, W.-T.; Shen, K.-P. Removal of VOCs from Gaseous Streams in a High-Voidage Rotating Packed Bed. *J. Chem. Eng. Jpn.* **2004**, *37*, 1471–1477. [CrossRef]
70. Liu, L. VOC removal in rotating packed bed: ANN model vs empirical model. *Alex. Eng. J.* **2021**, *61*, 4507–4517. [CrossRef]
71. Liu, Z.-W.; Liang, F.-N.; Liu, Y.-Z. Artificial neural network modeling of biosorption process using agricultural wastes in a rotating packed bed. *Appl. Therm. Eng.* **2018**, *140*, 95–101. [CrossRef]
72. Li, W.; Wei, S.; Jiao, W.; Qi, G.; Liu, Y. Modelling of adsorption in rotating packed bed using artificial neural networks (ANN). *Chem. Eng. Res. Des.* **2016**, *114*, 89–95. [CrossRef]
73. Singh, S.P.; Wilson, J.H.; Counce, R.M.; Lucero, A.J.; Reed, G.D.; Ashworth, R.A.; Elliott, M.G. Removal of volatile organic compounds from groundwater using a rotary air stripper. *Ind. Eng. Chem. Res.* **1992**, *31*, 574–580. [CrossRef]
74. Pirdashti, M.; Curteanu, S.; Kamangar, M.H.; Hassim, M.H.; Khatami, M.A. Artificial neural networks: Applications in chemical engineering. *Rev. Chem. Eng.* **2013**, *29*, 205–239. [CrossRef]

Disclaimer/Publisher's Note: The statements, opinions and data contained in all publications are solely those of the individual author(s) and contributor(s) and not of MDPI and/or the editor(s). MDPI and/or the editor(s) disclaim responsibility for any injury to people or property resulting from any ideas, methods, instructions or products referred to in the content.

Article

Effect of Steam on Carbonation of CaO in Ca-Looping

Ruzhan Bai [1], Na Li [1,*], Quansheng Liu [1], Shenna Chen [2], Qi Liu [3] and Xing Zhou [2,4,*]

[1] College of Chemical Engineering, Inner Mongolia University of Technology, Huhhot 010051, China; 20211800161@imut.edu.cn (R.B.)
[2] Hebei Key Laboratory of Inorganic Nanomaterials, School of Chemistry and Materials Science, Hebei Normal University, Shijiazhuang 050024, China
[3] College of Chemistry and Chemical Engineering, Taiyuan University of Technology, Taiyuan 030024, China
[4] College of Zhongran, Hebei Normal University, Shijiazhuang 050024, China
* Correspondence: nali87@imut.edu.cn (N.L.); zhoux@hebtu.edu.cn (X.Z.)

Abstract: Ca-looping is an effective way to capture CO_2 from coal-fired power plants. However, there are still issues that require further study. One of these issues is the effect of steam on the Ca-looping process. In this paper, a self-made thermogravimetric analyzer that can achieve rapid heating and cooling is used to measure the change of sample weight under constant temperature conditions. The parameters of the Ca-looping are studied in detail, including the addition of water vapor alone in the calcination or carbonation stage and the calcination/carbonation reaction temperatures for both calcination and carbonation stages with water vapor. Steam has a positive overall effect on CO_2 capture in the Ca-looping process. When steam is present in both calcination and carbonation processes, it increases the decomposition rate of $CaCO_3$ and enhances the subsequent carbonation conversion of CaO. However, when steam was present only in the calcination process, there was lower CaO carbonation conversion in the following carbonation process. In contrast, when steam was present in the carbonation stage, CO_2 capture was improved. Sample characterizations after the reaction showed that although water vapor had a negative effect on the pore structure, adding water vapor increased the diffusion coefficient of CO_2 and the carbonation conversion rate of CaO.

Keywords: Ca-looping; steam; carbonation conversion; catalysis

Citation: Bai, R.; Li, N.; Liu, Q.; Chen, S.; Liu, Q.; Zhou, X. Effect of Steam on Carbonation of CaO in Ca-Looping. *Molecules* **2023**, *28*, 4910. https://doi.org/10.3390/molecules28134910

Academic Editors: Zilong Liu, Meixia Shan and Yakang Jin

Received: 23 April 2023
Revised: 31 May 2023
Accepted: 14 June 2023
Published: 22 June 2023

Copyright: © 2023 by the authors. Licensee MDPI, Basel, Switzerland. This article is an open access article distributed under the terms and conditions of the Creative Commons Attribution (CC BY) license (https://creativecommons.org/licenses/by/4.0/).

1. Introduction

It is believed that climate change is mainly caused by the increased CO_2 concentration in the atmosphere [1]. Fossil fuel combustion systems, such as coal-fired power plants, are one of the major fixed sources of CO_2 emissions [2,3]. Ca-looping is a high-temperature and low-cost technology that is under study for capturing CO_2 from fossil fuel combustion [4,5].

Ca-looping systems use calcium carbonates that are typically derived from limestone or dolomite and are regenerable, abundant, and cheap sorbents [6,7].

Ca-looping is based on the reversible reaction described in Equation (1). The forward reaction is known as carbonation, and the reverse reaction is known as calcination. Calcination is an endothermic process which readily goes to completion under a wide range of conditions [8,9].

$$CaO + CO_2 \underset{\text{calcination}}{\overset{\text{carbonation}}{\rightleftarrows}} CaCO_3 \qquad (1)$$

Ca-looping can be achieved using a double fluidized bed system [10,11]. The solid adsorbent is continuously circulated between two interconnected fluidized bed reactors. Carbon dioxide (CO_2) in flue gas is absorbed by calcium oxide (CaO) in a carbonizer at about 650 °C. Calcium carbonate ($CaCO_3$) is then transferred to the regeneration reactor (calciner) [12].

In the calcinator, calcium carbonate is decomposed into calcium oxide (CaO) and carbon dioxide at 900 °C. The regenerated CaO is returned to the carbonizer, leaving a pure carbon dioxide stream, resulting in simultaneous regeneration of the adsorbent (CaO → CaCO$_3$ → CaO) [13]. Under actual conditions, the energy required to regenerate the adsorbent (calcined carbonate) is provided by firing coal or biomass using oxy-fuel technology to avoid dilution of the CO$_2$ stream with N$_2$ from air [14].

It has been widely reported [15,16] that the conversion ratio of CaO to CaCO$_3$ decreases with an increase in calcination/carbonation cycles. This phenomenon is considered to be related to the decreased active surface of adsorbent caused by sintering [17]. It seems that sintering plays a principal role in the decreased reactivity of the Ca-based sorbents. Sintering reduces the surface area of the sorbents and the number of active sites [18,19].

There are many factors affecting the sintering of Ca-based sorbent, including reaction temperature, reaction duration, cycling numbers, and reaction atmosphere [20,21]. For the reaction atmosphere, water vapor plays an important role [22]. On the one hand, it has been reported that water vapor may accelerate the sintering of CaO [23]; on the other hand, water vapor can promote both reaction rate and conversion between CaO and CO$_2$. Wang et al. [24] studied the effect that water vapor has on the carbonation of CaO in oxy-fuel CFB combustion conditions. Water vapor significantly improves absorption of carbon dioxide [25]. Donat et al. [26] investigated the influence that steam has on the reactivity of four kinds of limestone in an atmospheric pressure bench-scale bubbling fluidized bed (BFB) reactor that was made of a quartz tube and using periodically changing temperatures. In a word, water vapor is one of the factors that affect sintering in the Ca-looping process, and its role requires further study.

The effect of steam atmosphere on the Ca-looping calcination/carbonate process in laboratory tests still faces some challenges. A fluidized bed reactor is very good at simulating actual working conditions of Ca-looping cycles, but it is difficult to obtain reaction kinetics data and to evaluate the influences that individual operation parameters have. TGA is commonly used to measure calcination/carbonation kinetics [27–29]. Therefore, more researchers have used TGA to conduct Ca-looping experiments. However, the heating rate of TGA is usually set as 10–20 K/min, and diffusion resistance is serious, which is inconsistent with the factor that the Ca sorbent is cycling between the high-temperature calcination reactor (about 900 °C) and low-temperature carbonation reactor (about 650 °C) [30].

In this work, a custom-made thermogravimetric analyzer system with a large sample scope (100–300 mg) is used, which allows fast temperature and atmosphere switching when considering the water vapor atmosphere. On the one hand, it can simulate the sudden temperature change of the CaO adsorber in the calciner and carbonation reactor to the maximum extent; on the other hand, it can switch the temperature and atmosphere as quickly as 5–15 s and thus can exclude the long temperature/atmosphere switching time in the conventional TGA and the influence of the residence time on the sintering. In turn, the effect of water vapor in the calcination and/or carbonation stages, including conversion, kinetics, and surface morphology, can be studied more accurately. In the course of this paper, the temperature rise and fall rates of the samples are fully considered. In addition, the effect of CO$_2$ diffusion in the N$_2$ and N$_2$-H$_2$O mixture atmosphere on CO$_2$ diffusion was calculated. The results of this work will help in producing more effective designs and control strategies of the Ca-looping process.

2. Results and Discussion

2.1. Effects of Steam during Both Calcination and Carbonation

A multicycle test of eight cycles was performed using BD limestone. Figure 1 shows the results that were obtained with two different experimental methods. As mentioned in the Introduction, with an increase in the cycle number, the amount of newly formed CaCO$_3$ decreases and the time required for its decomposition decreases [14].

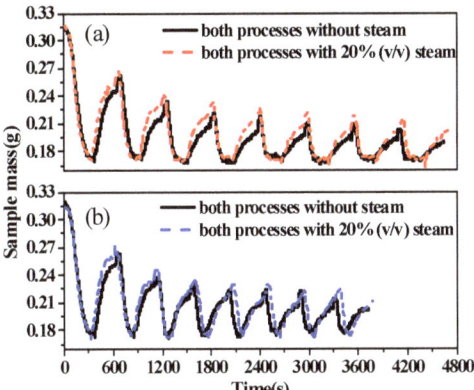

Figure 1. Sample weight vs. time for eight cycles of calcination/carbonation. (**a**) All calcination and carbonation periods lasted 300 s except the first calcination period, which was 360 s. (**b**) Calcination was completed and sent immediately to the carbonation furnace for carbonation, and all the carbonation periods were 300 s.

In Figure 1a, both calcination and carbonation time were fixed; the calcination period was 360 s and carbonation time was 300 s, whether calcination was completed or not. The calcination time in Figure 1b is adjusted with the increasing cycling number; once calcination was completed, the calcination was immediately terminated, and then, carbonation was carried out. On this basis, the influence of water vapor is also considered. Under the two different experimental methods, the carbonation conversion ratios were different. Generally, the conversion ratios in (a) were slightly lower than those in (b). This was obviously a result of sintering because of the excessive calcination time. Many studies used testing procedures that were similar to method (a). However, method (b) is considered more reasonable and was used in this study.

In addition, several characteristics are noted in Figure 1. Steam had a positive effect on CO_2 capture in Ca-looping. Steam increased the decomposition rate of $CaCO_3$ and enhanced the subsequent carbonation conversion of CaO. With 20% steam, both the reaction rate and calcium conversion ratio improved during all eight of the carbonation cycles compared to the results that were obtained without steam.

Figure 2 shows the conversion rate of carbonations in Figure 1b. Steam had a positive influence on carbonation, and this effect was more pronounced at higher steam concentrations and increasing cycle numbers.

2.2. Steam Addition during Calcination or Carbonation

In terms of the eight calcination/carbonation cycles experiments, further work is needed to determine the role of water vapor in each of the calcination and carbonation processes. Thus, a new test method is suggested here: steam is present in either calcination or carbonation but not in both reaction periods. In addition, the steam concentrations (0%, 10%, and 20%) are also considered. The carbonation conversions of CaO from BD limestone are shown in Figure 3.

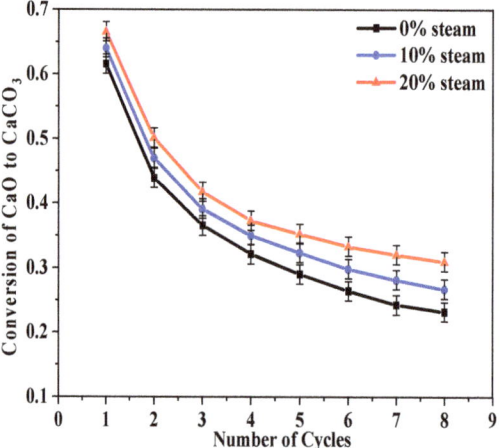

Figure 2. Influence of steam when it was present during both the calcination and carbonation stages on sorbent reactivity of BD limestone: calcination (80% CO_2 + 0%/10%/20% H_2O + O_2 balance, 900 °C); carbonation (15% CO_2 + 0%/10%/20% H_2O + N_2 balance, 650 °C).

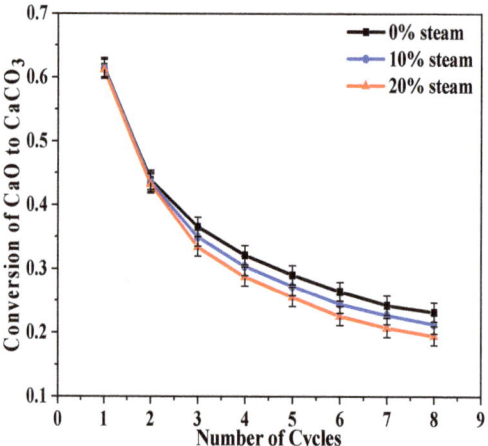

Figure 3. Influences of steam on conversion of CaO calcined from BD limestone: calcination (80% CO_2 + 0%/10%/20% H_2O + O_2 balance, 900 °C); carbonation (15% CO_2 + N_2 balance, 650 °C).

There was little difference in carbonation conversion for the CaO that calcined with steam until the third cycle. For the first carbonation cycle, the conversion was almost the same (~0.63) for all three of the calcination conditions. The differences became obvious at higher cycle numbers. For example, in the eighth cycle, the conversion ratio was 0.23 for CaO that was calcined without steam, 0.21 with 10% steam, and 0.19 with 20% steam. It can be concluded from the results shown in Figure 3 that the presence of steam in the calcination stage led to lower carbonation conversion of CaO, and when there is more steam, the sintering is more severe, which is consistent with previous studies [31] in showing that steam can accelerate the sintering of CaO and result in decreased surface area and porosity.

The first, fourth, and eighth carbonation cycles were recorded to help understand the effect of steam addition on carbonation reaction, and the results are shown in Figure 4. For the first cycle, there was almost no difference between CaO that was calcined with 0% and 20% steam. However, after four cycles, the differences became significant, especially after

50 s. Obviously, the addition of water vapor in the calcination shows a negative effect on the subsequent carbonation.

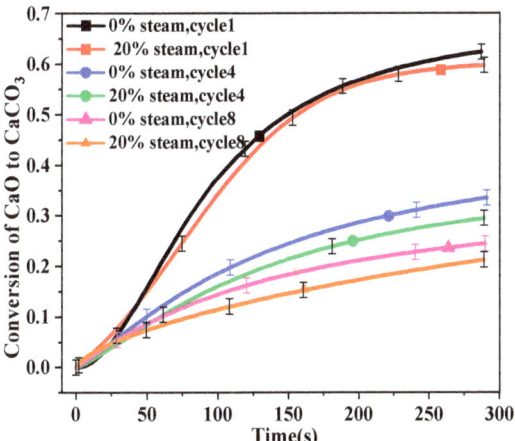

Figure 4. Conversion vs. time curves of BD limestone: calcination (80% CO_2 + 0%/20% H_2O + O_2 balance, 900 °C); carbonation (15% CO_2 balance N_2, 650 °C).

The effect of adding water vapor exclusively during the carbonation process is shown in Figure 5. For all eight cycles, the presence of steam in the carbonation stage improved the CO_2 capture capacity. The relative magnitude of the effect became more significant with an increase in the cycle number. For cycles 1 and 2, the differences in the carbonation conversion were minor at different steam concentrations. For the eighth cycle, the carbonation conversion was 0.33 with 20% steam, which is about 44% higher than the value of 0.23 without water vapor. Water vapor has a positive effect on the carbonation of CaO.

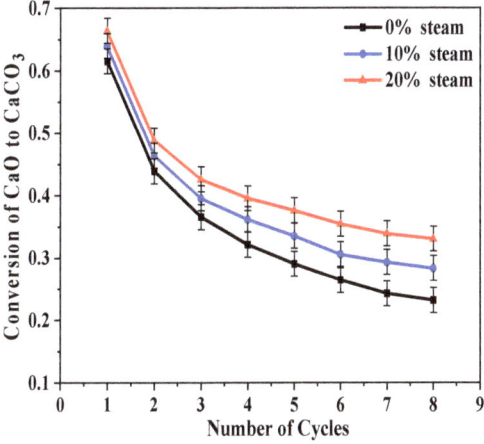

Figure 5. Influences that adding steam only during the carbonation stage has on sorbent reactivity of BD limestone: calcination (80% CO_2, O_2 balance, 900 °C); carbonation (15% CO_2 + 0%/10%/20% H_2O + N_2 balance, 650 °C).

Figure 6 shows the carbonation of CaO versus time for cycles 1, 4, and 8 when water vapor was only added in the carbonation stage. The effects were obvious from the start to the end. In this work, a sample was quickly moved into the furnace for calcination when the set temperature was reached and maintained. The phenomenon that is observed in Figure 6 may be easily explained by the catalytic effect of steam in the fast chemical control stage, as proposed by Wang [24] and Yang [32].

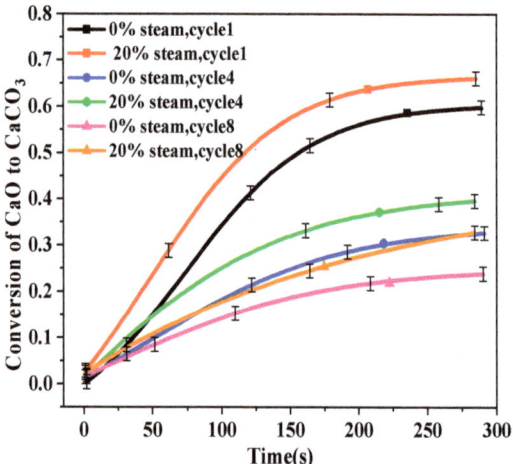

Figure 6. Conversion vs. time curves of different cycle numbers of BD limestone: calcination without water vapor (80% CO_2 + O_2 balance, 900 °C); only carbonation with vapor (15% CO_2 + 0%/20% H_2O + N_2 balance, 650 °C).

2.3. Reaction Temperature

The reaction temperature is one of the most important parameters for Ca-looping [14,33]. Calcination at 900 °C and carbonation at 650 °C were used in Sections 2.1 and 2.2. To further investigate the effects of temperature, more experiments were conducted. One set was performed at a carbonation temperature of 650 °C and calcination at 950 °C and 1000 °C. The second set of tests were conducted at a fixed calcination temperature of 900 °C and carbonation at 700 °C and 750 °C. The results are shown in Figures 6 and 7.

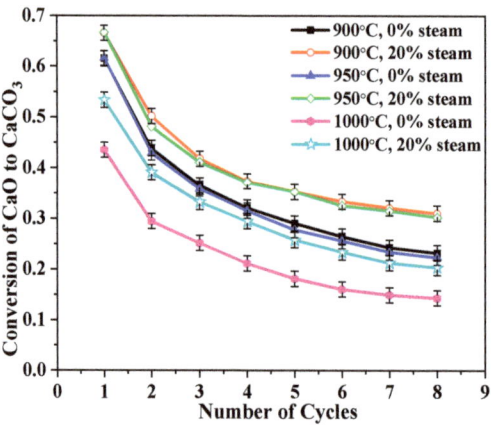

Figure 7. Influences of calcination temperature on sorbent reactivity of BD limestone: calcination (80% CO_2 + 0%/20% H_2O + O_2 balance) at different temperatures; carbonation (15% CO_2 + 0%/20% H_2O + N_2 balance) at 650 °C.

Several observations are made from Figure 7. The first is that for all three of the calcination temperatures, the conversion of CaO was improved by the presence of steam. The levels of improvement were very similar. The second observation is that the CaO conversion was almost the same (about 0.61) regardless of whether it was with 20% steam or without at 900 °C and 950 °C. However, the CaO carbonation conversion ratio decreased sharply at 1000 °C, and the conversion was about 0.45. The carbonation conversion ratio of CaO that was calcined at 900 °C and 950 °C for the first cycle was about 61%, but it was only 45% for the CaO that was calcined at 1000 °C.

Although adding steam can compensate for some of the lost CO_2 capacity, the carbonation conversion was still lower than that of the CaO that was calcined at the two lower temperatures even without added steam. A high calcination temperature led to sorbent reactivity decay in two ways: ion migration in sorbent crystal structure and sintering. Sintering decreased the sorbent surface area, and it is known that maximum conversions depend on the surface area of sorbents [34].

The initial calcination of BD limestone was compared in Figure 8. The calcination time at 900 °C was almost double that at 950 °C. It is known that the sintering of CaO is related both to temperature and to calcination time. Although higher temperature might have caused a higher rate of sintering, the shorter calcination time at 950 °C might reduce the overall sintering effect on CaO. This might be why the CaO calcined at 900 °C and 950 °C had similar CO_2 capture ability, as shown in Figure 9.

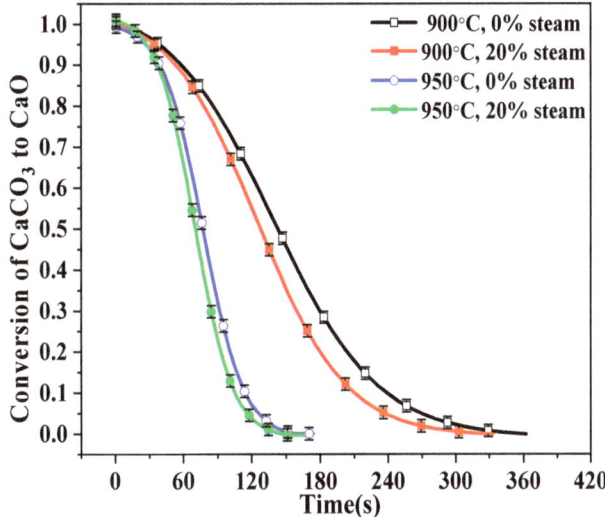

Figure 8. Decomposition curves of BD limestone calcination at 900 °C and 950 °C in 80% CO_2 + 0%/20% H_2O + O_2 balance.

Figure 9 showed carbonation conversions versus cycle number at 650, 700, and 750 °C for BD limestone. At 650 °C and 700 °C, 20% steam improved the carbonation conversion of CaO compared to the sample without steam, and it seems that 700 °C produces better conversion than 650 °C. However, when the carbonation temperature was increased to 750 °C, the carbonation conversion decreased greatly, even for the first cycle.

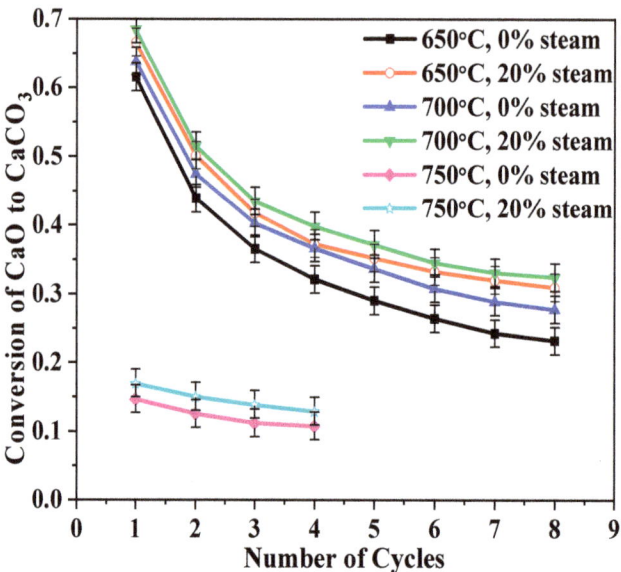

Figure 9. Influences of carbonation temperature on sorbent reactivity of BD limestone: calcination (900 °C, 80% CO_2 + 0%/20% H_2O + O_2 balance); carbonation (15% CO_2 + 0%/20% H_2O + N_2 balance).

As also seen in Figure 9, steam improved the conversion of CaO to $CaCO_3$ at all three carbonation temperatures. However, the magnitude of improvement was different, and the improvement was less pronounced at higher temperatures. A probable reason for this is that steam plays a catalytic role in the carbonation process. As a necessary process for a catalyst, steam must be adsorbed onto the surface of CaO. However, higher temperature had a negative effect on the adsorption process [18]. Hence, with an increase in the carbonation temperature, the catalytic effect of steam decreased.

To confirm this hypothesis, more carbonation experiments were conducted at different temperatures. Figure 10, which shows carbonation conversion versus time curves of CaO at four temperatures, clearly shows that steam had an obvious effect on the CaO carbonation rate during the initial fast reaction stage but not during the following slow diffusion-controlled reaction stage. In addition, steam has a greater catalytic effect on the carbonation reaction at lower temperatures. Linden et al. [35] found that a partial pressure of steam had a positive effect on the conversion of CaO to $CaCO_3$ in the temperature range of 400~550 °C. Figure 10 also shows that the reaction was quite slow in the temperature range of 550~600 °C, although CaO achieved a reasonable conversion ratio over much longer time periods. Given that the volume of CO_2-containing flue gas from coal-fired power plants is very large, a slower reaction rate to achieve high CO_2 capture efficiencies requires the construction of very bulky adsorber units to considerably extend the contact time of CaO and CO_2. Thus, from the results in Figure 10, it is determined that 650~700 °C is the most suitable temperature for carbonation for BD limestone.

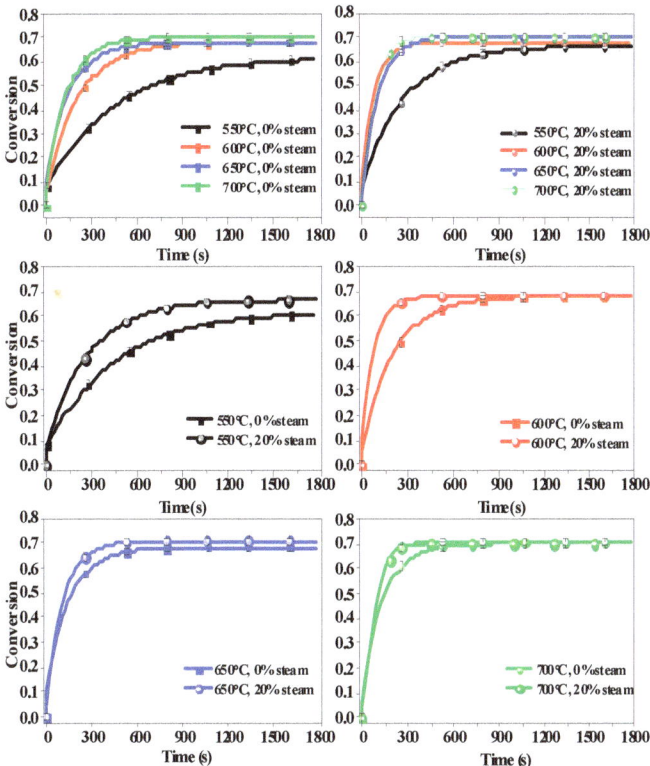

Figure 10. CaO carbonation conversion vs. time at different temperatures. (CaO was derived from BD limestone, calcination was at 900 °C in 80% CO_2 + O_2 balance, and carbonation was in 15% CO_2 + 0%/20% H_2O +N_2 balance.)

2.4. Pore Structure of CaO in Looping and Gas Diffusion Coefficient

In Section 2.1–2.3, steam has important influences on the carbonation ability of CaO. The surface morphology [36] and pore characteristics [37] of CaO are important factors that affect its carbonation in the looping cycle. Hence, comparing the surface morphology and pore characteristics of CaO that was calcined with different concentrations of steam may be helpful to understand the influence of steam. Figure 11 shows SEM images of the samples under each condition. In Figure 11a, $CaCO_3$ is dense and nonporous. After calcination, the CaO shown in Figure 11b has a rich pore structure. After calcination in an atmosphere containing water vapor, in Figure 11c, the particle size of CaO increased, and some small particles grew. After carbonization in an atmosphere with water vapor addition, the surface of $CaCO_3$ (in Figure 11d) is more compact, which corresponds to higher carbonation conversion.

The pore structures of CaO samples that were calcined in typical conditions were tested using N_2 absorption, and the results are shown in Table 1.

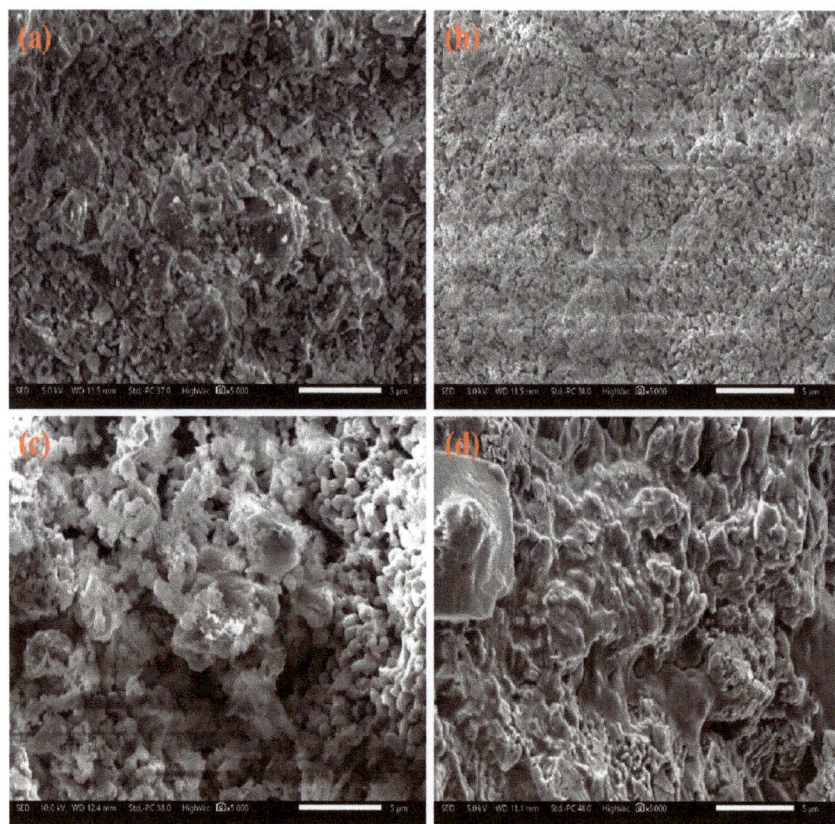

Figure 11. SEM images of samples under different conditions: (**a**) CaCO$_3$, (**b**) CaO, (**c**) CaO (calcined with water), and (**d**) CaCO$_3$ (calcined in water).

Table 1. Specific surface area and pore volume of CaO after first, third, and eighth calcinations.

	Surface Area m^2/g			Pore Volume mm^3/g		
	1st	3rd	8th	1st	3rd	8th
$\phi_{H_2O,cal} = \phi_{H_2O,car} = 0\%$	8.38	4.45	3.08	28.28	11.45	5.65
$\phi_{H_2O,cal} = \phi_{H_2O,car} = 20\%$	7.86	4.24	3.07	24.12	11.13	5.32

Table 1 shows that regardless of whether or not there is water vapor, the specific surface area and pore volume of CaO decrease with increasing cycles. For example, without water vapor, the specific surface area of CaO decreased from 8.38 to 3.08 m^2/g after the eighth cycle. Another observation is that there is very little difference in the CaO pore structure with or without vapor. However, in Section 2, it was clear that water vapor has a positive effect on CaO carbonation. Hence, the fact that the pore structures were similar with or without water vapor may indicate that the presence of steam predominantly plays a catalytic role in this case.

It was shown that steam did not have a significant influence on the preference of a Ca-based sorbent to adsorb CO$_2$; thus, more explanations are needed to account for this observation. Mass transfer is a key step in these reactions, and Figure 12 shows the effective diffusion coefficient of CO$_2$ (De) in a N$_2$ and N$_2$ + H$_2$O mixture. The De of CO$_2$

also increased with temperature and steam addition. It is clear that adding steam could improve the diffusion of CO_2 from the gas phase to the solid sorbent.

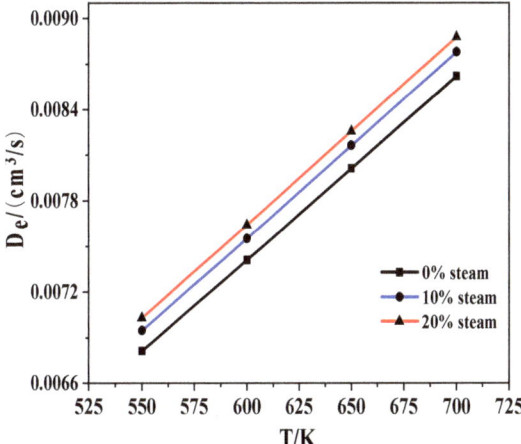

Figure 12. D_e of CO_2 vs. temperature in the presence of steam.

On the basis of the surface morphology, pore structure, and D_e of CO_2, a hypothesis is suggested here that there are three ways for steam to have an effect on the calcination and carbonation of a sorbent in looping. These three are high-temperature sintering, water vapor increasing CO_2 diffusion ability, and the catalytic effect of H_2O. The sintering of CaO that was accelerated by steam during calcination decreases its carbonation ability, and this negative effect should decrease with more cycles. The catalytic effect of steam has little relationship with its pore structure. Thus, the carbonation ability of CaO that undergoes severe sintering should be almost the same regardless of whether the sintering is with steam or not.

3. Experiment

3.1. Sample Preparation

In this study, natural limestone (BD) was used as a raw material. After crushing and screening, BD with a particle size of 150~250 μm was selected. The samples were dried at 105 °C for 4 h. Analysis results of BD limestone are given in Table 2.

Table 2. Composition of limestone (wt.%).

Compound	SiO_2	Al_2O_3	Fe_2O_3	TiO_2	P_2O_5	CaO	MgO	SO_3	Na_2O	K_2O	LOI
BD (wt%)	2.13	1.31	<0.55	<0.03	<0.03	53.22	1.48	<0.10	<0.20	0.13	41.02

3.2. Experimental Parameters

The limestone samples underwent calcination and carbonation in two separate tube furnaces, and the sample weight was measured continuously using a detection system. The experimental setup is shown in Figure 13.

First, two tube furnaces were heated to the test temperature, with a reaction gas flow rate of 1200 mL/min. At 1200 mL/min, mass transfer is not the limiting factor of the reaction, based on a previous experiment.

Second, a limestone sample (about 300 mg) was loaded into a quartz boat (110 mm long, 15 mm wide, and 12 mm deep).

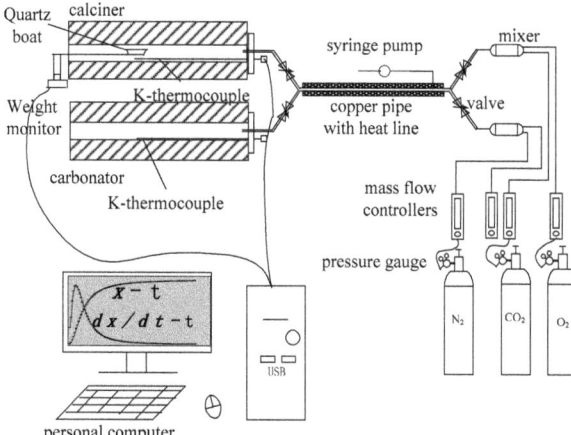

Figure 13. Tube furnace experimental system.

Third, when the furnaces reached the test temperature and stabilized, the limestone sample was quickly moved into the calciner. When the weight loss stopped (signaling the end of the calcination period), the sample was quickly moved into the carbonation furnace to carbonate with the synthetic flue gas. After carbonation, the sample was returned to the calcination furnace.

Finally, the above procedure was repeated for 8 complete calcination/carbonation cycles. The time needed to move the sample holder between the furnaces was less than 2 s.

The steam was added through a steam-generating unit, which has a programmable injection pump that feeds water into the evaporator. The temperature of the evaporator was maintained at about 220 °C. As a result, when the liquid water enters the evaporator, it is immediately converted into steam.

The CO_2 uptake capacity of the samples is calculated from mass changes using Equation (2):

$$X_N = \frac{m_{carb}^N - m_{cal}^N}{\beta m_0} \frac{M_{CaO}}{M_{CO_2}} \qquad (2)$$

where X_N is the carbonation conversion of the sample, N is the cycle number, m_0 is the initial mass (g) of the sample, m_{carb}^N is the mass (g) of the recarbonated sample after the Nth cycle, m_{cal}^N is the mass (g) of the calcined sample after the Nth cycle, and β is the content of CaO in the initial sample. M_{CaO} and M_{CO_2} are the molar weights (g/mol) of CaO and CO_2, respectively.

Diffusivity of CO_2 (D_e) was then calculated using the following formulas [38]:
D is the intrinsic diffusion coefficient:

$$D = \frac{1}{1/D_{CO_2,N_2-H_2O} + 1/D_K} \qquad (3)$$

D_{CO_2,N_2-H_2O} is the effective diffusion coefficient of CO_2 in the mixture of N_2 and H_2O:

$$D_{CO_2,N_2-H_2O} = \left(1 - y_{CO_2,N_2-H_2O}\right) / \sum_{i=2}^{k}(y_i/D_{1i}) \qquad (4)$$

D_K is the Knudsen diffusion coefficient:

$$D_K = 97 r_K \sqrt{\frac{T}{M_{CO_2}}} \qquad (5)$$

De is the effective diffusion coefficient, which is primarily controlled by Knudsen diffusion:

$$D_e = D\frac{\varepsilon_p}{\tau} \quad (6)$$

$$\varepsilon_p = V_p \rho_p \quad (7)$$

where ε_i is the porosity of the particle, V_p is the quality of the catalyst pore volume (kg/m^3) for the unit, ρ_p is the density (kg/m^3) of the sorbent particles, and τ is the labyrinth factor, which has a value of 3 here.

4. Conclusions

In this paper, the effects of water vapor on the calcination of calcium-based sorbents and the carbonation of calcium-based carbon dioxide sorbents were studied using a self-made thermogravimetric analyzer that can achieve rapid cooling. The results show that when water vapor is present in both the calcination and carbonation periods, it has an overall positive effect on CO_2 capture in the Ca-looping process. Water vapor increases the calcination rate of $CaCO_3$ and enhances the subsequent carbonation conversion of calcined CaO. However, the presence of water vapor in the calcination of limestone only leads to the deactivation of CaO, and this can result in lower carbonation conversion of CaO. The presence of water vapor in the carbonation stage improved CO_2 capture for all eight of the cycles tested. There are only insignificant differences in the carbonation conversion ratio for CaO samples that were calcined at 900 °C and 950 °C. This is likely caused by the combined effect of the calcination rate and the sintering time. It seems that carbonation at 700 °C has a better conversion ratio than that at 650 °C, and the carbonation conversion was much worse at 750 °C. Water vapor accelerates the sintering of CaO during calcination. On the other hand, the effective diffusion coefficient De of CO_2 increases with the addition of water vapor and an increase in temperature.

Author Contributions: Conceptualization, N.L. and X.Z.; Methodology, S.C. and Q.L. (Qi Liu); Resources, Q.L. (Quansheng Liu); Data curation, R.B., N.L. and X.Z.; Writing—original draft, R.B.; Writing—review & editing, N.L. and X.Z.; Supervision, S.C. All authors have read and agreed to the published version of the manuscript.

Funding: This project is supported by the National Natural Science Foundation of China (21868021), Ordos Science and Technology Plan Project (2021YY-Gong-19-49), Higher Educational Scientific Research Projects of Inner Mongolia Autonomous Region (JY20220021), Research Project of Colleges and Universities in Hebei Province (QN2021090), and Doctoral Fund of Hebei Normal University (L2021034).

Institutional Review Board Statement: Not applicable.

Informed Consent Statement: Not applicable.

Data Availability Statement: The data presented in this study are available on request from the corresponding author.

Conflicts of Interest: The authors declare no conflict of interest.

Sample Availability: Samples are available from the authors.

References

1. Stocker, T.F.; Quin, D. IPCC 2013: Climate Change 2013—The Physical Science Basis. In *Contribution of Working Group I to the Fifth Assessment Report of the Intergovernmental Panel on Climate Change*; Cambridge University Press: Cambridge, UK, 2013.
2. Norbarzad, M.J.; Tahmasebpoor, M.; Heidari, M.; Pevida, C. Theoretical and experimental study on the fluidity performance of hard-to-fluidize carbon nanotubes-based CO_2 capture sorbents. *Front. Chem. Sci. Eng.* **2022**, *16*, 1460–1475. [CrossRef]
3. Troya, J.A.; Jiménez, P.S.E.; Perejón, A.; Moreno, V.; Valverde, J.M.; Maqueda, P.A.P. Kinetics and cyclability of limestone ($CaCO_3$) in presence of steam during calcination in the CaL scheme for thermochemical energy storage. *Chem. Eng. J.* **2021**, *417*, 129194. [CrossRef]
4. Han, R.; Wang, Y.; Xing, S.; Pang, C.; Hao, Y.; Song, C.; Liu, Q. Progress in reducing calcination reaction temperature of Calcium-Looping CO_2 capture technology: A critical review. *Chem. Eng. J.* **2022**, *450*, 137952. [CrossRef]

5. Chen, J.; Duan, L.; Sun, S. Review on the development of sorbents for calcium looping. *Energy Fuels* **2020**, *34*, 7806–7836. [CrossRef]
6. Imani, M.; TPeridas, G.; Mordick Schmidt, B. Improvement in cyclic CO_2 capture performance and fluidization behavior of eggshell-derived $CaCO_3$ particles modified with acetic acid used in calcium looping process. *J. CO2 Util.* **2022**, *65*, 102207. [CrossRef]
7. Peridas, G.; Mordick Schmidt, B. The role of carbon capture and storage in the race to carbon neutrality. *Electr. J.* **2021**, *34*, 106996. [CrossRef]
8. Heidari, M.; Tahmasebpoor, M.; Mousavi, S.B.; Pevida, C. CO_2 capture activity of a novel CaO adsorbent stabilized with $(ZrO_2+Al_2O_3+CeO_2)$-based additive under mild and realistic calcium looping conditions. *J. CO2 Util.* **2021**, *53*, 101747. [CrossRef]
9. Mousavi, S.B.; Heidari, M.; Rahmani, F.; Sene, R.A.; Clough, P.T.; Ozmen, S. Highly robust ZrO_2-stabilized CaO nanoadsorbent prepared via a facile one-pot MWCNT-template method for CO_2 capture under realistic calcium looping conditions. *J. Clean. Prod.* **2023**, *384*, 135579. [CrossRef]
10. Shimizu, T.; Hirama, T.; Hosoda, H.; Kitano, K.; Inagaki, M.; Tejima, K. A Twin Fluid-Bed Reactor for Removal of CO_2 from Combustion Processes. *Chem. Eng. Res. Des.* **1999**, *77*, 62–68. [CrossRef]
11. Moreno, J.; Hornberger, M.; Schmid, M.; Scheffknecht, G. Oxy-Fuel Combustion of Hard Coal, Wheat Straw and Solid Recovered Fuel in a 200 kWth Calcium Looping CFB Calciner. *Energies* **2021**, *14*, 2162. [CrossRef]
12. Xu, H.; Shi, B. Design and System Evaluation of Mixed Waste Plastic Gasification Process Based on Integrated Gasification Combined Cycle System. *Processes* **2022**, *10*, 499. [CrossRef]
13. Heidari, M.; Tahmasebpoor, M.; Antzaras, A.; Lemonidou, A.A. CO_2 capture and flfluidity performance of CaO-based sorbents: Effect of Zr, Al and Ce additives in tri-, bi- and mono-metallic confifigurations. *Process Saf. Environ. Prot.* **2020**, *14*, 349–365. [CrossRef]
14. Heidari, M.; Mousavi, S.B.; Rahmani, F.; Clough, P.T.; Ozmen, S. The novel Carbon Nanotube-assisted development of highly porous $CaZrO_3$-CaO xerogel with boosted sorption activity towards high-temperature cyclic CO_2 capture. *Energy Convers. Manag.* **2022**, *274*, 116461. [CrossRef]
15. Chen, J.; Duan, L.; Sun, S. Accurate control of cage-Like CaO hollow microspheres for enhanced CO_2 capture in calcium looping via a template-assisted synthesis approach. *Environ. Sci. Technol.* **2019**, *53*, 2249–2259. [CrossRef]
16. Tian, S.C.; Yan, F.; Zhang, Z.T.; Jiang, J.G. Calcium-looping reforming of methane realizes in situ CO_2 utilization with improved energy efficiency. *Sci. Adv.* **2019**, *5*, eaav5077. [CrossRef] [PubMed]
17. Labus, K. Comparison of the Properties of Natural Sorbents for the Calcium Looping Process. *Materials* **2021**, *14*, 548. [CrossRef]
18. Wang, N.N.; Feng, Y.C.; Guo, X. Atomistic mechanisms study of the carbonation reaction of CaO for high temperature CO_2 capture. *Appl. Surf. Sci.* **2020**, *532*, 147425. [CrossRef]
19. Bian, Z.G.; Li, Y.J.; Zhang, C.X.; Zhao, J.L.; Wang, Z.Y.; Liu, W.Q. $CaO/Ca(OH)_2$ heat storage performance of hollow nanostructured CaO-based material from Ca-looping cycles for CO_2 capture. *Fuel Process. Technol.* **2021**, *217*, 106834. [CrossRef]
20. Criado, Y.A.; Arias, B.; Abanades, J.C. Effect of the Carbonation Temperature on the CO_2 Carrying Capacity of CaO. *Ind. Eng. Chem. Res.* **2018**, *57*, 12595–12599. [CrossRef]
21. Wang, H.; Guo, S.; Liu, D.Y.; Guo, Y.Z.; Gao, D.Y.; Sun, S.Z. A dynamic study on the impacts of water vapor and impurities on limestone calcination and cao sulfurization processes in a microfluidized bed reactor analyzer. *Energy Fuels* **2016**, *30*, 4625–4634. [CrossRef]
22. Guo, H.X.; Yan, S.L.; Zhao, Y.G.; Ma, X.B.; Wang, S.P. Influence of water vapor on cyclic CO_2 capture performance in both carbonation and decarbonation stages for Ca-Al mixed oxide. *Chem. Eng. J.* **2019**, *359*, 542–551. [CrossRef]
23. Dong, J.; Tang, Y.; Nzihou, A.; Weiss-Hortala, E. Effect of steam addition during carbonation, calcination or hydration on re-activation of CaO sorbent for CO_2 capture. *J. CO2 Util.* **2020**, *39*, 101167. [CrossRef]
24. Wang, C.B.; Jia, L.F.; Tan, Y.W.; Anthony, E.J. Carbonation of fly ash in oxy-fuel CFB combustion. *Fuel* **2008**, *87*, 1108–1114. [CrossRef]
25. Champagne, S.; Lu, D.Y.; Macchi, A.; Symonds, R.T.; Anthiny, E.J. Influence of Steam Injection during Calcination on the Reactivity of CaO-Based Sorbent for Carbon Capture. *Ind. Eng. Chem. Res.* **2013**, *52*, 2241–2246. [CrossRef]
26. Donat, F.; Florin, N.H.; Anthony, E.J.; Fennell, P.S. Influence of high-temperature steam on the reactivity of CaO sorbent for CO capture. *Environ. Sci. Technol.* **2012**, *46*, 1262–1269. [CrossRef]
27. Ar, I.; Dogu, G. Calcination kinetics of high purity limestones. *Chem. Eng. J.* **2001**, *2*, 131–137. [CrossRef]
28. Ramezani, M.; Tremain, P.; Doroodchi, E.; Moghtaderi, B. Determination of carbonation/calcination reaction kinetics of a limestone sorbent in low CO_2 partial pressures using TGA experiments. *Energy Procedia* **2017**, *114*, 259–270. [CrossRef]
29. Valverde, J.M. On the negative activation energy for limestone calcination at high temperatures nearby equilibrium. *Chem. Eng. Sci.* **2015**, *132*, 169–177. [CrossRef]
30. Li, D.; Wang, Y.; Li, Z.S. Limestone Calcination Kinetics in Microfluidized Bed Thermogravimetric Analysis (MFB-TGA) for Calcium Looping. *Catalysts* **2022**, *12*, 1661. [CrossRef]
31. Iliuta, I.; Radfarnia, H.R.; Iliuta, M.C. Hydrogen Production by Sorption-Enhanced Steam Glycerol Reforming: Sorption Kinetics and Reactor Simulation. *AIChE J.* **2013**, *59*, 2105–2118. [CrossRef]
32. Yang, S.; Xiao, Y. Steam Catalysis in CaO Carbonation under Low Steam Partial Pressure. *Ind. Eng. Chem. Res.* **2008**, *47*, 4043–4048. [CrossRef]

33. Li, Z.S.; Fan, F.; Tang, X.Y.; Cai, N.S. Effect of Temperature on the Carbonation Reaction of CaO with CO_2. *Energy Fuels* **2012**, *26*, 2473–2482. [CrossRef]
34. Sun, P.; Grace, J.R.; Lim, J.C.; Anthony, E. A discrete-pore-size-distribution-based gas–solid model and its application to the reaction. *Chem. Eng. Sci.* **2008**, *63*, 57–70. [CrossRef]
35. Lindén, I.; Backman, P.; Brink, A. Influence of Water Vapor on Carbonation of CaO in the Temperature Range 400–550 °C. *Ind. Eng. Chem. Res.* **2011**, *50*, 14115–14120. [CrossRef]
36. Silakhori, M.; Jafarian, M.; Alfonso, C.; Saw, W.; Venkataraman, M.; Lipinski, W.; Nathan, G. Effects of steam on the kinetics of calcium carbonate calcination. *Chem. Eng. Sci.* **2021**, *246*, 116987. [CrossRef]
37. Yang, J.; Ma, L.P.; Liu, H.P.; Yi, W.; Keomounlath, B.; Dai, Q.X. Thermodynamics and kinetics analysis of Ca-looping for CO_2 capture: Application of carbide slag. *Fuel* **2019**, *242*, 1–11. [CrossRef]
38. Fedunik-Hofaman, L.; Bayon, A.; Donne, S.W. Comparative Kinetic Analysis of $CaCO_3$/CaO Reaction System for Energy Storage and Carbon Capture. *Appl. Sci.* **2019**, *9*, 4601. [CrossRef]

Disclaimer/Publisher's Note: The statements, opinions and data contained in all publications are solely those of the individual author(s) and contributor(s) and not of MDPI and/or the editor(s). MDPI and/or the editor(s) disclaim responsibility for any injury to people or property resulting from any ideas, methods, instructions or products referred to in the content.

Article

Microbial Preparations Combined with Humic Substances Improve the Quality of Tree Planting Material Needed for Reforestation to Increase Carbon Sequestration

Aleksey Nazarov [1,*], Sergey Chetverikov [2], Darya Chetverikova [2], Iren Tuktarova [1], Ruslan Ivanov [2], Ruslan Urazgildin [2], Ivan Garankov [1] and Guzel Kudoyarova [2]

[1] Department of Environment and Rational Use of Natural Resources, Faculty of Business Ecosystem and Creative Technologies, Ufa State Petroleum Technological University, ul. Kosmonavtov 1, Ufa 450064, Russia; himcenter@mail.ru (I.G.)

[2] Ufa Institute of Biology, Ufa Federal Research Centre, RAS, Prospekt Oktyabrya 69, Ufa 450054, Russia; che-kov@mail.ru (S.C.); belka-strelka8031@yandex.ru (D.C.); ivanovirs@mail.ru (R.I.); guzel@anrb.ru (G.K.)

* Correspondence: nazarovam1501@gmail.com

Abstract: Restoring forests in areas where they once stood is an important step towards increasing carbon sequestration. However, reforestation requires an increase in current levels of seedling production in the tree nurseries. The purpose of this work was to study the effectiveness of preparations based on bacteria and humic substances (HSs) to stimulate the growth of tree seedlings in a nursery. Two selected strains of *Pseudomonas* and humic substances were used to treat pine and poplar plants. The treatment of seedlings was carried out during their transplantation and after it, and the effects of treatment on shoot elongation, shoot and root mass were evaluated. Treatments with both bacterial strains enhanced the growth of poplar and pine shoots and roots, which was explained by their ability to synthesize auxins. *P. protegens* DA1.2 proved to be more effective than *P.* sp. 4CH. The treatment of plants with humic substances increased the nitrogen balance index and the content of chlorophyll in the leaves of poplar seedlings, which can elevate carbon storage due to the higher rate of photosynthesis. In addition, the combination of humic substances with *P. protegens* DA1.2 increased shoot biomass accumulation in newly transplanted pine plants, which indicates the possibility of using this combination in plant transplantation. The increase in length and weight of shoots and roots serves as an indicator of the improvement in the quality of planting material, which is necessary for successful reforestation to increase capture of carbon dioxide.

Keywords: decarbonization; woody plantations; seedling growth; *Pseudomonas* species; humic sustances

Citation: Nazarov, A.; Chetverikov, S.; Chetverikova, D.; Tuktarova, I.; Ivanov, R.; Urazgildin, R.; Garankov, I.; Kudoyarova, G. Microbial Preparations Combined with Humic Substances Improve the Quality of Tree Planting Material Needed for Reforestation to Increase Carbon Sequestration. *Sustainability* 2023, *15*, 7709. https://doi.org/10.3390/su15097709

Academic Editors: Zilong Liu, Meixia Shan and Yakang Jin

Received: 15 March 2023
Revised: 6 May 2023
Accepted: 6 May 2023
Published: 8 May 2023

Copyright: © 2023 by the authors. Licensee MDPI, Basel, Switzerland. This article is an open access article distributed under the terms and conditions of the Creative Commons Attribution (CC BY) license (https://creativecommons.org/licenses/by/4.0/).

1. Introduction

Deforestation is one of the largest anthropogenic sources of increased GHG emissions in the world [1]. Planting trees has the potential to increase the capacity of forests to sequester carbon [2] and reforestation helps mitigate climate [3]. However, most of the forest trees face severe problems in successful regeneration [4] and young pine seedlings experience slow growth in the first years after planting [5]. Meanwhile, reforestation requires increasing the number and quality of tree seedlings produced in nurseries. Improving cultivation of tree seedlings for reforestation has attracted the attention of researchers [6]. The nursery stage of tree planting is a very important phase for the later success of field plantations and has been shown to need critical care to ensure their long-term productivity [7]. A recent report highlights that carbon sequestration resulting from tree planting and reforestation depends on nursery management [8].

One of the ways to solve the problem of obtaining high-quality tree seedlings for woody plantations is the introduction of modern physiologically active preparations based on bacteria and humic substances into the technology of their cultivation. Their use is considered an environmentally friendly mechanism of plant growth promotion compared to large amounts of fertilizers and is becoming more widespread in the agricultural industry [9]. It is well known that rhizospheric bacteria [10–12] as well as humate substances (HSs, products of degradation of organic matter extracted from brown coal, peat, and other sources) [13–15] have a positive effect on plant growth when introduced separately. However, there are significantly fewer reports of their effects on trees than on herbaceous plants. Nevertheless, it was shown that humates increased the growth of trees and the yield of oranges and grapefruits [16], while humic acids enhanced the growth of rubber tree planting materials [17]. Rhizosphere microbes enhanced the growth of teak seedlings (Tectona grandis) [7], and bacteria of the *Pseudomonas fluorescens* strain promoted the growth of aspen seedlings under nutrient stress [18]. Plant-growth-promoting (PGP) rhizobacteria have a beneficial effect on plants, including trees, by increasing the availability of soil nutrients to the plant and the production of metabolites such as plant hormones [19]. Reports on agricultural crops have shown that the combined effect of humates and bacteria turned out to be more effective than the use of each of them separately [20,21]. However, there are practically no works devoted to the combined use of plant growth promoting rhizobacteria (PGPR) and humic substances for tree nursery management. In this regard, the aim of this work was to study the effectiveness of preparations based on bacteria and HSs for stimulating plant growth as a means of improving the quality of tree planting material and more sustainable production of seedlings for reforestation in order to optimize carbon sequestration. The novelty of this work lies in the study of the complex effect of humates and microorganisms on the growth of tree seedlings, which, as far as we know, has not been carried out before. We hypothesized that the combined use of bacteria and humic substances may be more effective in stimulating the growth of tree seedlings in nurseries than the use of each separately.

2. Materials and Methods

2.1. Bacterial Strain and Cultural Media

We used two strains of Gram-negative bacteria from the collection of microorganisms of the Ufa Institute of Biology isolated from natural sources: *Pseudomonas protegens* DA1.2 (deposited in All-Russian Collection of Microorganisms B-3542D) described in the article by Chetverikov et al. [22] and *Pseudomonas* sp. 4CH (deposited in the collection of microorganisms UIB-57). These bacterial strains were chosen for treating tree seedlings since in previous experiments their combination with humates stimulated the growth of herbaceous plants (wheat) [20]. Bacteria were cultivated in Erlenmeyer flasks with King's B medium (2% peptone, 1% glycerol, 0.15% K_2HPO_4, 0.15% $MgSO_4*7H_2O$) on an Innova 40R shaker (New Brunswick, NJ, USA) (160 rpm) for 48 h at 28 °C. The number of cells in cultures was measured by applying serial dilutions to King B medium with agar-agar (15 g L^{-1}) and then counting the number of colony-forming units (CFU). The bacterial culture was diluted with sterile water to give a solution for treatment of plants. The ability to mobilize phosphates was assessed by measuring the size of transparent zones on Pikovskaya medium, and acetylene reduction assay was used as a measure of bacterial nitrogenase activity as described [23].

2.2. Taxonomic Affiliation of Bacterial Strain

For the taxonomic affiliation of the 4CH bacterial strain, the nucleotide sequence of 16S rRNA gene was determined. Total DNA from bacterial colonies was isolated using the RIBO-sorb reagent kit (Amplisens®, Central Research Institute of Epidemiology, Moscow, Russia) according to the manufacturer's recommendations. Amplification of the 16S rRNA gene fragment was performed using universal primers: 27F (5'-AGAGTTTGATCTGGCTCAG-3') and 1492R (5'-ACGGTACCTTGTTACGACTT-3') [24]. 16S rRNA sequencing was performed using the BigDye Terminator sequencing kit (Applied Biosystems, Thermo Fisher Scientific Inc., Waltham, MA, USA) using a Genetic

Analyzer 3500 xL (Applied Biosystems, Thermo Fisher Scientific Inc., Waltham, MA, USA). To sequester cycle-sequencing reaction components, we used the BigDye® XTerminator™ purification kit (Applied Biosystems, Thermo Fisher Scientific Inc., Waltham, MA, USA). The search for 16S rRNA nucleotide sequences similar to the corresponding sequences of the studied strains was performed in the GenBank sequence database using the BLAST software package (http://www.ncbi.nlm.nih.gov/blast accessed on 2 February 2023).

2.3. Extraction of Humic Substances

The source of humic substances was the brown coal from the Tyulganskoe deposit in the Orenburg region of the Russian Federation. Coal was mixed with 0.1 M KOH in a ratio of 1:10 and HSs were extracted for two hours with stirring at 1500 rpm. The precipitate was removed by centrifugation at 12,000 rpm for 10 min.

2.4. Plant Growth Conditions and Treatments

We used generally accepted technology for growing tree seedlings in nurseries, while the combined treatment of tree seedlings with bacteria and humates was used for the first time in this study. We modified the technology that was previously successfully used on wheat plants [20]. Experiments were carried out in the tree nursery of the Bashkir Agrarian University (54°80′ N, 55°84′ E, 170 m a.s.l., Ufa region of Bashkortostan, Russian Federation). For the experiment, we used cuttings of Bashkir pyramidal poplar (a hybrid of Italian pyramidal poplar and black poplar (*P. italica pyramidalis* × *P. nigra*), obtained in the late 1930s at the Bashkir forest experimental station). These tree species were chosen for the present experiments as they are often used for reforestation and urban greening in many regions. In the first ten days of April, with the beginning of intensive snowmelt, poplar branches of last year's generation were selected, from which 20 cm long cuttings were separated and laid for monthly stratification. Before treatment with preparations based on bacteria and HSs, the cuttings were placed in water for the germination of the first roots. The cuttings were planted in the ground for 2/3 of their length at an angle of 45° to the ground surface.

Before planting, poplar seedlings were soaked in 2 L of water, to which 50 mL of bacterial suspension ((4 ± 0.5) 10^9 CFU mL^{-1}) and HSs (2 g L^{-1}) were added singly or in combination. Seedlings were watered 2 times each month with 2 L of the mixture of bacteria and HSs of the same concentration. Control plants were treated with the same amount of water without additives. Three months after planting the poplar seedlings, the length of lateral shoots was measured.

Two-year-old seedlings of *Pinus sylvestris* L. from the nursery were planted at a distance of 50 cm from each other. Before planting they were soaked in suspension of bacteria, humates, or their mixture and then watered with the same solutions as described above. In parallel, pine seedlings transplanted from the nursery 1 and 2 years prior to the present experiments (denoted below as 3- and 4-years old seedlings, respectively) were watered in the same way with bacteria and HSs (singly or in mixture) 3 times. Pine growth rate was assessed by the change in the length of the main and side shoots within 4 months. Shoots and roots of two-year-old pine seedlings were sampled after the end of seedling growth in the current year. Roots taken at the same time were washed with tap water and both shoots and roots were dried in a ventilated oven at 60 °C for 48 h to measure their dry mass.

2.5. Analysis of the Content of Pigments

The content of chlorophyll (a + b), flavonoids, and nitrogen balance index (NBI) [24] in the leaves was measured using a DUALEX SCIENTIFIC+ device (FORCE-A, Paris, France) according to the manufacturer's recommendations.

2.6. IAA (Auxin, Indoleacetic Acid) Assay in Bacterial Cultural Media

On the second day of cultivating bacteria, immunoassay of culture media was performed. IAA was partitioned from culture media of bacteria with diethyl ether as described [25]. Briefly, 1 ml of bacterial culture media was diluted with distilled water and acidified with HCl to pH 2.5 to extract IAA with diethyl ether. Then, hormones were partitioned from diethyl ether into $NaHCO_3$ solution and re-extracted with diethyl ether from the acidified aqueous phase. IAA analysis was carried out with enzyme-linked immunosorbent assay using specific antibodies against IAA as described [26]. The reliability of the method is due to specificity of antibodies to auxins and the use of an extraction method that makes it possible to efficiently extract hormones while reducing the amount of impurities by decreasing the volume of extractants at each stage of solvent partitioning. The efficiency of purification of IAA prior to immunoassay was confirmed by the study of chromatographic distribution, which showed that the peaks of immunoreactivity coincided only with the positions of the IAA standards.

2.7. Statistics

The data were processed using Statistica version 10 (Statsoft, Moscow, Russia) and are presented in the tables and figures as means ± standard errors. The statistical significance of differences between the mean values was assessed using analysis of variance followed by Duncan's test ($p < 0.05$). In the figures, mean values that are statistically different from each other are indicated by different letters. The number of replications (n) is provided in the figure legends.

3. Results

To identify the 4CH strain, the nucleotide sequence (1401 bp) of the 16S rRNA gene was determined; it was deposited in the GenBank database as OQ381088. Its comparison with other known sequences made it possible to attribute the studied microorganism to *P. chlororaphis* with a high degree of probability (99.93% similarity with the type strain of *P. chlororaphis* subsp. *aureofaciens* NBRC 3521(T) was found). Based on the data on the nucleotide sequence of the 16S rRNA gene, a phylogenetic tree was constructed to identify relationships with other species of the genus *Pseudomonas* (Figure 1).

Studied bacterial strains showed nitrogenase activity and the ability to solubilize phosphates and synthesize auxins (Table 1).

Table 1. Properties of plant-growth-promoting bacteria.

Strain	Nitrogenase Activity, nmol C_2H_4 h^{-1} mL^{-1}	Synthesis of Auxins, ng mL^{-1}	Phosphates Solubilization, mm
Pseudomonas protegens DA1.2	20.8 ± 0.3	870 ± 44	18 ± 2
Pseudomonas sp. 4CH	20.0 ± 0.2	837 ± 55	15 ± 2

Measurement of the shoot length 4 months after transplanting of 2-year-old pine seedlings showed that the rate of shoot elongation was significantly accelerated by treatment with *P. protegens* DA1.2 and its combination with HSs (Figure 2). The rates of shoot elongation in plants treated with *Pseudomonas* sp. 4CH and its combination with HSs were intermediate between control plants and plants treated with *P. protegens* DA1.2. The mean increase in shoot length of plants treated with HSs in combination with *Pseudomonas protegens* DA1.2 was significantly different from that of plants treated with HSs alone, while there was no difference between plants treated with HSs or these bacteria alone (Figure 2).

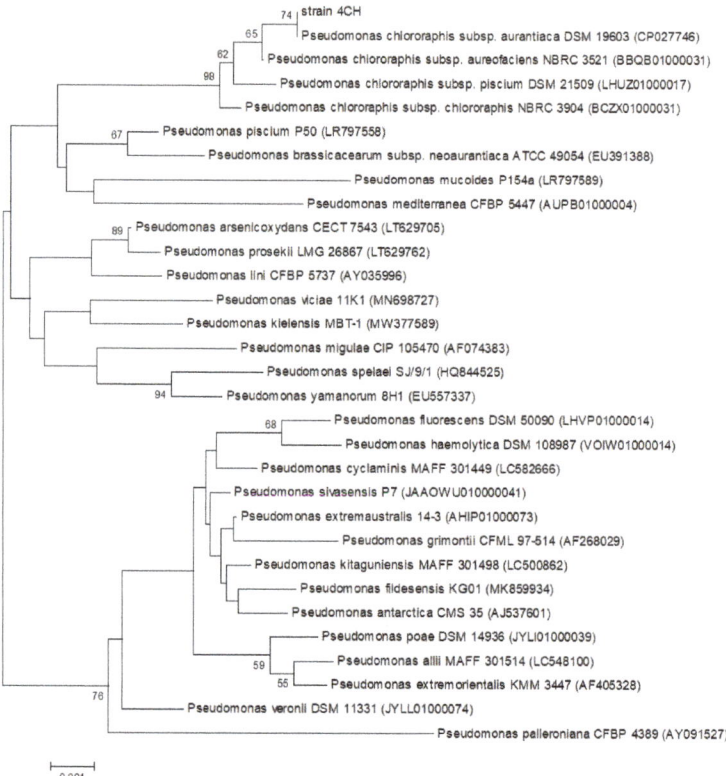

Figure 1. Phylogenetic position of the strain *Pseudomonas* sp. 4CH according to the analysis of the nucleotide sequence of the 16S rRNA gene (the evolutionary distance corresponding to 1 nucleotide change in every 1000 is shown on a scale, the numbers are the statistical significance of the branching order determined with bootstrap analysis, and the indicator values above 50% are shown).

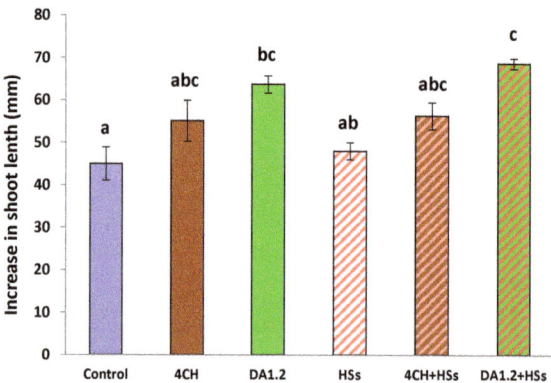

Figure 2. The increase in shoot length (averaged values for the main and lateral shoots) of 2-year-old pine seedlings 4 months after their transplantation and triple watering with the bacterial suspensions of *Pseudomonas* sp. 4CH, *Pseudomonas protegens* DA1.2, and humic substances (HSs) applied alone or in combination (4CH + HSs and DA1.2 + HSs). Means that are statistically different from each other are marked with different letters, $p \leq 0.05$, $n = 10$ (ANOVA followed by Duncan's test).

When pine seedlings were treated with both bacterial preparations, the mass of shoots or roots increased compared to the control (Figure 3). Humates did not affect the accumulation of root mass, but increased the mass of shoots when applied alone or in combination with any of the bacteria. The mass of shoots of plants treated with any bacterium in combination with HS was larger than in plants treated only with the corresponding bacteria.

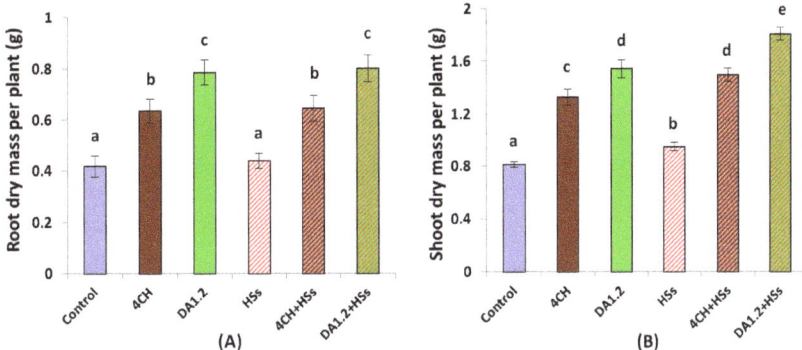

Figure 3. Dry mass of roots (**A**) and shoots (**B**) of 2-year-old pine seedlings sampled after the end of seedling growth in the current year. Mean values that are statistically different from each other are marked with different letters, $p \leq 0.05$, $n = 10$ (ANOVA followed by Duncan's test).

Bacterial treatment of pine seedlings transplanted three years before the present experiments accelerated elongation of their shoots compared with the control plants (Figure 4). The rate of shoot elongation of plants treated with HSs did not differ from that in the control or in plants treated with *Pseudomonas* sp. 4CH, while in the case of combination of *Pseudomonas* sp. 4CH and HSs the increase in shoot length was significantly greater than in plants treated with HSs alone.

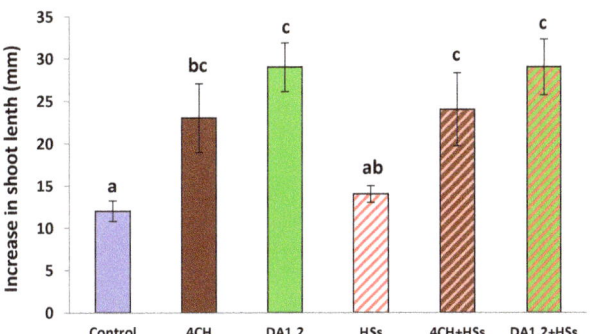

Figure 4. The increase in the shoot length (averaged values for the main and side shoots) of 3-year-old pine seedlings in 4 months after triple watering with suspensions of bacteria *Pseudomonas* sp. 4CH, *Pseudomonas protegens* DA1.2, and humic substances (HSs) used alone or in combination (4CH + HSs and DA1.2 + HSs). Mean values that are statistically different from each other are marked with different letters, $p \leq 0.05$, $n = 10$ (ANOVA followed by Duncan's test).

The treatment of 4-year-old seedlings with bacteria and HS separately and in combination accelerated the elongation of their shoots (Figure 5). However, the effect of HSs applied alone was significantly lower than that of bacterial treatments used alone or in combination with HSs.

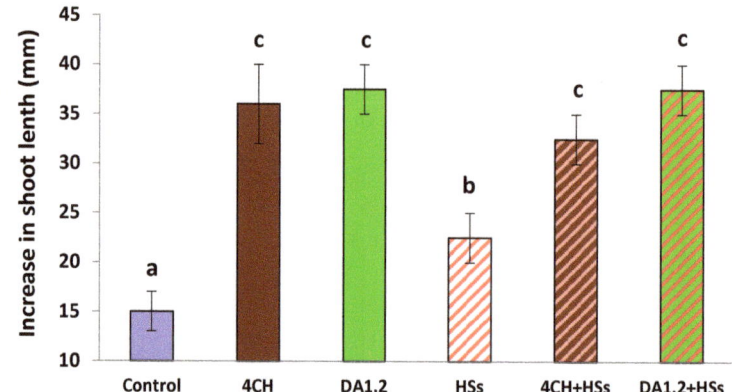

Figure 5. The increase in the shoot length (averaged values for the main and side shoots) of 4-years-old pine seedlings in 4 months after triple watering with the suspensions of bacteria *Pseudomonas* sp. 4CH, *Pseudomonas protegens* DA1.2, and humic substances (HSs) used alone or in combination (4CH + HSs and DA1.2 + HSs). Mean values that are statistically different from each other are marked with different letters, $p \leq 0.05$, $n = 10$ (ANOVA followed by Duncan's test).

Statistically significant increase in the shoot length of poplar plants compared to the control was found only when they were treated with *P. protegens* DA1.2 applied alone (Figure 6).

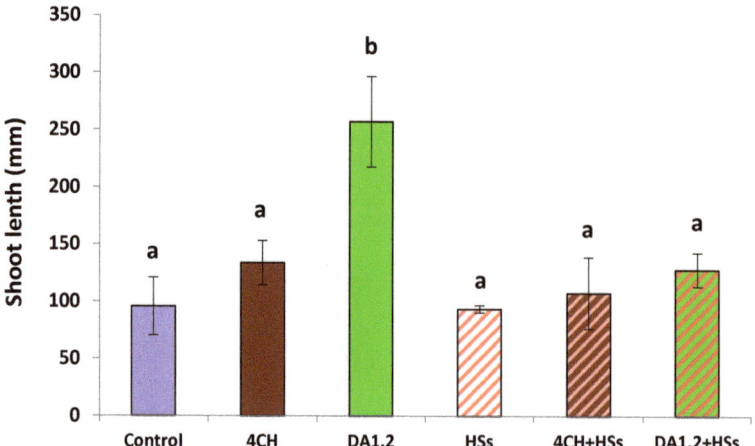

Figure 6. Length of shoots of the poplar plants after watering with bacterial suspensions *Pseudomonas* sp. 4CH, *Pseudomonas protegens* DA1.2, and humic substances (HSs) used alone or in combination (4CH + HSs and DA1.2 + HSs). Mean values that are statistically different from each other are marked with different letters, $p \leq 0.05$, $n = 6$ (ANOVA followed by Duncan's test).

None of the treatment options increased the content of flavonoids in poplar plants (Figure 7A). The content of chlorophyll and NBI increased only when plants were treated with HSs. (Figure 7B,C).

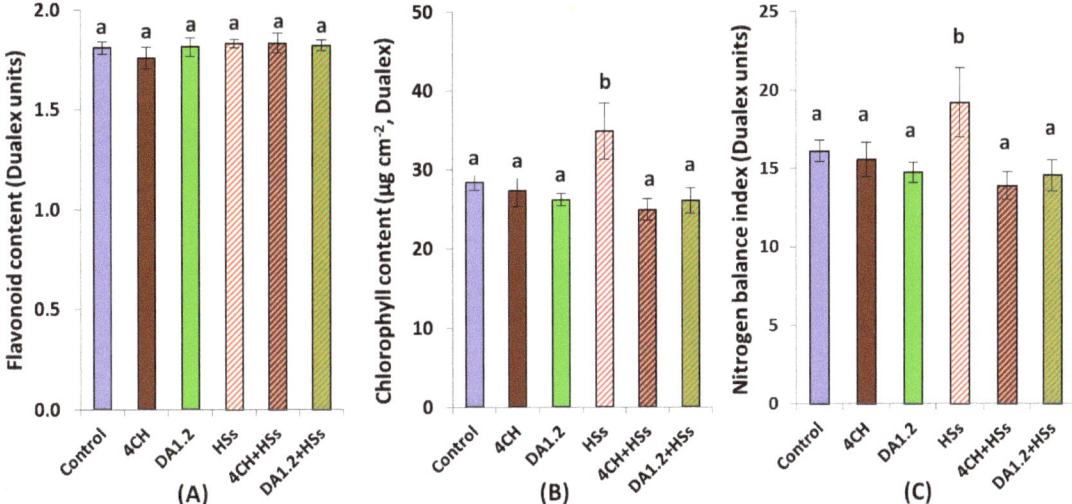

Figure 7. Content of flavonoids (**A**), chlorophyll (**B**), and nitrogen balance index (NBI) (**C**) of the poplar plants after watering with suspensions of bacteria *Pseudomonas* sp. (4CH), *Pseudomonas protegens* DA1.2, and humic substances (HSs) used alone or in combination (4CH + HSs and DA1.2 + HSs). Mean values that are statistically different from each other are marked with different letters, $p \leq 0.05$, $n = 15$ (ANOVA followed by Duncan's test).

4. Discussion

We were able to demonstrate that both bacterial treatments increased elongation of pine shoots (Figures 2 and 4). In addition, larger mass of shoots and roots was found in bacteria-treated plants (Figure 3). The use of microorganisms to increase plant growth and productivity is an important practice required for agriculture [27,28]. During the past decades, the use of PGPR for sustainable agriculture has greatly increased [29]. However, there are significantly fewer reports of their effects on trees than on herbaceous plants. Our data are consistent with a report that showed increased growth of tree seedlings under the influence of rhizospheric microbes as a sustainable way to optimize forestry [7]. Inoculation of *Pinus taeda* seedlings with *Bacillus subtilis* increased root and shoot biomass [30]. Shoot and root growth of *Swietenia macrophylla* was also stimulated by *Bacillus* spp. under nursery conditions [31].

In the present experiments, bacterial effects can be explained by the ability of the used bacterial strains to synthesize auxins, their nitrogenase activity and ability to solubilize phosphates (Table 1). Auxins are known to stimulate cell extension [32] and auxin production by these bacterial strains was detected in the present (Table 1) and previous [20] experiments. The capacity of PGPR to synthesize auxins is considered one of the most important mechanisms through which microbes regulate plant growth [10,14]. Increased root biomass found in the present experiments is consistent with the data of other researchers. Thus, inoculation with *Azospirillum brasilense* and *Pseudomonas geniculata* strains increased the root mass and length of *Linum usitatissimum* [10]. Stimulation of root growth by bacteria contributes to increased water and nutrient uptake by plants, thereby promoting their growth [10,14]. The increase in length and weight of shoots and roots found in the present experiments serves as an indicator of the improvement in the quality of planting material [33], which is necessary for successful reforestation to mitigate climate change and capture carbon dioxide [3].

The increment in shoot length was greatest in 2-year-old pine seedlings, transplanted just before the present experiments (about 45–70 mm (Figure 2) versus no more than 35 mm in older plants (Figures 4 and 5)). Bacterial preparations were most effective on 4-year-old pine seedlings transplanted 2 years before the present experiments. In this case, treatment with both bacterial strains led to a 2.5-fold increase in the length of shoots compared with the control (Figure 5). Bacterial treatments were less effective in recently transplanted plants (Figure 2): they only resulted in an increase in the length of the shoots by about 1.5 times compared to the control in the case of *P. protegens* DA1.2, while in plants treated with *Pseudomonas* sp. 4CH the rate of shoot elongation of 2-year-old pine seedlings was close to the control. In most cases, *P. protegens* DA1.2 bacteria were more effective than *Pseudomonas* sp. 4CH. Thus, the effectiveness of the action of bacteria on shoot elongation depended on the type of microorganisms and the age of the plants.

Bacterial treatment of poplar plants was less effective than in the case of pine plants, and only the treatment with *P. protegens* DA1.2 led to a statistically significant increase in the length of poplar shoots compared with the control (Figure 6).

The effect of HSs was lower than that of bacterial treatments when each was applied separately. Humates increased the biomass of shoots, but not roots. This can be explained by the presence of cytokinin-like substances in humates [34], and it is known that these hormones stimulate the growth of shoots and inhibit root growth [35]. Our data are consistent with reports indicating an increase in the growth rate of orange and grape trees by humates [16] and an acceleration of the growth of rubber planting materials with foliar application of humic acid [17].

The additive effect of the combination of bacteria and HSs was less pronounced than in previous experiments with herbaceous plants [20]. Nevertheless, the shoots of 2-year-old pine plants treated during their transplantation were significantly heavier in the case of combination of HSs with either *P. protegens* DA1.2 or *Pseudomonas* sp. 4CH compared with corresponding bacterial treatments applied alone. Thus, the combination of HSs and these bacteria may be recommended for use during transplantation of pine plants.

HSs increased concentrations of chlorophyll and NBI (nitrogen balance index) [24] in the poplar leaves (Figure 7B,C). Humic substances stimulated the uptake of nitrate by roots and the accumulation of the anion at the leaf level of maize [36]. In our previous experiments we found an increased accumulation of total nitrogen in the shoots of wheat plants that had received organomineral fertilizers with humates [15]. The chlorophyll molecule contains nitrogen, which makes availability of this element an important factor in the development of the photosynthetic apparatus. Therefore, an increase in NBI and chlorophyll in leaves of HSs-treated poplar plants is likely to contribute to enhancing photosynthesis and improving carbon accumulation by the plants [37].

In order to increase the production of tree seedlings, nurseries are currently using large amounts of fertilizers that can lead to environmental pollution [38]. Furthermore, this technology produces individual trees which are unbalanced in size and more likely to suffer infections from phytopathogenic fungi [33]. Bacteria and humic substances may be important for plant nutrition by increasing N and P uptake by the plants without addition of excessive amounts of fertilizers.

The results of our research show that the use of bacteria and humic substances improves the quality of tree planting material for reforestation, while reforestation is a means of increasing carbon sequestration. The combination of bacteria and humic substances can work better than either of them alone.

5. Conclusions

We were first to study the combined effects of bacteria and humates on the growth of tree seedlings. Our studies have shown the ability of bacterial preparations to accelerate the growth of shoots of poplar and pine plants. *P. protegens* DA1.2 proved to be more effective than *P.* sp. 4CH, which indicates the prospects for further search for more effective strains. The treatment of plants with humic substances increased nitrogen balance index and chlorophyll content in the leaves of poplar seedlings, which is likely to increase carbon storage due to increased photosynthesis. In addition, combination of HSs with *P. protegens*

DA1.2 increased shoot biomass accumulation of recently transplanted pine plants, which suggests the possibility of using this combination during plant transplantation.

The nursery stage is believed to be a very important phase for the later success of field plantations. Nevertheless, further studies are needed to confirm the long-term positive effects of humates and bacteria on the behavior of trees in field plantations. Study of the effects of inoculating the rhizosphere of seedlings of other tree species with various bacterial strains and treating them with humates is needed to find an effective combination to achieve successful reforestation as a means of increasing the carbon storage capacity of forests under various climatic conditions. Nevertheless, the data obtained in the present study demonstrate promising prospects and the feasibility of such a study.

Author Contributions: Conceptualization, A.N., I.T. and S.C.; methodology, R.U. and I.T.; software, R.I.; formal analysis, D.C.; investigation, D.C. and I.G.; resources, A.N.; data curation, D.C.; writing—original draft preparation, G.K.; writing—review and editing, A.N. and S.C.; visualization, R.I.; supervision, A.N. and S.C.; project administration, A.N.; funding acquisition, A.N. All authors have read and agreed to the published version of the manuscript.

Funding: This research was performed within the state assignment framework of the Ministry of Science and Higher Education of the Russian Federation «Program for the creation and functioning of a carbon polygons» on the territory Bashkortostan Republic «Eurasian carbon polygon» for 2022–2023 (Publication number: FEUR-2022-0001).

Institutional Review Board Statement: Not applicable.

Informed Consent Statement: Not applicable.

Data Availability Statement: Not applicable.

Conflicts of Interest: The authors declare no conflict of interest.

Abbreviations

HSs	humate substances
GDP	gross domestic product
GHG	greenhouse gas
PGPR	plant growth promoting rhizobacteria
NBI	nitrogen balance index
IAA	Indole-3-acetic acid (auxin)

References

1. Pearson, T.R.H.; Brown, S.; Murray, L.; Sidman, G. Greenhouse gas emissions from tropical forest degradation: An underestimated source. *Carbon Balance Manag.* **2017**, *12*, 3. [CrossRef] [PubMed]
2. Grant, M.; Domke, G.M.; Oswalt, S.N.; Walters, B.F.; Morin, R.S. Tree planting has the potential to increase carbon sequestration capacity of forests in the United States. *Proc. Natl. Acad. Sci. USA* **2020**, *117*, 40, 24649–24651. [CrossRef]
3. Fargione, J.; Haase, D.L.; Burney, O.T.; Kildisheva, O.A.; Edge, G.; Cook-Patton, S.C.; Chapman, T.; Rempel, A.; Hurteau, M.D.; Davis, K.T.; et al. Challenges to the reforestation pipeline in the United States. *Front. For. Glob. Chang.* **2021**, *4*, 629198. [CrossRef]
4. Ahangar, M.A.; Dar, G.H.; Bhat, Z.A. Growth response and nutrient uptake of blue pine (Pinus wallichiana) seedlings inoculated with rhizosphere microorganisms under temperate nursery conditions of Kashmir. *Ann. For. Res.* **2012**, *55*, 217–227. [CrossRef]
5. Xu, Y.; Zhang, Y.; Li, Y.; Li, G.; Liu, D.; Zhao, M.; Cai, N. Growth promotion of Yunnan pine early seedlings in response to foliar application of IAA and IBA. *Int. J. Mol. Sci.* **2012**, *13*, 6507–6520. [CrossRef]
6. Dumroese, R.K.; Landis, T.D.; Pinto, J.R.; Haase, D.L.; Wilkinson, K.W.; Davis, A.S. Meeting forest restoration challenges: Using the target plant concept. *Reforesta* **2016**, *1*, 37–52. [CrossRef]
7. Chaiya, L.; Gavinlertvatana, P.; Teaumroong, N.; Pathom-aree, W.; Chaiyasen, A.; Sungthong, R.; Lumyong, S. Enhancing Teak (*Tectona grandis*) seedling growth by rhizosphere microbes: A sustainable way to optimize agroforestry. *Microorganisms* **2021**, *9*, 1990. [CrossRef]
8. Moser, R.L.; Windmuller-Campione, M.A.; Russell, M.B. Natural resource manager perceptions of forest carbon management and carbon market participation in Minnesota. *Forests* **2022**, *13*, 1949. [CrossRef]
9. Shah, A.; Nazari, M.; Antar, M.; Msimbira, L.A.; Naamala, J.; Lyu, D.; Rabileh, M.; Zajonc, J.; Smith, D.L. PGPR in agriculture: A sustainable approach to increasing climate change resilience. *Front. Sustain. Food Syst.* **2021**, *5*, 667546. [CrossRef]

10. Omer, A.M.; Osman, M.S.; Badawy, A.A. Inoculation with *Azospirillum brasilense* and/or *Pseudomonas geniculata* reinforces flax (*Linum usitatissimum*) growth by improving physiological activities under saline soil conditions. *Bot. Stud.* **2022**, *63*, 15. [CrossRef]
11. Abdel Latef, A.A.H.; Omer, A.M.; Badawy, A.A.; Osman, M.S.; Ragaey, M.M. Strategy of salt tolerance and interactive impact of *Azotobacter chroococcum* and/or *Alcaligenes faecalis* inoculation on canola (*Brassica napus* L.) Plants Grown Saline Soil. *Plants* **2021**, *10*, 110. [CrossRef] [PubMed]
12. Kudoyarova, G.; Arkhipova, T.; Korshunova, T.; Bakaeva, M.; Loginov, O.; Dodd, I.C. Phytohormone mediation of interactions between plants and non-symbiotic growth promoting bacteria under edaphic stresses. *Front. Plant Sci.* **2019**, *10*, 1368. [CrossRef]
13. Canellas, L.P.; Olivares, F.L.; Aguiar, N.O.; Jones, D.L.; Nebbioso, A.; Mazzei, P. Humic and fulvic acids as biostimulants in horticulture. *Sci. Hortic.* **2015**, *196*, 15–27. [CrossRef]
14. Olaetxea, M.; de Hita, D.; Garcia, C.A.; Fuentes, M.; Baigorri, R.; Mora, V.; Garnica, M.; Urrutia, O.; Erro, J.; Zamarreño, A.M.; et al. Hypothetical framework integrating the main mechanisms involved in the promoting action of rhizospheric humic substances on plant root and shoot-growth. *Appl. Soil Ecol.* **2018**, *123*, 521–537. [CrossRef]
15. Nazarov, A.M.; Garankov, I.N.; Tuktarova, I.O.; Salmanova, E.R.; Arkhipova, T.N.; Ivanov, I.I.; Feoktistova, A.V.; Prostyakova, Z.G.; Kudoyarova, G.R. Hormone balance and shoot growth in wheat (*Triticum durum* Desf.) plants as influenced by sodium humates of the granulated organic fertilizer. *Sel'skokhozyaistvennaya Biol.* **2020**, *55*, 945–955. [CrossRef]
16. Alva, A.K.; Obreza, T.A. By-product iron-humate increases tree growth and fruit production of orange and grapefruit. *HortScience* **1998**, *33*, 71–74. [CrossRef]
17. Cahyo, A.N.; Ardika, R.; Saputra, J.; Wijaya, T. Acceleration on the growth of rubber planting materials by using foliar application of humic acid. *J. Agric. Sci.* **2014**, *36*, 112–119. [CrossRef]
18. Shinde, S.; Cumming, J.R.; Collart, F.R.; Noirot, P.H.; Larsen, P.E. *Pseudomonas fluorescens* transportome is linked to strain-specific plant growth promotion in aspen seedlings under nutrient stress. *Front. Plant Sci.* **2017**, *8*, 348. [CrossRef]
19. Noirot-Gros, M.-F.; Shinde, S.V.; Akins, C.; Johnson, J.L.; Zerbs, S.; Wilton, R.; Kemner, K.M.; Noirot, P.; Babnigg, G. Functional imaging of microbial interactions with tree roots using a microfluidics setup. *Front. Plant Sci.* **2020**, *11*, 408. [CrossRef]
20. Feoktistova, A.; Bakaeva, M.; Timergalin, M.; Chetverikova, D.; Kendjieva, A.; Rameev, T.; Hkudaygulov, G.; Nazarov, A.; Kudoyarova, G.; Chetverikov, S. Effects of humic substances on the growth of *Pseudomonas plecoglossicida* 2,4-d and wheat plants inoculated with this strain. *Microorganisms* **2022**, *10*, 1066. [CrossRef]
21. Alharbi, K.; Rashwan, E.; Hafez, E.; Omara, A.E.-D.; Mohamed, H.H.; Alshaal, T. Potassium humate and plant growth-promoting microbes jointly mitigate water deficit stress in soybean cultivated in salt-affected soil. *Plants* **2022**, *11*, 3016. [CrossRef]
22. Chetverikov, S.P.; Chetverikova, D.V.; Bakaeva, M.D.; Kenjieva, A.A.; Starikov, S.N.; Sultangazin, Z.R. A promising herbicide-resistant bacterial strain of *Pseudomonas protegens* for stimulation of the growth of agricultural cereal grains. *Appl. Biochem. Microbiol.* **2021**, *57*, 110–116. [CrossRef]
23. Bakaeva, M.; Kuzina, E.; Vysotskaya, L.; Kudoyarova, G.; Arkhipova, T.; Rafikova, G.; Chetverikov, S.; Korshunova, T.; Chetverikova, D.; Loginov, O. Capacity of *Pseudomonas* strains to degrade hydrocarbons, produce auxins and maintain plant growth under normal conditions and in the presence of petroleum contaminants. *Plants* **2020**, *9*, 379. [CrossRef] [PubMed]
24. Zhang, K.; Liu, X.; Ma, Y.; Zhang, R.; Cao, Q.; Zhu, Y.; Cao, W.; Tian, Y. A comparative assessment of measures of leaf nitrogen in rice using two leaf-clip meters. *Sensors* **2019**, *20*, 175. [CrossRef] [PubMed]
25. Veselov, D.S.; Sharipova, G.V.; Veselov, S.U.; Kudoyarova, G.R. The effects of NaCl treatment on water relations, growth and ABA content in barley cultivars differing in drought tolerance. *J. Plant Growth Regul.* **2008**, *27*, 380–386. [CrossRef]
26. Arkhipova, T.; Martynenko, E.; Sharipova, G.; Kuzmina, L.; Ivanov, I.; Garipova, M.; Kudoyarova, G. Effects of plant growth promoting rhizobacteria on the content of abscisic acid and salt resistance of wheat plants. *Plants* **2020**, *9*, 1429. [CrossRef]
27. Backer, R.; Rokem, J.S.; Ilangumaran, G.; Lamont, J.; Praslickova, D.; Ricci, E.; Subramanian, S.; Smith, D.L. Plant growth-promoting rhizobacteria: Context, mechanisms of action, and roadmap to commercialization of biostimulants for sustainable agriculture. *Front. Plant Sci.* **2018**, *9*, 1473. [CrossRef]
28. Ruzzi, M.; Aroca, R. Plant growth-promoting rhizobacteria act as biostimulants in horticulture. *Sci. Hortic.* **2015**, *196*, 124–134. [CrossRef]
29. Das, A.J.; Kumar, M.; Kumar, R. Plant growth promoting rhizobacteria (PGPR): An alternative of chemical fertilizer for sustainable, Environment friendly agriculture. *Res. J. Agric. Fores. Sci.* **2013**, *1*, 21–23.
30. Shekhawat, S.; Alessa, N.; Rathore, H.; Sharma, K. A green approach—Cost optimization for a manufacturing supply chain with MFIFO warehouse dispatching policy and inspection policy. *Sustainability* **2022**, *14*, 14664. [CrossRef]
31. Trujillo-Elisea, F.I.; Labrín-Sotomayor, N.Y.; Becerra-Lucio, P.A.; Becerra-Lucio, A.A.; Martínez-Heredia, J.E.; Chávez-Bárcenas, A.T.; Peña-Ramírez, Y.J. Plant growth and microbiota structural effects of *Rhizobacteria* inoculation on mahogany (*Swietenia macrophylla* King [*Meliaceae*]) under nursery conditions. *Forests* **2022**, *13*, 1742. [CrossRef]
32. Spaepen, S.; Vanderleyden, J. Auxin and plant-microbe interactions. *Cold Spring Harb. Perspect. Biol.* **2011**, *3*, a001438. [CrossRef] [PubMed]
33. Otero, M.; Salcedo, I.; Txarterina, K.; González-Murua, C.; Duñabeitia, M.K. Quality assessment of *Pinus radiata* production under sustainable nursery management based on compost tea. *J. Plant Nutr. Soil Sci.* **2019**, *3*, 356–366. [CrossRef]
34. Pizzeghello, D.; Francioso, O.; Ertani, A.; Muscolo, A.; Nardi, S. Isopentenyladenosine and cytokinin-like activity of different humic substances. *J. Geochem. Explor.* **2013**, *129*, 70–75. [CrossRef]
35. Werner, T.; Nehnevajova, E.; Köllmer, I.; Novak, O.; Strnad, M.; Krämer, U.; Schmülling, T. Root-specific reduction of cytokinin causes enhanced root growth, drought tolerance, and leaf mineral enrichment in Arabidopsis and tobacco. *Plant Cell* **2010**, *22*, 3905–3920. [CrossRef]

36. Quaggiotti, S.; Ruperti, B.; Pizzeghello, D.; Francioso, O.; Tugnoli, V.; Nardi, S. Effect of low molecular size humic substances on nitrate uptake and expression of genes involved in nitrate transport in maize (*Zea mays* L.). *J. Exp. Bot.* **2004**, *55*, 803–813. [CrossRef]
37. Hebat-Allah, A.A.; Alshammari, S.O.; Abd El-Sadek, M.E.; Kenawy, S.K.M.; Badawy, A.A. The promotive effect of putrescine on growth, biochemical constituents, and yield of wheat (*Triticum aestivum* L.) plants under water stress. *Agriculture* **2023**, *13*, 587. [CrossRef]
38. Aslantaş, R.; Çakmakçi, R.; Şahin, F. Effect of plant growth promoting rhizobacteria on young apple tree growth and fruit yield under orchard conditions. *Sci. Hortic.* **2007**, *111*, 371–377. [CrossRef]

Disclaimer/Publisher's Note: The statements, opinions and data contained in all publications are solely those of the individual author(s) and contributor(s) and not of MDPI and/or the editor(s). MDPI and/or the editor(s) disclaim responsibility for any injury to people or property resulting from any ideas, methods, instructions or products referred to in the content.

Review

A Systematic Review of Syngas Bioconversion to Value-Added Products from 2012 to 2022

Marta Pacheco [1,2], Patrícia Moura [1] and Carla Silva [2,*]

[1] LNEG, Laboratório Nacional de Energia e Geologia, Unidade de Bioenergia e Biorrefinarias, 1649-038 Lisboa, Portugal
[2] Universidade de Lisboa, Faculdade de Ciências, Instituto Dom Luiz, 1749-016 Lisboa, Portugal
* Correspondence: camsilva@ciencias.ulisboa.pt

Abstract: Synthesis gas (syngas) fermentation is a biological carbon fixation process through which carboxydotrophic acetogenic bacteria convert CO, CO_2, and H_2 into platform chemicals. To obtain an accurate overview of the syngas fermentation research and innovation from 2012 to 2022, a systematic search was performed on Web of Science and The Lens, focusing on academic publications and patents that were published or granted during this period. Overall, the research focus was centered on process optimization, the genetic manipulation of microorganisms, and bioreactor design, in order to increase the plethora of fermentation products and expand their possible applications. Most of the published research was initially funded and developed in the United States of America. However, over the years, European countries have become the major contributors to syngas fermentation research, followed by China. Syngas fermentation seems to be developing at "two-speeds", with a small number of companies controlling the technology that is needed for large-scale applications, while academia still focuses on low technology readiness level (TRL) research. This systematic review also showed that the fermentation of raw syngas, the effects of syngas impurities on acetogen viability and product distribution, and the process integration of gasification and fermentation are currently underdeveloped research topics, in which an investment is needed to achieve technological breakthroughs.

Keywords: synthesis gas; syngas fermentation; carboxylic acids; alcohols; Web of Science; The Lens

Citation: Pacheco, M.; Moura, P.; Silva, C. A Systematic Review of Syngas Bioconversion to Value-Added Products from 2012 to 2022. *Energies* **2023**, *16*, 3241. https://doi.org/10.3390/en16073241

Academic Editors: Zilong Liu, Meixia Shan and Yakang Jin

Received: 6 March 2023
Revised: 29 March 2023
Accepted: 31 March 2023
Published: 4 April 2023

Copyright: © 2023 by the authors. Licensee MDPI, Basel, Switzerland. This article is an open access article distributed under the terms and conditions of the Creative Commons Attribution (CC BY) license (https:// creativecommons.org/licenses/by/ 4.0/).

1. Introduction

The pressing global concerns over atmospheric carbon dioxide (CO_2) levels, climate change, and waste recycling have served as major incentives to pursuing commodities that are derived from alternative non-fossil raw materials. The escalating urgency of these issues has spurred researchers to explore both established and emerging carbon sequestration technologies to meet the demand for carbon-based fuels and chemicals, while drastically reducing the emissions from fossil-derived sources [1]. These challenges are in line with the United Nations Sustainable Development Goals (UNSDG) for 2030, namely the Sustainable Cities and Communities (Goal 11), Responsible Production and Consumption (Goal 12), and Climate Change mitigation (Goal 13) [2].

A way to comply with this demand is through the reutilization of carbon that is trapped in hard-to-treat residues. This can be accomplished by the application of thermochemical technologies, such as gasification and pyrolysis, that facilitate the conversion of recalcitrant and heterogeneous forms of residual carbon, such as those that are contained in municipal solid waste (MSW) or lignocellulosic waste biomass, into simpler forms, as in the case of synthesis gas or syngas [3,4]. This gaseous product is mainly composed of carbon monoxide (CO), CO_2, and hydrogen (H_2), but can also contain, in lower amounts, methane (CH_4), short chain hydrocarbons (C_nH_m), hydrogen sulfide (H_2S), ammonia (NH_3), and hydrogen cyanide (HCN). Syngas can be easily transformed into heat and power by direct burning, but it can also be converted to liquid fuels, traditionally through the Fischer–Tropsch process (FTP) [5,6]. FTP is a catalytic process that converts a mixture of H_2 and CO into aliphatic products. However, like many other catalytic processes, it presents several downsides. Notably, it requires high energy inputs, due to the high temperature and pressure that are required to kickstart the reactions, and expensive rare metal catalysts composed of critical raw materials, which tend to become easily fouled with the contaminants that are present in the syngas [6,7].

Syngas fermentation can be the biological alternative to FTP. It consists of the use of microorganisms, which are designated carboxydotrophic acetogens, to assimilate the gaseous carbon forms, i.e., CO and CO_2, in the absence of oxygen (O_2), converting them into microbial biomass, carboxylic acids, and alcohols that can be used as platform chemicals, or as feedstock for liquid fuel production [8,9]. This conversion can be performed in a wide range of pressures and temperatures, since there are strains of acetogens that grow under mild conditions, normally at 37 °C, while others thrive in thermophilic environments, growing at temperatures higher than 50 °C [9]. Syngas fermentation is not only applicable to syngas itself, but virtually to any gaseous mixture that is composed mainly of CO, CO_2, and H_2, such as industrial off-gases from steel mills, including Linz–Donawitz gas, and the cement industry [10–13].

Research on syngas fermentation goes as far back as the 1970s, but has seen an exponential increase in interest from 2008 onward, nowadays being a promising research field for bioenergy and climate action [6,14]. Over the years, there have been a multitude of review publications on the subject, focusing on different approaches, such as revealing the most promising mesophilic and thermophilic microbial catalysts, bioreactor configurations that facilitate the gas–liquid mass transfer and increase the carbon fixation yields, genetic manipulation to develop more robust acetogens or to support the specific metabolic pathways that increase the product yield, or the optimization of the fermentation parameters for a more efficient bioconversion process [7,15–18]. In 2022, Calvo et al., performed a much-needed systematic review of the research into syngas fermentation, yet their focus was on scientific publications and collaborations, pointing towards the most fruitful areas of publication and future collaborative research interests [6]. However, none of these reviews performed a holistic, integrated evaluation between research and final application.

The end goal of research and innovation is to gain knowledge, to improve our understanding of the universe, and to allow for the progress of society. However, knowledge by itself has little value; it is only through sharing that it can fulfill its role. This is mainly achieved through scientific publications and intellectual property (IP) protection via patent applications, both of which facilitate innovation and improve knowledge [19]. This review aims to fill this information gap by crossing the data from both scientific and patent databases. Taking into consideration the particular case of syngas fermentation, a series of questions concerning the advances from 2012 to 2022 were addressed:

- What is being researched in syngas fermentation and what are the differences to what is being patented?
- Which value-added product(s) resulting from syngas fermentation is/are the main focus of this research?
- Has syngas fermentation technology been commercialized? What is the TRL of the research vs. implementation?
- What should be the main focus of syngas fermentation scientific research going forward?

Having these questions in mind, the objective of this systematic review was to obtain a thorough overview of the research and patent publications from 1 January 2012 to 31 December 2022 in the field of syngas fermentation. This work aims to understand the real development level of the research on syngas fermentation and its technological advances, pointing not only to the immediate research needs and bottlenecks, but also targeting the interaction between academia and industry to identify opportunities for innovation and investment.

2. Materials and Methods

A systematic search methodology was followed, as represented in Figure 1, based on the method that was developed by Calvo et al., 2022 [6], using two databases: the Web of Science™ (WoS) (www.webofscience.com/, accessed on 5 March 2023) for the scientific publications and The Lens (LENS) (www.lens.org/, accessed on 5 March 2023) for the patent publications. From the datasets that were obtained, Bibliometrix and Microsoft® Excel® were used to categorize the research papers and patent publications over the 2012 to 2022 period [20]. The entries for the scientific publications were then sorted by topic, product focus, and the funding country of origin, while patents were sorted by jurisdiction, leading assignees with the highest number of patents published, and leading owners.

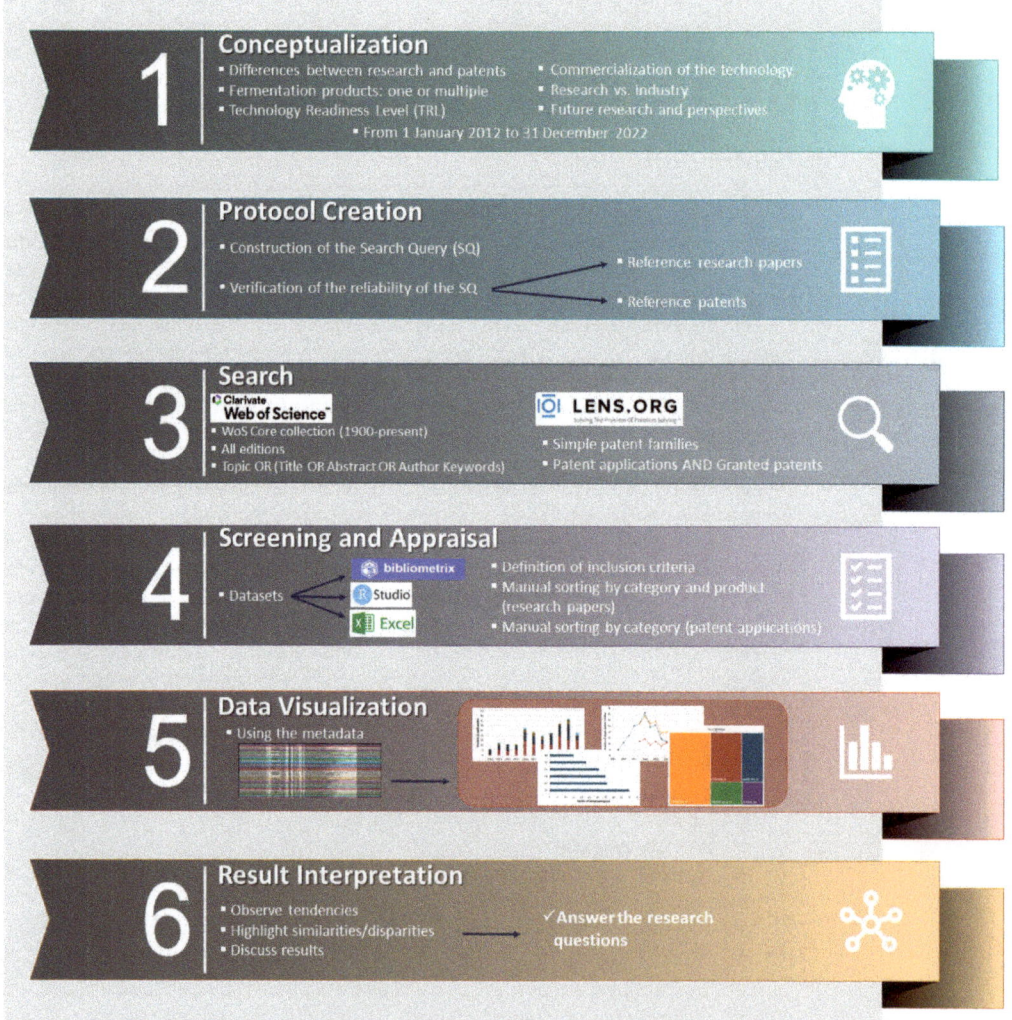

Figure 1. Flowchart of the systematic search methodology followed for the systematic review of syngas bioconversion to value-added products.

2.1. Databases and Search Query

To obtain an accurate overview of the scientific publications and patents that were written in English from 1 January 2012 to 31 December 2022, WoS and LENS were chosen as the search engines for the systematic review. The Web of Science Core Collection (1900-present) database, with all of its available editions, was used for the WoS search engine. The search was performed by "Topic" (title, abstract, author keywords, and Keywords Plus) OR "Title OR Abstract OR Author keywords". This search method was adopted because many reference publications on syngas fermentation were left out when only the search by "Topic" was performed, including life cycle assessment/techno-economic analysis (LCA/TEA) publications and some important reviews. The datasets that were obtained from the WoS were downloaded in BibTeX format and fed to bibliometrix, which is an R-Tool for comprehensive science mapping analyses [20] through biblioshiny [21] in RStudio ver. 2022.07.1+554 open-source software [22]. The data on the number of researchers (in full-time equivalent) per million inhabitants were obtained from UNESCO [23], and the total annual number of publications was obtained from the WoS Core Collection in all topics, by searching "PY = year" (from 2012 to 2022).

In the LENS database, the search was filtered by simple patent families and document type, focusing on patent applications and granted patents. Simple patent families are a collection of patent documents that are considered to cover a single invention. This allows for a simplified overview of a group of patents with the same priorities and the same technical content, facilitating data screening.

The use of a simple search query (SQ), namely "(syngas fermentation) OR (synthesis gas fermentation)", in LENS resulted in a reliable number of relevant results. However, the same search in the WoS resulted in data with many off-topic publications, while other publications of interest were absent. As such, an augmented SQ model, drawing inspiration from the one that was employed by Calvo et al., (2022) was implemented, and is provided for reference in Appendix A.

It was necessary to include not only the word syngas, but also all the possible variations that represent syngas, such as gas components, producer gas, and synthesis gas. Additionally, it was crucial to exclude not only products that were obtained from the catalysis of syngas, but also the FTP processing of syngas. Finally, "NOT ((. . .) "syngas from" OR "synthesis gas from")" was also removed from the query, as some publications that used raw syngas from various sources, like syngas from the industry, were excluded.

2.2. Data Screening and Visualization

To facilitate the data screening and outlier exclusion, namely the publications or patents that did not have syngas fermentation as their main focus, the datasets were converted to .csv files for easy manipulation and validation in Excel® (Microsoft® Excel® for Microsoft 365 MSO ver. 2209 Build 16. 0. 15629. 20200). The publications and patents were verified and categorized manually, through title, abstract, and claim verification. Several inclusion criteria were selected in order to normalize the data screening, as follows:

Criterion 1: A fermentation process using a gaseous carbon source that is composed of CO or CO_2+H_2, or a mixture of the three.

Criterion 2: The utilization of acetogenic microorganisms, mixed cultures of acetogenic microorganisms, or mixed cultures of acetogenic microorganisms with other types of microorganisms + *Criterion 1*.

Criterion 3: The production of acids, alcohols, or a mixture of both + Criterion 1.

Criterion 4: The genetic modification of microorganisms to optimize and/or increase syngas conversion and/or the metabolite production from syngas.

2.2.1. Research Publications

Biblioshiny enabled the distribution of publications by year, funding institution, and document type. Publications were automatically sorted by: book chapter—research article, book chapter—review, conference proceedings—research article, conference proceedings—

review, and research article and review, based on the information that was contained in the metadata [21]. The publications that were retrieved from the dataset were further divided into 10 different topics based on defined classification rules, as follows:

Bioreactor design: Publications that included "bioreactor", "continuous fermentation", "batch", "reactor", or "bubble column" in the title. Publications that referred to the performance of "fermentation in bioreactor", or that used a physical support for the biocatalyst, such as nanoparticles or biochar, in the definition of the "objective of the work" that was included in the abstract.

Raw syngas: Publications that included "producer gas", "furnace gas", "industrial", or "biomass-derived" in the title. Publications that referred to the utilization of syngas from industrial or thermochemical processes in the abstract.

Culture medium optimization: Publications that included "medium" or "supplement" in the title. Publications that referred to the supplementation of components into the culture medium in the definition of the "objective of the work" that was included in the abstract.

LCA/TEA: Publications that included "life cycle assessment", "techno-environmental assessment", "sustainability assessment", "techno-economical assessment", or "financial assessment" in the title. Publications that referred the performance of a "life cycle assessment", "sustainability analysis and/or assessment", "emission analysis and/or assessment", or a "techno-economical assessment" in the definition of the "objective of the work" that was included in the abstract.

Modeling: Publications that included "modeling/model", "simulation", or "algorithm" in the title. Publications that referred to the performance of a "predicting model" or "simulation" in the definition of the "objective of the work" that was included in the abstract.

Molecular biology: Publications that included "engineered", "mutation/mutagenesis", "expression", "gene replacement", "genome and/or genomic", or "metabolic engineering" in the title. Publications that referred to the performance of the genetic manipulation, manipulation of pathways or enzymes, the development of engineered acetogenic strains, or a description of the pathways and/or novel enzymes in the definition of the "objective of the work" that was included in the abstract.

New analytical techniques: Publications that included "technique" or "method" in the title. Publications that referred to the performance of an analysis using a new sensor or a new analytical technique in the definition of the "objective of the work" that was included in the abstract.

New microorganisms: Publications that included "novel" or "new" before the word "acetogen", or before the genus and species name of a microorganism. Publications that referred to the description of new syngas-converting microorganisms in the definition of the "objective of the work" that was included in the abstract.

Process optimization: Publications that included "improvement", "production", "formation", "effect", "tolerance", "integration", "optimization", or "enhanced" in the title. Publications that referred to the performance of a process optimization or integration in the definition of the "objective of the work" that was included in the abstract.

Review: Publications that were bibliographic or process reviews. Publications that compared technologies.

The publications in the dataset were also sorted by the main product of the syngas fermentation, in a total of 19 different categories: *Acetate and Acetic acid, Acetone, Butyrate and Butyric acid, Butanol, Isobutanol, 2,3-butanediol, Ethanol, Ethylene, Hexanoate and Hexanoic acid, Malate and Malic acid, Methane, Methanol, Mevalonate and Mevalonic acid, n-caproate and n-caproic acid, n-caprylate and n-caprylic acid, Poly-3-hidroxibutyrate, Propionate and Propionic acid, Isopropanol,* and *Succinate and Succinic acid.* The publications that focused on combinations of products, e.g., mixtures of acids, mixtures of alcohols, or mixtures of acids and alcohols, were categorized as "multiple". The TRL of the scientific publications of the 5 leading publishing authors, and of the patents from the 2 leading inventors who also authored scientific publications, was assessed by an abstract and materials and methods screening, following the TRL definition of the European Commission [24].

2.2.2. Patents

A procedure similar to the one that was performed for the scientific publications was also carried out for the patents. The dataset that was returned by the SQ was thoroughly assessed to select only the patents that were related to or usable for syngas fermentation. Patents can have two statuses: published or granted. Granted patents hand the assignees a monopoly over the patented invention for a set timeframe. Published patent applications do not confer these rights to the assignees of the invention, but simply state that the patented invention might be protected in the future [25]. For this study, the filtered dataset was further sorted by publication year, grant year, the jurisdiction of the publication or grant, and by the 5 leading assignees with published and/or granted patents, for comparison. To provide a topic overview of the submitted patents, a classification into five topics was performed, using a methodology similar to the one that was followed for the scientific publications. The classification rules were as follows:

Genetic engineering: Patents that described genetically modified microorganisms or enzymes that performed/assisted with syngas/CO/CO_2+H_2 fermentation.

New microorganisms: Patents that described a novel syngas/CO/CO_2+H_2 fermenting microorganism.

Parts: Patents that described a part or object that could be used in a syngas/CO/CO_2+H_2 fermentation process.

Process: Patents that described a production process using syngas/CO/CO_2+H_2 fermentation with acetogenic microorganisms.

Product: Patents that described a product composition with one or more precursors that were obtained through syngas/CO/CO_2+H_2 fermentation with acetogenic microorganisms.

Multiple: Patents that described combinations of the previous topics.

The sorted scientific publications, patents, and TRL classification datasets can be found in the Supplementary Materials (File S1).

3. Results and Discussion

3.1. Scientific Publications

The SQ for the WoS search engine retrieved a total of 312 results, of which, 254 fit the selection criteria (see Section 2.2). The number of publications per year and the document types are summarized in Figure 2.

From 2012 to 2022, research articles represented the majority of the publications (with an average of 19.4 publications per year), followed by reviews (an average of 2.9 publications per year), conference proceedings (an average of 0.5 publications per year), and both types of book chapters (research articles and reviews), which were less represented, with averages of 0.2 and 0.1 publications per year, respectively. It is important to note that the number of conference proceedings publications is most likely underestimated, since only the WoS-indexed proceedings were accounted for. In addition, many conference publications might not have been written in English, which would also exclude them from the search. In the period that is under analysis, a significant increase in the number of publications was only observed from 2016 onwards. In fact, from 2012 to 2015, the number of yearly publications on the subject were few and varied only slightly in number, with a maximum of 15 results matching the search criteria in 2014 (Figure 2). The relevant publications in this period, i.e., those with the highest number of total citations, addressed the production of ethanol, butanol, and hexanol from syngas, the use of mixed cultures in membrane biofilm reactors, and the upgrading of ethanol to n-caproate [26–29]. Important reviews on syngas fermentation were also published by Bengelsdorf et al., in 2013, with 129 citations, and by Latif et al., in 2014, with 125 citations, providing an overview on the state of the art and showing the most promising syngas fermentation applications at the time [8,30].

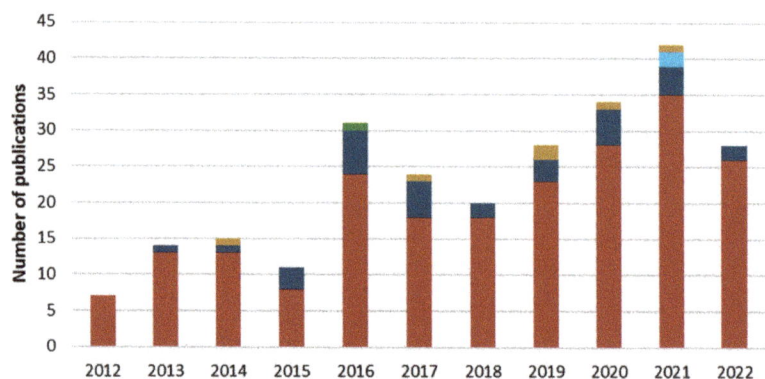

Figure 2. Number of scientific publications per year on the subject "syngas fermentation", sorted by document type, from 2012 to 2022. Color coding is as follows: ■ book chapter (research article); ■ book chapter (review); ■ conference proceedings (research article); ■ research article; ■ and review.

In 2016, the number of scientific publications doubled in value from 2014, reaching a new maximum of 31 publications, comprising research articles, reviews, and one research article that was published as book chapter. This suggests that, in 2016, the interest level for syngas fermentation technology increased, with the release of 24 original article publications, a value that was only achieved again in 2019. From 2016 onward, the number of yearly publications was considerably higher compared to the 2012 to 2015 period. Even in 2018, when the number of publications reached a lower point, with 20 publications, this was still 33% above the number that was reported for 2014 (Figure 2). From 2018 to 2021, the number of yearly publications kept increasing, with an average rate of 28% per year. In 2021, the year with the highest number of publications, a total of 42 publications were published, with 35 being original articles, 4 being reviews, 2 being reviews in book chapters, and 1 being a conference proceeding. The number of publications experienced a decrease of 33% in 2022 compared to the previous year, which had 28 publications. The publications in 2022 equaled the total of 2019 and corresponded to 26 original research articles and 2 reviews [31,32].

The evaluation of the number of publications that is presented in Figure 2 could be biased by two variables: the increasing number of researchers worldwide and the increasing number of total publications in the WoS Core Collection. Figure 3a,b shows the variation in the total number of publications in the WoS Core Collection, the number of researchers per million of the world's inhabitants, the number of yearly scientific publications that are related to "syngas fermentation", and their respective value per capita (considering researchers at a full-time equivalent) and per total publications in the WoS Core Collection during the analyzed period.

Within the analyzed period, the yearly increase in researchers per million inhabitants was rather constant (Figure 3a), occurring at an approximate rate of 2.8% per year. Similarly, from 2012 to 2021, the total number of publications in the WoS Core Collection increased approximately 4.8% per year, with 2015 having an increase of 15%. In 2022, for the first time within the analyzed period, there was a decrease in the number of total publications, to a value that was 14% lower than that of 2021. This deceleration in scientific production might be due to the COVID-19 pandemic and the lockdown measures that were applied during 2020 and 2021, with the temporary closure of research institutes all over the world and/or the conscription of laboratory equipment and research teams to respond to the pandemic emergency, with its effects only being felt in the subsequent years. From Figure 3b, it can be observed that the steady increase in the number of researchers and the variation in the number of publications over the analyzed period exerted little effect on the observed trend for the syngas fermentation publications. However, these variables might help to explain the periods of great variation in the data, as in the cases of 2015 and 2022. A more in-depth analysis of the yearly data seems to indicate that 2015 presented a lower number of publications regarding syngas fermentation, as decreases of 27% and 36% were registered in the number of syngas fermentation publications and the syngas fermentation publications per total publications, respectively, even though the total number of the publications in the WoS database increased in that year. As for 2022, while the total publications in the WoS Core Collection decreased by 14%, both the number of syngas fermentation publications and the number of syngas fermentation publications per total publications showed more substantial decreases of 33% and 23%, respectively. Compared to other research areas, these numbers reveal that syngas fermentation is still a niche field of research. The occurrence of interest peaks appears to be a common occurrence and might be related to intermittent funding incentives. However, the fact that there was a constantly growing trend in the syngas fermentation scientific publications over the analyzed time period ensures that this is a topic of interest and a focus for new research.

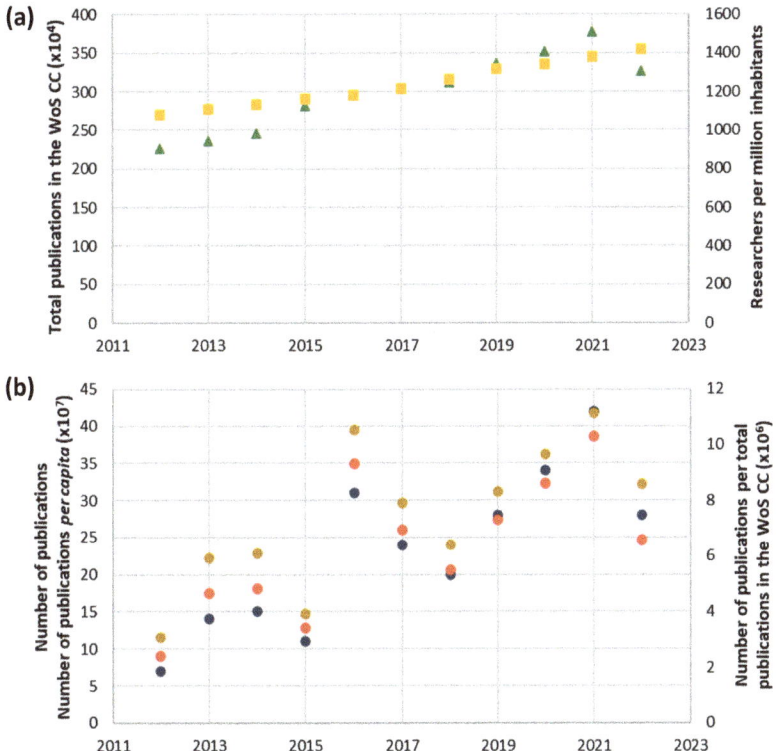

Figure 3. (a) Variation of the total number of publications in the WoS Core Collection and the number of researchers per million inhabitants in the world from 2012 to 2022, and (b) number of scientific publications per year related to "syngas fermentation" and its value per capita and per total publications in the WoS Core Collection. Color coding is as follows: (a) ■ number of researchers per million inhabitants; and ▲ total publications in the WoS Core Collection; and (b) ● number of scientific publications in "Syngas fermentation" (main yy axis); ● number of scientific publications in "Syngas fermentation" per capita (main yy axis); ● and number of scientific publications in "Syngas fermentation" per total publications in the WoS Core Collection (secondary yy axis).

In order to refine the information on the specific topics within the syngas fermentation research field along the studied period, the publications were categorized into different topics, according to the criteria that were previously selected (Section 2.2.1.). The relative percentages for each topic, considering the total publications per year to be 100%, are shown in Figure 4.

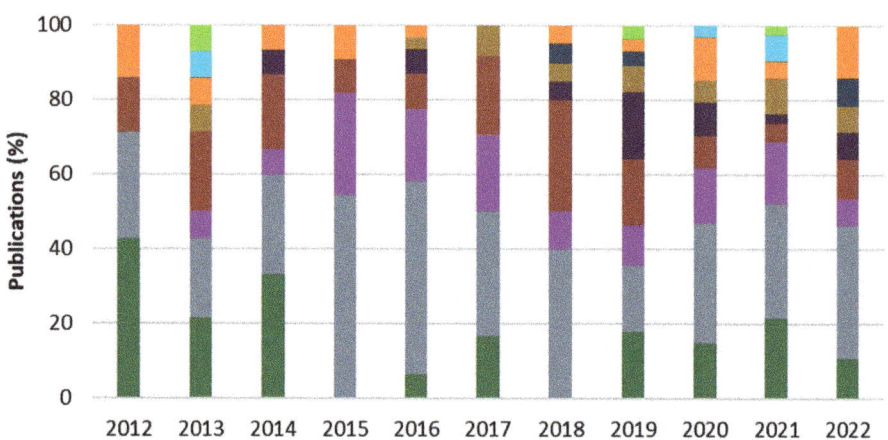

Figure 4. Distribution per year of the scientific publications about "syngas fermentation" sorted by topic, from 2012 to 2022. Color coding is as follows: ■ bioreactor design; ■ culture medium optimization; ■ LCA/TEA; ■modeling; ■ molecular biology; ■ new analytical techniques; ■ new microorganisms; ■ process optimization; ■ raw syngas; and ■ review.

Over the 2012–2022 period, not only was there an increase in the number of total publications, but there was also a diversification of the research topics. Process optimization was the most published topic, accounting for 34% of all the publications over the studied period, followed by molecular biology at 16%. The relevant publications on process optimization have focused mainly on the production of longer-chain products, such as n-caprylate, n-caproate, and hexanol, which are higher-value commodity chemicals when compared to shorter-chain products [28,33,34]. From 2012 to 2014, molecular biology played an important role in syngas fermentation research. An example of this is the scientific publications of Berzin et al., 2012 and 2013, and Köpke et al., 2014, on the development of recombinant strains for the selective production of non-traditional alcohols, such as acetone, butanol, and 2,3-butanediol [35–37]. During that 3-year period, researchers also focused on process optimization, as well as new bioreactor configurations that would allow for higher productivities and easy scalability [27,38,39]. An increase in scientific reviews was observed from 2015 onwards, which summarized the research that was performed, with a focus on the evaluation of the future technological and economic viability of syngas fermentation technology, and how it could be integrated into an industrial setting using biomass-derived syngas or industrial off-gases [40,41]. In 2016, the second most cited paper in the 2012–2022 period was published. This was a review of gas fermentation as a flexible platform for the commercial scale production of low carbon fuels and chemicals from waste and renewable feedstocks, by Liew et al., with 208 citations [15]. From 2016 onward, the publications that were related to process optimization presented the highest increase in the studied period, with 2.6 times more publications than in 2015. The publications on process optimization, which showed a slight growth tendency since 2012, increased visibly in 2015 and 2016, remaining the topic with the highest number of publications until the end of the studied period. In 2016, Marcellin et al., published the most cited research publication on process modeling, with 85 citations, where a systematic platform that allowed for the development of energy efficient pathways for the production of chemicals and advanced fuels via C1 fermentation, using complete omics technologies that were augmented with genetic tools and a genome-scale mathematical model, was presented [42].

From 2017 to 2019, the research focus shifted yet again to experiments on bioreactors, envisioning an easy scalability with the addition of novel support materials for the biocatalysts, such as biochar and nanoparticles, and new reactor configurations that would tackle one of the biggest limitations of these syngas fermentation processes, namely the gas–liquid mass transfer [43–47]. In 2018, a decrease in the overall publications was observed. However, the most cited publication in the studied period was from this year, a scientific paper by Haas et al., in Nature Catalysis (280 total citations) that focused on a solar-powered CO_2 and H_2O reduction to syngas, followed by its fermentation to alcohols [48]. Articles describing newly isolated syngas-fermenting microorganisms and strains appeared for the first time in 2018, reappearing in 2019 and 2022. This suggests that the search for more robust isolates capable of performing syngas fermentation and producing interesting new products was challenging, with only four publications in 11 years [49–52]. The desire to advance syngas fermentation to an implementable technology was substantiated by the highest number of publications on the subject being in 2020 and 2021. LCA/TEA analyses and studies using syngas impurities or raw syngas, i.e., the syngas from thermochemical processes and industrial off-gases, have tried to demonstrate the potential of this technology for the future production of biofuels and platform chemicals [11,12,53–61]. Research article publications with a focus on process optimization, including culture media optimization, process or genetic modeling, systems integration, and molecular biology, all saw a visible increase in these two years, with the publication of important studies, such as Heffernan et al., 2020 (process optimization), Nangle et al., 2020 (molecular biology), and Roy et al., 2021 (raw syngas) [61–63].

From 2012 to 2022, the research focus was also centered on the multiple value-added products that can be produced by syngas-fermenting microorganisms. Such products, mainly carboxylic acids and alcohols, are pivotal as platform chemicals for the reduction in fossil fuel use and to the energy transition effort. Figure 5 is a representation of the top four most referred to products that are obtained from syngas fermentation, which were sorted from the publications on the subject during the studied period, representing more than 90% of the total publications.

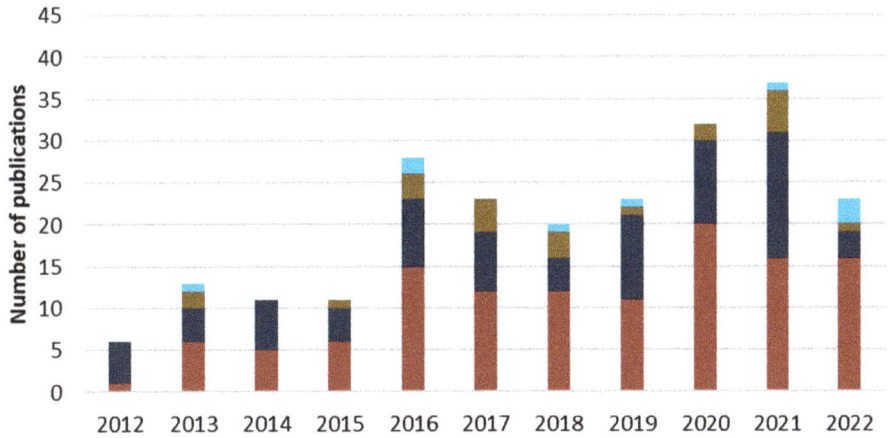

Figure 5. Scientific publications per year related to "syngas fermentation", sorted by top 5 value-added products produced from syngas, from 2012 to 2022. Color coding is as follows: ■ acetic acid; ■ butanol; ■ ethanol; and ■ multiple.

From 2012 to 2022, the research focused mainly on the production of multiple products from syngas fermentation. Combinations of different alcohols, such as ethanol, acetone, hexanol, propanol, and butanol or 2,3-butanediol, and carboxylic acids, such as acetic, butyric, propionic, n-caproic, n-caprylic, formic, malic, and mevalonic acids, and other value-added products, such as fatty acid methyl esters (FAMEs) or polyhydroxyalkanoates (PHAs), were the main focus of this research [64,65]. This reflects the effort to maximize the value that syngas fermentation could represent in the replacement of fossil-based chemicals and to increase the plasticity of a future syngas biorefinery [27,29,34,49]. The production of long-chain value-added products is a means of increasing carbon capture efficiency, especially in cases where it is coupled with other CO_2 capture technologies and a catalytic transformation to syngas [66]. In 2022 there seems to have been a shift of the research focus to the production of butanol [67–69], whereas the publications focusing on ethanol production were lowered considerably. This shift could be justified by the fact that butanol is considered to be an advanced biofuel that offers numerous advantages over ethanol. Specifically, it has a similar calorific value, octane value, and air–fuel ratio to gasoline, is less corrosive than ethanol, and can be transported through pipelines. Butanol is also immiscible with water (H_2O) and can be used pure or blended with gasoline, without requiring engine modifications [70]. Moreover, butanol can be upgraded to 2,3-butanediol, which is also utilized as a sustainable aviation fuel (SAF) with remarkable results [71].

During the analyzed period, the authors with the highest number of publications on syngas fermentation were Kennes C. and Veiga M.C. from the University of A Coruña, with 16 publications [68], followed by Atiyeh H.K. from Ohio State University, with 14 publications [72], and Angenent L.T. from the University of Tübingen and Weuster-Botz D. from the Technical University of Munich, with 10 research articles each [73,74]. An analysis of the research conducted by these authors revealed that the leading research on syngas fermentation was mainly focused on TRL 2 and 3, with only one publication by Pardo-Planas O. and Atiyeh H.K., among others, modeling a TRL 7 installation using the data from TRL 3 assays [75].

In order to comply with ever-restricting emission regulations and mitigate climate change, governments around the world have been actively funding research into alternative biotechnological processes for the production of sustainable biofuels, or for reducing or cutting emissions, creating funding programs and packages specifically for this purpose. The funding information that was present in the dataset was evaluated in terms of its country or region of origin. Funded publications represented 91% of the total published works, 228 out of 250, and among these, 31 publications were supported by private funding, such as business angels, family and friends, private companies, or foundations. The analyzed publications were funded by the national funds of 25 different countries and the European Union (EU). A distinction was established depending on the origin of the funding. Communitarian funds were classified as EU, whereas national funds were categorized as the individual country, regardless of whether they originated from an EU member state. The results for the six countries or regions with the highest number of funded contributions (above 10 publications) are presented in Figure 6.

The United States of America (US) holds the highest number of funded publications, with a total of 50 in the analyzed dataset. Germany was also a major contributor to this syngas fermentation research, with 36 publications, of which 23 were funded by national funds from the German Federal Ministry of Education and Research, 8 with national funds in collaboration with private companies and foundations, and only 5 were funded by joint national and EU funds. Germany had its highest numbers of outputs in 2016 and 2022, with seven funded publications. China was a relatively recent contributor to the syngas fermentation research field, with a total of 35 funded publications during the studied period, reaching its highest number of publications in 2020.

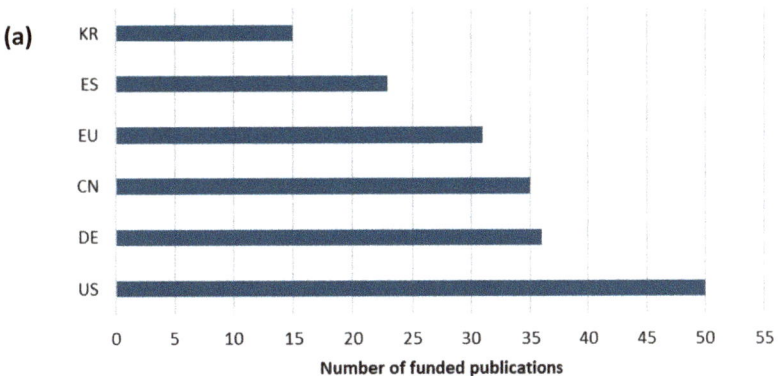

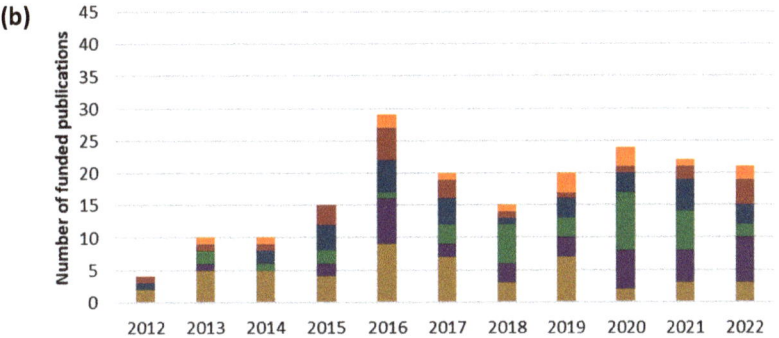

Figure 6. (a) Number of funded scientific publications per funding institution's country or region from 2012 to 2022, and (b) number of scientific publications funded per year per funding institution's country or region. Color coding is as follows: ■ Republic of Korea (KR); ■ Spain (ES); ■ European Union (EU); ■ China (CN); ■Germany (DE); and ■ United States of America (US).

A total of 31 publications acknowledged the funding of EU communitarian funds. The highest number of publications that were funded by the EU was achieved in 2016 and 2021, coinciding with the years when European projects like SYNPOL (Biopolymers from syngas fermentation—2012 to 2016, [76]), SYNTOBU (Biological production of butanol from syngas—2013 to 2017, [77]), AMBITION (Advanced biofuel production with energy system integration—2016 to 2019, [78]), and BIOCONCO$_2$ (BIOtechnological processes based on microbial platforms for the CONversion of CO$_2$ from the iron and steel industry into commodities for chemicals and plastics—2018 to 2022, [79]) passed their midterm or were near their ending date. Spain appeared as the fifth-largest investor in syngas fermentation, with 23 funded publications, 4 of which were funded solely by the Spanish Ministry of Economy and Innovation or by regional governments. Spain and the US were the only countries that had publications every year in the analyzed period.

The Republic of Korea has funded 15 publications on syngas fermentation, mainly through the C1 gas refinery program [80]. This program was funded by the National Research Foundation of Korea and the Ministry of Education, Science and Technology in 2015, focusing on the development of core C1 gas refinery technologies with the economic feasibility to decrease the fossil fuel dependency of Korean industries [81]. Accordingly, a substantial increase in the number of scientific publications that were funded by the Republic of Korea was observed since 2016, with the highest number of these funded publications being achieved in 2019 and 2020.

In terms of external collaborations, Germany had the highest number of collaborations with countries and companies that were outside of Europe, with eight co-publications with partners from various countries such as Australia, Canada, Brazil, and the US. The countries with the fewest collaborations were China and the Republic of Korea, both with only one publication that was in collaboration with another country or private company. The publications that were funded by EU communitarian funds tended to rely on collaborations, mainly with countries from within the EU or from the European Economic Area, such as Norway, for example.

3.2. Patents

The SQ that was used in LENS returned a dataset with 345 simple patent families, from which 249 were selected as being related to or appliable within syngas fermentation. Of these, 103 were granted and 88 are currently active. Figure 7 shows the distribution of the simple patent families by filing, publishing, and granting year (Figure 7a) and their distribution by topic (Figure 7b).

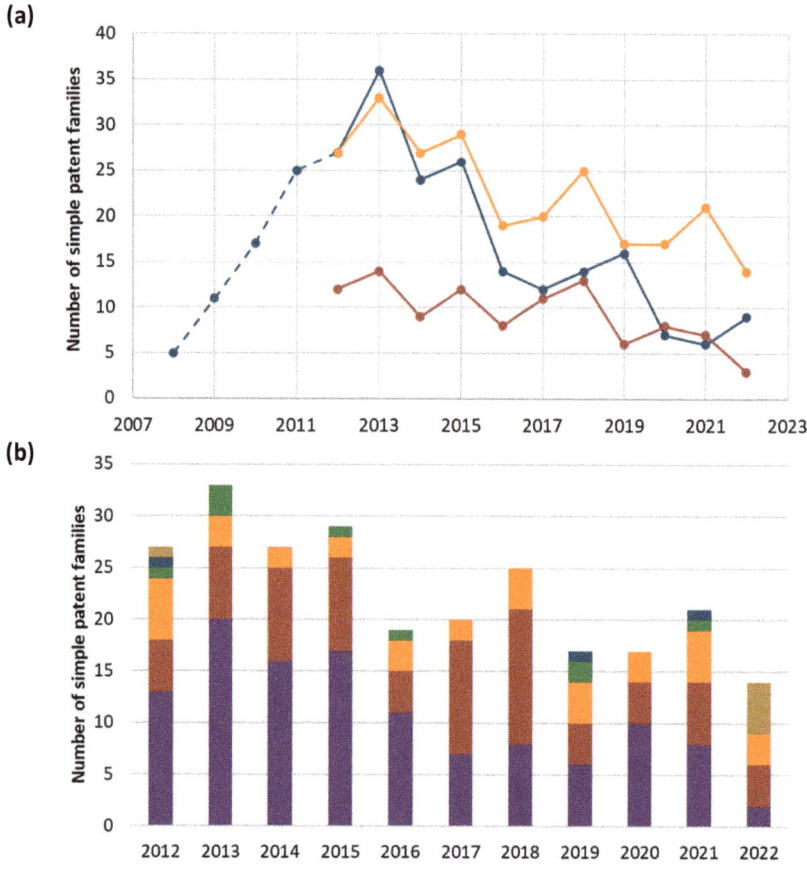

Figure 7. (a) Number of simple patent families filed, published and granted from 2012 to 2022, and (b) number of simple patent families from 2012 to 2022, organized by topic. Color coding is as follows: (a) – ● – filed before 2012; ——— filed after 2012; ——— granted; and ——— published; and (b) ■ genetic engineering; ■ multiple; ■ new microorganisms; ■ parts; ■ process; and ■ products.

From the data presented in Figure 7a, it can be seen that there was a growing trend in the patent filing related to syngas fermentation until 2013, with a maximum of 36 filed patents. However, a decrease in this patent filing occurred from 2013 onwards, with occasional peaks in 2015 and 2019. It is important to note that this decline in patent filing over the analyzed period may have been a result of underestimation, due to the delay between the filing and publication dates. While the average time between the patent filing and the publication varies between 18 months to 3 years, this actual time frame can vary widely depending on several factors, including the jurisdiction in which the patent was filed, the complexity of the invention, and the level of scrutiny that is required to assess the patent application. Filing information is only publicly available after publication, and the publishing process can take several years, as can be seen from the patents that were published during the analyzed period and those that had filing dates prior to 2012, going back to 2008. Nonetheless, it is observable that 2013 was indeed a year of technological progress in syngas fermentation, with 33 published patents. The year with the highest number of granted patents was 2018, with an increase of 63% compared to 2016.

The topic that was related to the syngas fermentation process had the highest representation, with 118 patents being published. The years of 2013 and 2015 saw the highest number of patents describing different syngas fermentation processes, with 20 and 17 patents, respectively. The patents describing multiple topics, namely combinations of these syngas fermentation processes with descriptions of parts, such as novel bioreactors or propeller shapes, or with new and/or engineered microorganisms, were the second most frequent type of published patent among the 249 patents that were analyzed, with 76 published patents and a peak of 13 patents in 2018. Genetic engineering was also an important patent topic, with 37 patent applications, while parts, product descriptions, and new syngas-fermenting microorganisms appeared with 9, 6, and 3 patent applications, respectively. Process-focused patents had their publishing peak between 2012 and 2016. Between 2017 and 2018, the inventors' focus seemed to have shifted to multiple-topic patents. The highest number of patents focusing on genetic engineering for the production of novel products with acetogenic microorganisms occurred in 2012, but this topic has been a constant focus during the period that was studied, with at least two patents published every year. Parts, products, and new syngas-fermenting microorganisms were sporadic topics, with patents about product composition appearing more prominently in 2022.

In order to access the global distribution of these syngas fermentation patents, the dataset was organized by jurisdiction, i.e., the country, territory, or organization of the patent publishing or granting. Figure 8 shows the results that were obtained for each jurisdiction, organized by patent topic.

From Figure 8, it is possible to observe that the US published the highest number of patents on syngas fermentation, 216, followed by the World Intellectual Property Organization (WO) with 189 patents. China (CN) and Europe (EP) were in the third and fourth positions, with similar numbers of published patents, 140 and 139, respectively. Canada, South Korea, Japan, Brazil, and Australia published 105, 86, 83, 80, and 76 patents, respectively. The remaining countries had less than 50 published patents between 2012 and 2022, with some having as few as 1 published patent. The US, CN, and EP jurisdictions are often preferred by assignees due to their large market potential, strong legal systems with strong IP enforcement, efficient patent offices, and easy access to innovation, while WO allows for a time saving, cost-effective option for achieving protection in multiple countries, justifying the increased patent publication values under these jurisdictions [82].

Thematically, the process-related patents were the majority within most jurisdictions, followed by multiple thematic patents and genetic engineering. The multiple thematic patents described either the apparatus and methods for the cultivation of microorganisms with syngas (under the topics "parts + process") or novel and/or engineered microorganisms and cultivation methods with syngas (under the topics "new microorganisms and/or genetic engineering + process"). The patents that were related to parts and new microorganisms were most common in the 12 leading publishing jurisdictions, while the product-related patents were exclusive to the US and WO jurisdictions.

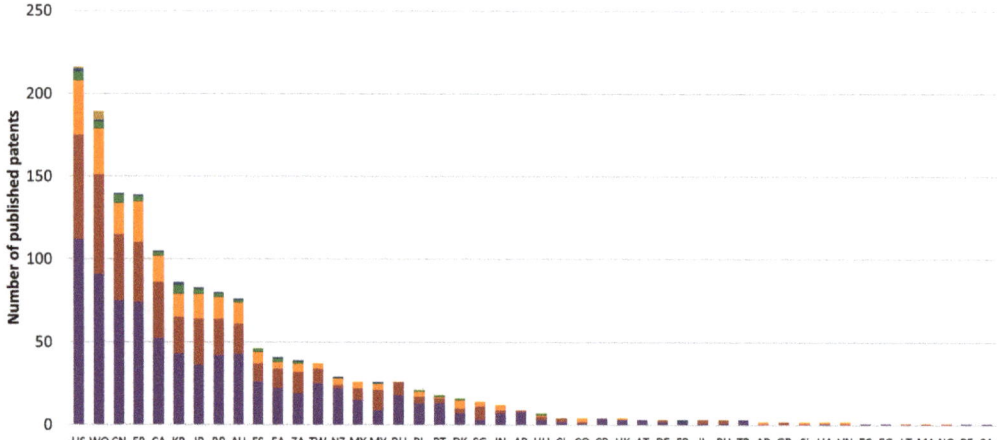

Figure 8. Number of patents related to syngas fermentation published between 2012 and 2022 by jurisdiction and sorted by topic. Color coding is as follows: ■ genetic engineering; ■ multiple; ■ new microorganisms; ■ parts; ■ process; and ■ products. Country or organization abbreviation is as follows: WO—World Intellectual Property Organization; EP—European Union Intellectual Property Office; EA—Eurasian Patent Organization; and AP—African Regional Intellectual Property Organization; the remaining country coding is set by the Alpha-2 (ISO 3166) code.

Patents can be submitted by either individual assignees, universities, or companies as a means of protecting IP. Patents are especially important for companies, since they are proof of a company's innovative strength, facilitating their product marketing and investor/capital attraction. From the dataset, the five leading assignees with the highest number of patents published were selected, namely Lanzatech, Coskata, Genomatica, Ineos Bio, and Evonik, representing 56% of the total published patents. This top five was composed in its entirety by companies that performed syngas fermentation to bioethanol or to other products of interest, such as 1,2-butanediol and 2-hydroxyisobutyric acid, showing the influence of bioenergy in the syngas fermentation field. Figure 9a,b depicts the number of patents published, the leading patent owners, and the yearly distribution by the topic of the patents that were published by the top companies.

During the 2012–2022 period, five companies appeared as the leading investors/patent assignees in syngas fermentation technologies: LanzaTech, Coskata, INEOS Bio, Evonik, and Genomatica (Figure 9a).

LanzaTech was the company with the highest number of published patents and is currently the only fully successful company to commercialize its gas fermentation technologies. The company was founded in 2005 in New Zealand and moved their headquarters to the US soon after. Since then, it has been investing in gas fermentation, first to ethanol, and now to a multitude of platform chemicals that are used to make products such as PolyEthylene Terephthalate (PET) bottles [81]. The company has commercial plants that are fully operating from steel mill gases in China, India, and Belgium, and many more that are scheduled for construction all over the world in the next few years, using not only steel mill gases as feedstock, but also gasification–fermentation integrated technologies for the use of different types of biomasses, such as MSW and agricultural residues [16]. Over the last 11 years, the company has been applying for patents on a regular basis, with five patents being published in 2022. LanzaTech's syngas fermentation patents were mainly related to genetic engineering and process methodology.

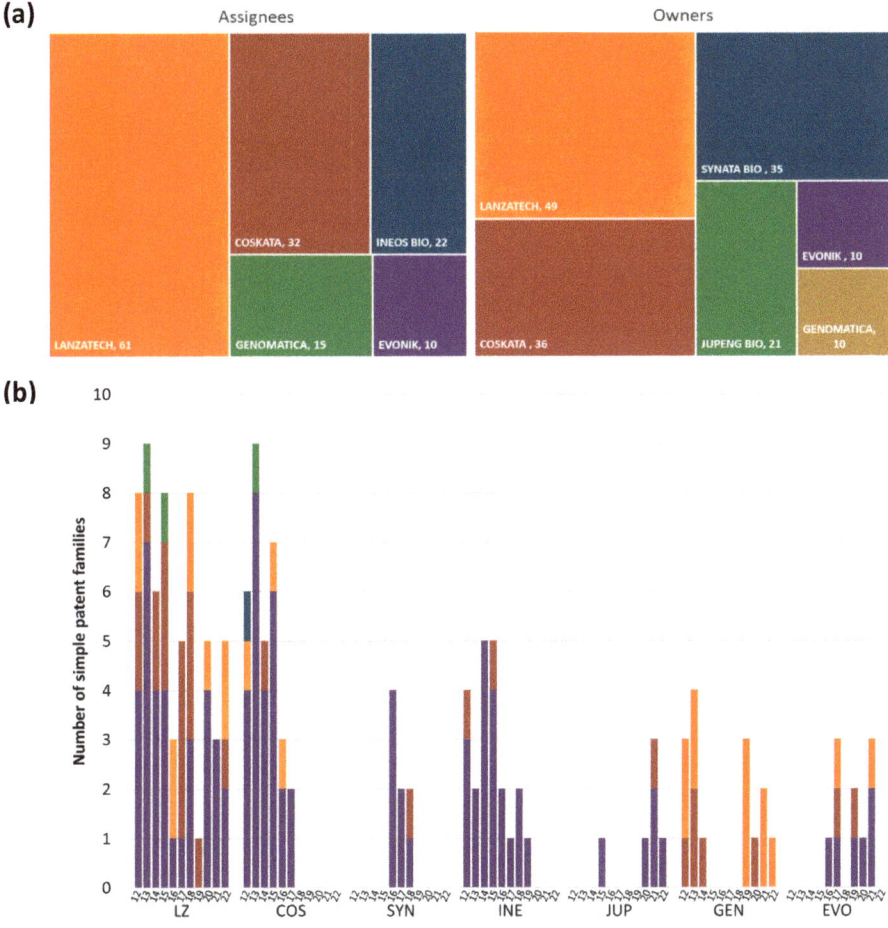

Figure 9. (a) Number of simple patent families published by the 5 leading assignees (left) and 6 leading patent owners (right), and (b) number of patents published by the 7 leading assignees from 2012 to 2022, sorted by topic. Color coding is as follows: (b) ■ genetic engineering; ■ multiple; ■ new microorganisms; ■ parts; and ■ process. LZ—LanzaTech; COS—Coskata; SYN—Synata Bio; INE—INEOS Bio; JUP—Jupeng Bio; GEN—Genomatica; and EVO—Evonik.

Coskata appeared as the second most important patent publisher and holder. However, as shown in Figure 9b, these patents were published only from 2012 to 2016. Coskata declared bankruptcy in 2015, and its technology was bought that year by Synata Bio, a company that is still active nowadays, having published patents that were related to syngas fermentation until 2018 [83]. It focused mainly on single-step, feedstock-flexible fermentation processes that converted syngas through fermentation into ethanol. In 2016, Synata Bio bought the Hugoton (Kansas, US) cellulosic ethanol plant from the bankrupt company Abengoa, adapting it for the fermentation of syngas to ethanol; however, it was not able to keep the plant operational, selling it in 2021 [84]. Nowadays, Synata Bio focuses on cellulosic bioethanol production, having reconverted their syngas-fermenting pilot plants to that end.

INEOS Bio, the third-biggest patent holder within the analyzed period, was a branch of the multinational chemicals manufacturer INEOS, which was focused on the production of advanced biofuels. The company published several patents at a constant rate until 2019 and was the proprietor of a semi-commercial plant in Vero Beach (Florida, US), which produced bioethanol at a commercial level from MSW syngas. However, in 2016, the plants that were owned by INEOS Bio were put on sale and the company was bought by Jupeng Bio. Jupeng Bio, another bioenergy company, is the proprietor of an integrated gasification and fermentation process that was demonstrated at the pilot level in 2017 [85]. The company published patents that were related to syngas fermentation in 2020 and 2021, namely the design of parts and process description.

Much like INEOS Bio, Evonik is a company that is focused on specialty chemicals and is based in Germany. This company uses syngas to produce 2-hydroxyisobutyric acid and other value-added platform chemicals. In 2020, in partnership with Siemens, it commissioned a pilot plant to convert H_2O and CO_2 into CO and H_2, which could be converted into value-added products by the Evonik fermenters in Marl, Germany. Evonik's patents focused on process description and the development of genetically engineered microorganisms for the production of various molecules, such as acetone, N-acetyl homoserine, or diverse alcohols.

Finally, Genomatica is a US company that is focused on replacing fossil-based chemicals with more sustainable building blocks, particularly on the production of 2,3-butanediol from sugarcane. Its portfolio of syngas fermentation patents is mainly composed of the genetic engineering of syngas-fermenting microorganisms for the production of isobutanol, aniline, and other valuable platform chemicals, although it does not actively produce them from syngas fermentation.

During the analyzed period, LanzaTech appeared as the dominant company in terms of patent publishing and ownership, accounting for 25% and 20% of the total published patents, respectively. From Figure 9b, it is possible to observe that LanzaTech was the only company that was consistently publishing patents during the analyzed period. The company mainly focused on the syngas fermentation process, including microbial cultivation methods, the production of multiple products of interest, and patents that were related to the genetic engineering of syngas-fermenting microorganisms and enzyme optimization. From 2012 to 2016, Coskata emerged as LanzaTech's most direct competitor, owning and publishing patents that were related to process description, genetic engineering, and multiple topics. Furthermore, it was granted a patent for a novel syngas-fermenting clostridium strain, *Clostridium coskatii*, in 2012, which is still in force at the time of this study. Synata Bio published some patents with Coskata during 2016 and 2017, but their interest in syngas fermentation seemed to wane after 2018. INEOS Bio was the third-biggest contributor to the syngas fermentation patents, with 22 patents being published from 2012 to 2019. Ineos Bio patents mainly focused on process description and process and apparatus description for a syngas fermentation to ethylene. Jupeng Bio appeared as one of the largest patent owners in this study, albeit with inconsistent patent publishing, with one patent published in 2015 and the remaining patents being published in the last three years of this study, indicating the relative novelty in the syngas fermentation market. Genomatica also published inconsistently during the analyzed period, having a hiatus between 2015 and 2018. The patents that were published by Genomatica mainly focused on the genetic engineering of syngas-fermenting strains for the production of multiple platform chemicals, their maintenance, and their product separation processes. As for Evonik, they have been publishing patents that are centered on process description, genetic engineering, and multiple topics since 2016.

3.3. Overview

In analyzing the results that were obtained in this systematic review, it was possible to observe an inverse trend between the scientific and patent publications from 2012 to 2022. The years 2013, 2016, and 2021 were especially relevant, as these were years of increased scientific and patent publications, indicating a possible technological breakthrough. Patent filing related to syngas fermentation saw a fast-paced increase from 2005 to 2013 (as can be verified by a quick search on LENS), much as was observed for syngas fermentation scientific publications [6]. The years of 2012 and 2013 marked the installation and first runs of many of the pilot plants that have operated over the last 11 years, and companies started to develop optimized processes and strains [7]. However, only in 2016 did the syngas fermentation research field see a turn from articles with a focus on fundamental science to articles with a more industrial and commercial point of view. LCA and TEA evaluations of this technology also started to appear more frequently from 2016 onward, most probably fed with data from the pilot and demonstration facilities that were active at the time [7,16,18,31]. As for 2020 and 2021, it was possible to observe the resurgence of scientific publications that were related to raw syngas fermentation, and the study of the effects of syngas impurities, such as HCN and H_2S, which tend to hinder fermentation, lowering the process productivity [10–12,60,61,86,87]. This is a recent topic in syngas fermentation research, with only seven publications appearing during the studied period. However, research using raw syngas is of utmost importance, since some of the active syngas fermentation plants from the analyzed period were forcibly closed due to tar and syngas impurity accumulations, leading to "catastrophic" decreases in their production [7,88].

Another important observation from the obtained data is that the research tended to focus on the production of multiple products with long carbon chains from syngas fermentation. Research on the production of longer-chain value-added products is important, not only to maximize carbon fixation, but also in the transition from a long to a short carbon cycle economy. With a large product portfolio, the greater the chances are of increasing the versatility of the biorefinery, which could also be coupled with genetic engineering tools to optimize the productivities and customize the microorganism towards the customer's interest. LanzaTech is an example of a company with a large portfolio of these high-income products, that already have guaranteed partnerships with brands such as Danone, L'Oréal, or Zara, beyond the traditional bioethanol and SAF market [89].

In terms of the most contributing countries for syngas fermentation research, the US had the highest number of funded publications (Figure 6a) and the highest number of published patents (Figure 8). Except for LanzaTech (which was founded in New Zealand, but soon after moved its headquarters to the US) and Evonik, all the other five leading companies were founded in the US at the turn of the millennium, many by acquiring technology from the pioneering researchers in the gas fermentation field from several US universities at the time. However, from 2015 for the EU, and from 2017 for China, an increase in the investments into syngas fermentation research was observed, achieving higher publication levels than those from the US later in the time span (2020 and 2021). The enforcement of the EU Green Deal, the climate law, the push for stricter climate targets for 2030, the war between Russia and Ukraine, and the Chinese pollution crisis might be some of the reasons for such a sudden interest increase in these carbon mitigation technologies, which not only severely cut emissions, but also add value with the co-production of biofuels and bio-based chemicals that can help to substitute fossil-based ones [90]. Funding programs such as the C1 gas refinery program of KR and EU projects such as SYNPOL, AMBITION, BIOCONCO$_2$, and PYROCO$_2$ are examples of the investment from the world's governmental entities into syngas fermentation research [76,78–80,91]. The decrease in the number of Chinese publications in 2022 might be related to the strict COVID-19 policies that have been enforced, with China being one of the most affected countries in the world by the pandemic.

From the results that were obtained, it appears that syngas fermentation, as a technology, is developing at two different speeds. From 2012 to 2022, the research mainly focused on low TRLs, namely TRL 2 and 3. However, by crossing the data on the research authors and patent inventors, it was observed that the leading inventors co-authored a relatively high number of scientific publications. Köpke M. and Simpson S.D. (LanzaTech) were the inventors of 23 patents that are owned by LanzaTech and the co-authors of 9 research publications. Most publications that were authored by Köpke and Simpson are considered to have a TRL of 2 or 3, but both inventors co-authored an LCA research paper with data from LanzaTech's first pilot plant (TRL 7). As stated before, scientific publications with the fermentation of raw syngas or a focus on the effect of the impurities that are present in raw syngas have only recently been released. Researchers have tended to focus mainly on productivity increases, the effects of the different gaseous components on the product distribution, and how to tackle the inherent mass transfer challenges, but not on how to fully integrate gasification and fermentation in a more efficient, economically viable way [60,87]. On the other hand, LanzaTech is actively commercializing its technology, giving actual proof that it works with variable feedstocks and that it can be an added value for any carbon-producing industry, succeeding where several other companies of the same branch have failed.

This disparity between research and industry might be due to the resources and investments that are required for the construction and maintenance of a gasification and fermentation pilot plant. A large consortium with industrial partners and constant investment would be necessary to take the syngas fermentation technology from the bench scale research to higher TRLs, and this might be hard to achieve in the current socio-economical conjuncture. There are operational difficulties related to the syngas flow regulation between the gas source and fermentation vessel, and scale adjustments are required in order to fully integrate syngas fermentation into higher-TRL facilities, which might be hard to achieve without prior testing and optimization [3,7,16,92]. Furthermore, the most advanced research into syngas fermentation, belonging to the US, was bought by private companies, who tend to protect innovation through industrial secrets and not fully disclose their research. This has resulted in an observable decrease in US-funded publications in recent years, but not necessarily in patent publications, opening the way for competitors such as the EU and China to occupy the lead in novel syngas fermentation research and contribute to this "two-speed" phenomenon.

4. Conclusions

The relatively low number of patents and publications that have tackled syngas fermentation in the period of 2012 to 2022 reveals that this is still a niche research topic. In this review, the following research questions were answered:

- What is being researched in syngas fermentation and what are the differences to what is being patented?
- Initially, the research focus was centered on the genetic manipulation of microorganisms and most published research was funded/developed by the United States of America. Over the years, the research focus changed to process optimization, and European countries became the biggest contributors in terms of the number of publications. In terms of patent publications, five companies stood out, namely LanzaTech, Coskata, INEOS Bio, Genomatika, and Evonik. These companies published 56% of the total analyzed dataset of simple patent families, focusing on process description, the genetic engineering of syngas-fermenting microorganisms, and multiple-topic patents, i.e., patents that described combinations of these topics.
- Which value-added product(s) resulting from syngas fermentation is/are the main focus of the research?
- Ethanol was the most studied solo product, followed by acetic acid and butanol. The production of multiple fermentation products was the main focus of the research over the studied period. The production of multiple products is a way of increasing the applicability and plasticity of a future syngas biorefinery. However, the companies tended to focus mainly on the production of bioethanol from syngas, while the production of other alcohols or value-added products were a minority in the company's product portfolios, due to the inherent complexity that is associated with culturing engineered microorganisms.
- Has this syngas fermentation technology been commercialized? What is the TRL of the research vs. the implementation?
- LanzaTech was the only company that was able to achieve the commercialization of its syngas fermentation technology. It owns several demonstration and pilot plants in

- the US and China, and, in collaboration with the steel mill industry, has been able to sell several of these industrial facilities based on carbon capture for producing value-added products from syngas. These facilities can be adapted to consume the escaped gases from virtually any CO_2-emitting industry. The most recent of such ventures was the ArcelorMittal plant in Ghent, Belgium, which was launched in December 2022.
- The research on syngas fermentation focused mainly on low TRLs, between TRL 2 and TRL 3, while the companies operated at a TRL of 7 to 9. This is an indicator of a "two-speed" development between academia and industry. While a small number of companies based in the United States of America control the technology that is required for the large-scale application of syngas fermentation, academia is still focusing on low TRL research. This can be related, on the one hand, to the necessity for novelty or achievable disruption in the actual pool of knowledge, which can be easily achieved in small-scale assays, or, on the other hand, with the large investments that are necessary for large-scale infrastructures and operations, which are not easily accessible to research groups.
- What should be the main focus of syngas fermentation scientific research going forward?
- The focus of syngas fermentation research should be centered on raw syngas fermentation, the evaluation of syngas impurity effects on acetogens, and the process integration of gasification and fermentation technologies. Furthermore, in order to increase the carbon capture efficiency and further facilitate the transition from fossil fuels, the production of long-chain value-added products should be the focus of future research. Ideally, it would be necessary to transform syngas fermentation into a flexible and versatile technology that is applicable to any carbon-rich off-gas, offering a close to complete carbon fixation into products. These are topics that seem to lack research and such thematics are pivotal to advancing these syngas fermentation technologies to a more mature level.

Supplementary Materials: The following supporting information can be downloaded at: https://www.mdpi.com/article/10.3390/en16073241/s1, File S1: Scientific publication and patent datasets used in this systematic review.

Author Contributions: Conceptualization, M.P., P.M. and C.S.; methodology, M.P. and C.S.; validation, M.P., P.M. and C.S.; formal analysis, M.P.; data curation, M.P.; writing—original draft preparation, M.P.; writing—review and editing, M.P., P.M. and C.S.; supervision, P.M. and C.S.; funding acquisition, P.M. and C.S. All authors have read and agreed to the published version of the manuscript.

Funding: This work was funded by the Portuguese Fundação para a Ciência e a Tecnologia (FCT) I.P./MCTES through national funds (PIDDAC)—UIDB/50019/2020. M.P. was supported by FCT through PhD grant DFA/BD/6423/2020.

Data Availability Statement: The datasets generated in this study are available as Supplementary Material.

Conflicts of Interest: The authors declare no conflict of interest. The funders had no role in the design of the study; in the collection, analyses, or interpretation of data; in the writing of the manuscript; or in the decision to publish the results.

Appendix A

The search query that was used in this study was: ((TS = ((((syngas OR "synthesis gas" OR "producer gas" OR C1 OR "H2 and CO") NEAR/2 (convers * OR ferment * OR metabol *)) AND ((formate OR acet * OR propion * OR butyr * OR valer * OR capro * OR capryl * OR acid$ OR ethanol OR butanol OR propanol OR pentanol OR hexanol OR octanol OR alcohol$ OR chemical$) NEAR/2 (production OR synthesis))) NOT ("Fischer-tropsch synthesis" OR cataly * OR "CO hydrogenation"))) OR (TI = ((((syngas OR "synthesis gas" OR "producer gas" OR C1 OR "H2 and CO") NEAR/2 (convers * OR ferment * OR metabol *)) AND ((formate OR acet * OR propion * OR butyr * OR valer * OR capro * OR capryl

* OR acid$ OR ethanol OR butanol OR propanol OR pentanol OR hexanol OR octanol OR alcohol$ OR chemical$) NEAR/2 (production OR synthesis))) NOT ("Fischer-tropsch synthesis" OR cataly * OR "CO hydrogenation"))) OR (AB = ((((syngas OR "synthesis gas" OR "producer gas" OR C1 OR "H2 and CO") NEAR/2 (convers * OR ferment * OR metabol *)) AND ((formate OR acet * OR propion * OR butyr * OR valer * OR capro * OR capryl * OR acid$ OR ethanol OR butanol OR propanol OR pentanol OR hexanol OR octanol OR alcohol$ OR chemical$) NEAR/2 (production OR synthesis))) NOT ("Fischer-tropsch synthesis" OR cataly* OR "CO hydrogenation"))) OR (AK = ((((syngas OR "synthesis gas" OR "producer gas" OR C1 OR "H2 and CO") NEAR/2 (convers * OR ferment * OR metabol *)) AND ((formate OR acet * OR propion * OR butyr * OR valer * OR capro * OR capryl * OR acid$ OR ethanol OR butanol OR propanol OR pentanol OR hexanol OR octanol OR alcohol$ OR chemical$) NEAR/2 (production OR synthesis))) NOT ("Fischer-tropsch synthesis" OR cataly * OR "CO hydrogenation")))).

References

1. IEA. *International Energy Agency Global Energy Review 2021: Assessing the Effects of Economic Recoveries on Global Energy Demand and CO2 Emissions in 2021*; IEA: Paris, France, 2021.
2. United Nations. *The Sustainable Development Goals Report 2019*; United Nations: New York, NY, USA, 2019.
3. Chandolias, K.; Richards, T.; Taherzadeh, M.J. Chapter 5—Combined Gasification-Fermentation Process in Waste Biorefinery. In *Waste Biorefinery: Potential and Perspectives*; Elsevier: Amsterdam, The Netherlands, 2018; pp. 157–200, ISBN 9780444639929.
4. Ruan, R.; Ding, K.; Liu, S.; Peng, P.; Zhou, N.; He, A.; Chen, P.; Cheng, Y.; Wang, Y.; Liu, Y.; et al. Chapter 12—Gasification and Pyrolysis of Waste. In *Current Developments in Biotechnology and Bioengineering*; Kataki, R., Pandey, A., Khanal, S.K., Pant, D., Eds.; Elsevier: Amsterdam, The Netherlands, 2020; pp. 263–297. [CrossRef]
5. Darmawan, A.; Aziz, M. Chapter 2—Process and Products of Biomass Conversion Technology. In *Innovative Energy Conversion from Biomass Waste*; Elsevier: Amsterdam, The Netherlands, 2022; pp. 25–60. ISBN 978-0-323-85477-1.
6. Calvo, D.C.; Luna, H.J.; Arango, J.A.; Torres, C.I.; Rittmann, B.E. Determining Global Trends in Syngas Fermentation Research through a Bibliometric Analysis. *J. Environ. Manag.* **2022**, *307*, 114522. [CrossRef] [PubMed]
7. de Tissera, S.; Köpke, M.; Simpson, S.D.; Humphreys, C.; Minton, N.P.; Dürre, P. Syngas Biorefinery and Syngas Utilization. In *Advances in Biochemical Engineering/Biotechnology*; Springer Science and Business Media Deutschland GmbH: Berlin, Germany, 2019; Volume 166, pp. 247–280. [CrossRef]
8. Latif, H.; Zeidan, A.A.; Nielsen, A.T.; Zengler, K. Trash to Treasure: Production of Biofuels and Commodity Chemicals via Syngas Fermenting Microorganisms. *Curr. Opin. Biotechnol.* **2014**, *27*, 79–87. [CrossRef]
9. Debabov, V.G. Acetogens: Biochemistry, Bioenergetics, Genetics, and Biotechnological Potential. *Microbiology* **2021**, *90*, 273–297. [CrossRef]
10. Ramachandriya, K.D.; Wilkins, M.R.; Patil, K.N. Influence of Switchgrass Generated Producer Gas Pre-Adaptation on Growth and Product Distribution of Clostridium ragsdalei. *Biotechnol. Bioprocess Eng.* **2013**, *18*, 1201–1209. [CrossRef]
11. Pacheco, M.; Pinto, F.; Ortigueira, J.; Silva, C.; Gírio, F.; Moura, P. Lignin Syngas Bioconversion by Butyribacterium methylotrophicum: Advancing towards an Integrated Biorefinery. *Energies* **2021**, *14*, 7124. [CrossRef]
12. Novak, K.; Neuendorf, C.S.; Kofler, I.; Kieberger, N.; Klamt, S.; Pflügl, S. Blending Industrial Blast Furnace Gas with H_2 Enables Acetobacterium woodii to Efficiently Co-Utilize CO, CO_2 and H_2. *Bioresour. Technol.* **2021**, *323*, 124573. [CrossRef] [PubMed]
13. Zhang, L.; Shen, Q.; Pang, C.H.; Chao, W.; Tong, S.; Kow, K.W.; Lester, E.; Wu, T.; Shang, L.; Song, X.; et al. Life Cycle Assessment of Bio-Fermentation Ethanol Production and Its Influence in China's Steeling Industry. *J. Clean Prod.* **2023**, *397*, 136492. [CrossRef]
14. Diekert, G.B.; Thauer, R.K. Carbon Monoxide Oxidation by Clostridium thermoaceticum and Clostridium formicoaceticum. *J. Bacteriol.* **1978**, *136*, 597–606. [CrossRef]
15. Liew, F.; Martin, M.E.; Tappel, R.C.; Heijstra, B.D.; Mihalcea, C.; Köpke, M. Gas Fermentation-A Flexible Platform for Commercial Scale Production of Low-Carbon-Fuels and Chemicals from Waste and Renewable Feedstocks. *Front. Microbiol.* **2016**, *7*, 694. [CrossRef]
16. Frazão, C.J.R.; Walther, T. Syngas and Methanol-Based Biorefinery Concepts. *Chem. Ing. Tech.* **2020**, *92*, 1680–1699. [CrossRef]
17. Ciliberti, C.; Biundo, A.; Albergo, R.; Agrimi, G.; Braccio, G.; de Bari, I.; Pisano, I. Syngas Derived from Lignocellulosic Biomass Gasification as an Alternative Resource for Innovative Bioprocesses. *Processes* **2020**, *8*, 1567. [CrossRef]
18. Fackler, N.; Heijstra, B.D.; Rasor, B.J.; Brown, H.; Martin, J.; Ni, Z.; Shebek, K.M.; Rosin, R.R.; Simpson, S.D.; Tyo, K.E.; et al. Stepping on the Gas to a Circular Economy: Accelerating Development of Carbon-Negative Chemical Production from Gas Fermentation. *Annu. Rev. Chem. Biomol. Eng.* **2021**, *12*, 439–470. [CrossRef] [PubMed]
19. Park, M.; Leahey, E.; Funk, R.J. Papers and Patents Are Becoming Less Disruptive over Time. *Nature* **2023**, *613*, 138–144. [CrossRef] [PubMed]
20. Aria, M.; Cuccurullo, C. Bibliometrix: An R-Tool for Comprehensive Science Mapping Analysis. *J. Informetr.* **2017**, *11*, 959–975. [CrossRef]

21. Bibliometrix-Biblioshiny. Available online: https://www.bibliometrix.org/home/index.php/layout/biblioshiny (accessed on 20 December 2022).
22. Posit | The Open-Source Data Science Company. Available online: https://posit.co/ (accessed on 20 December 2022).
23. UNESCO Science, Technology and Innovation: 9.5.2 Researchers (in Full-Time Equivalent) per Million Inhabitants. Available online: http://data.uis.unesco.org/index.aspx?queryid=3685# (accessed on 20 December 2022).
24. European Comission. *Horizon 2020 Work Programme 2014–2015*; European Comission: Brussels, Belgium, 2015.
25. Patent Granted, vs. Published: Everything You Need to Know. Available online: https://www.upcounsel.com/patent-granted-vs-published (accessed on 20 December 2022).
26. Liu, K.; Atiyeh, H.K.; Tanner, R.S.; Wilkins, M.R.; Huhnke, R.L. Fermentative Production of Ethanol from Syngas Using Novel Moderately Alkaliphilic Strains of Alkalibaculum bacchi. *Bioresour. Technol.* 2012, 104, 336–341. [CrossRef]
27. Zhang, F.; Ding, J.; Zhang, Y.; Chen, M.; Ding, Z.-W.; van Loosdrecht, M.C.M.; Zeng, R.J. Fatty Acids Production from Hydrogen and Carbon Dioxide by Mixed Culture in the Membrane Biofilm Reactor. *Water Res.* 2013, 47, 6122–6129. [CrossRef]
28. Vasudevan, D.; Richter, H.; Angenent, L.T. Upgrading Dilute Ethanol from Syngas Fermentation to N-Caproate with Reactor Microbiomes. *Bioresour. Technol.* 2014, 151, 378–382. [CrossRef] [PubMed]
29. Phillips, J.R.; Atiyeh, H.K.; Tanner, R.S.; Torres, J.R.; Saxena, J.; Wilkins, M.R.; Huhnke, R.L. Butanol and Hexanol Production in Clostridium carboxidivorans Syngas Fermentation: Medium Development and Culture Techniques. *Bioresour. Technol.* 2015, 190, 114–121. [CrossRef]
30. Bengelsdorf, F.R.; Straub, M.; Dürre, P. Bacterial Synthesis Gas (Syngas) Fermentation. *Environ. Technol.* 2013, 34, 1639–1651. [CrossRef] [PubMed]
31. Liew, F.E.; Nogle, R.; Abdalla, T.; Rasor, B.J.; Canter, C.; Jensen, R.O.; Wang, L.; Strutz, J.; Chirania, P.; de Tissera, S.; et al. Carbon-Negative Production of Acetone and Isopropanol by Gas Fermentation at Industrial Pilot Scale. *Nat. Biotechnol.* 2022, 40, 335–344. [CrossRef] [PubMed]
32. Okolie, J.A.; Epelle, E.I.; Tabat, M.E.; Orivri, U.; Amenaghawon, A.N.; Okoye, P.U.; Gunes, B. Waste Biomass Valorization for the Production of Biofuels and Value-Added Products: A Comprehensive Review of Thermochemical, Biological and Integrated Processes. *Process Saf. Environ. Prot.* 2022, 159, 323–344. [CrossRef]
33. Kucek, L.A.; Spirito, C.M.; Angenent, L.T. High N-Caprylate Productivities and Specificities from Dilute Ethanol and Acetate: Chain Elongation with Microbiomes to Upgrade Products from Syngas Fermentation. *Energy Environ. Sci.* 2016, 9, 3482–3494. [CrossRef]
34. Diender, M.; Stams, A.J.M.; Sousa, D.Z. Production of Medium-Chain Fatty Acids and Higher Alcohols by a Synthetic Co-Culture Grown on Carbon Monoxide or Syngas. *Biotechnol. Biofuels* 2016, 9, 82. [CrossRef] [PubMed]
35. Berzin, V.; Kiriukhin, M.; Tyurin, M. Selective Production of Acetone during Continuous Synthesis Gas Fermentation by Engineered Biocatalyst Clostridium sp. MAceT113. *Lett. Appl. Microbiol.* 2012, 55, 149–154. [CrossRef]
36. Berzin, V.; Tyurin, M.; Kiriukhin, M. Selective N-Butanol Production by Clostridium sp. MTButOH1365 During Continuous Synthesis Gas Fermentation Due to Expression of Synthetic Thiolase, 3-Hydroxy Butyryl-CoA Dehydrogenase, Crotonase, Butyryl-CoA Dehydrogenase, Butyraldehyde Dehydrogenase, and NAD-Dependent Butanol Dehydrogenase. *Appl. Biochem. Biotechnol.* 2013, 169, 950–959. [CrossRef]
37. Köpke, M.; Gerth, M.L.; Maddock, D.J.; Mueller, A.P.; Liew, F.; Simpson, S.D.; Patrick, W.M. Reconstruction of an Acetogenic 2,3-Butanediol Pathway Involving a Novel NADPH-Dependent Primary-Secondary Alcohol Dehydrogenase. *Appl. Environ. Microbiol.* 2014, 80, 3394–3403. [CrossRef]
38. Mohammadi, M.; Younesi, H.; Najafpour, G.; Mohamed, A.R. Sustainable Ethanol Fermentation from Synthesis Gas by Clostridium ljungdahlii in a Continuous Stirred Tank Bioreactor. *J. Chem. Technol. Biotechnol.* 2012, 87, 837–843. [CrossRef]
39. Shen, Y.; Brown, R.; Wen, Z. Syngas Fermentation of Clostridium carboxidivorans P7 in a Hollow Fiber Membrane Biofilm Reactor: Evaluating the Mass Transfer Coefficient and Ethanol Production Performance. *Biochem. Eng. J.* 2014, 85, 21–29. [CrossRef]
40. Devarapalli, M.; Atiyeh, H.K. A Review of Conversion Processes for Bioethanol Production with a Focus on Syngas Fermentation. *Biofuel Res. J.* 2015, 2, 268–280. [CrossRef]
41. Dürre, P.; Eikmanns, B.J. C1-Carbon Sources for Chemical and Fuel Production by Microbial Gas Fermentation. *Curr. Opin. Biotechnol.* 2015, 35, 63–72. [CrossRef]
42. Marcellin, E.; Behrendorff, J.B.; Nagaraju, S.; DeTissera, S.; Segovia, S.; Palfreyman, R.W.; Daniell, J.; Licona-Cassani, C.; Quek, L.; Speight, R.; et al. Low Carbon Fuels and Commodity Chemicals from Waste Gases—Systematic Approach to Understand Energy Metabolism in a Model Acetogen. *Green Chem.* 2016, 18, 3020–3028. [CrossRef]
43. Shen, Y.; Brown, R.C.; Wen, Z. Syngas Fermentation by Clostridium carboxidivorans P7 in a Horizontal Rotating Packed Bed Biofilm Reactor with Enhanced Ethanol Production. *Appl. Energy* 2017, 187, 585–594. [CrossRef]
44. Sun, X.; Atiyeh, H.K.; Kumar, A.; Zhang, H. Enhanced Ethanol Production by Clostridium ragsdalei from Syngas by Incorporating Biochar in the Fermentation Medium. *Bioresour. Technol.* 2018, 247, 291–301. [CrossRef]
45. Jang, N.; Yasin, M.; Kang, H.; Lee, Y.; Park, G.W.; Park, S.; Chang, I.S. Bubble Coalescence Suppression Driven Carbon Monoxide (CO)-Water Mass Transfer Increase by Electrolyte Addition in a Hollow Fiber Membrane Bioreactor (HFMBR) for Microbial CO Conversion to Ethanol. *Bioresour. Technol.* 2018, 263, 375–384. [CrossRef]
46. Riegler, P.; Bieringer, E.; Chrusciel, T.; Stärz, M.; Löwe, H.; Weuster-Botz, D. Continuous Conversion of CO_2/H_2 with Clostridium aceticum in Biofilm Reactors. *Bioresour. Technol.* 2019, 291, 121760. [CrossRef]

47. Amulya, K.; Kopperi, H.; Venkata Mohan, S. Tunable Production of Succinic Acid at Elevated Pressures of CO_2 in a High Pressure Gas Fermentation Reactor. *Bioresour. Technol.* **2020**, *309*, 123327. [CrossRef] [PubMed]
48. Haas, T.; Krause, R.; Weber, R.; Demler, M.; Schmid, G. Technical Photosynthesis Involving CO_2 Electrolysis and Fermentation. *Nat. Catal.* **2018**, *1*, 32–39. [CrossRef]
49. Yang, C. Acetogen Communities in the Gut of Herbivores and Their Potential Role in Syngas Fermentation. *Fermentation* **2018**, *4*, 40. [CrossRef]
50. Lee, J.; Lee, J.W.; Chae, C.G.; Kwon, S.J.; Kim, Y.J.; Lee, J.-H.; Lee, H.S. Domestication of the Novel Alcohologenic Acetogen Clostridium sp. AWRP: From Isolation to Characterization for Syngas Fermentation. *Biotechnol. Biofuels* **2019**, *12*, 228. [CrossRef]
51. Kim, J.; Kim, K.-Y.; Ko, J.K.; Lee, S.-M.; Gong, G.; Kim, K.H.; Um, Y. Characterization of a Novel Acetogen Clostridium sp. JS66 for Production of Acids and Alcohols: Focusing on Hexanoic Acid Production from Syngas. *Biotechnol. Bioprocess Eng.* **2022**, *27*, 89–98. [CrossRef]
52. Pati, S.; Mohanty, M.K.; Mohapatra, S.; Samantaray, D. Bioethanol Production by Enterobacter hormaechei through Carbon Monoxide-rich Syngas Fermentation. *Int. J. Energy Res.* **2022**, *46*, 20096–20106. [CrossRef]
53. Benalcázar, E.A.; Noorman, H.; Filho, R.M.; Posada, J. Assessing the Sensitivity of Technical Performance of Three Ethanol Production Processes Based on the Fermentation of Steel Manufacturing Offgas, Syngas and a 3:1 Mixture Between H_2 and CO_2. *Comput. Aided Chem. Eng.* **2020**, *48*, 589–594. [CrossRef]
54. Okoro, O.V.; Faloye, F.D. Comparative Assessment of Thermo-Syngas Fermentative and Liquefaction Technologies as Waste Plastics Repurposing Strategies. *AgriEngineering* **2020**, *2*, 378–392. [CrossRef]
55. de Medeiros, E.M.; Noorman, H.; Maciel Filho, R.; Posada, J.A. Multi-Objective Sustainability Optimization of Biomass Residues to Ethanol via Gasification and Syngas Fermentation: Trade-Offs between Profitability, Energy Efficiency, and Carbon Emissions. *Fermentation* **2021**, *7*, 201. [CrossRef]
56. Berazneva, J.; Woolf, D.; Lee, D.R. Local Lignocellulosic Biofuel and Biochar Co-Production in Sub-Saharan Africa: The Role of Feedstock Provision in Economic Viability. *Energy Econ.* **2021**, *93*, 105031. [CrossRef]
57. Lee, U.; R Hawkins, T.; Yoo, E.; Wang, M.; Huang, Z.; Tao, L. Using Waste CO_2 from Corn Ethanol Biorefineries for Additional Ethanol Production: Life-cycle Analysis. *Biofuels Bioprod. Biorefining* **2021**, *15*, 468–480. [CrossRef]
58. Okolie, J.A.; Tabat, M.E.; Gunes, B.; Epelle, E.I.; Mukherjee, A.; Nanda, S.; Dalai, A.K. A Techno-Economic Assessment of Biomethane and Bioethanol Production from Crude Glycerol through Integrated Hydrothermal Gasification, Syngas Fermentation and Biomethanation. *Energy Convers. Manag. X* **2021**, *12*, 100131. [CrossRef]
59. Almeida Benalcázar, E.; Noorman, H.; Maciel Filho, R.; Posada, J.A. Decarbonizing Ethanol Production via Gas Fermentation: Impact of the $CO/H_2/CO_2$ Mix Source on Greenhouse Gas Emissions and Production Costs. *Comput. Chem. Eng.* **2022**, *159*, 107670. [CrossRef]
60. Infantes, A.; Kugel, M.; Raffelt, K.; Neumann, A. Side-by-side Comparison of Clean and Biomass-Derived, Impurity-containing Syngas as Substrate for Acetogenic Fermentation with Clostridium ljungdahlii. *Fermentation* **2020**, *6*, 84. [CrossRef]
61. Roy, M.; Yadav, R.; Chiranjeevi, P.; Patil, S.A. Direct Utilization of Industrial Carbon Dioxide with Low Impurities for Acetate Production via Microbial Electrosynthesis. *Bioresour. Technol.* **2021**, *320*, 124289. [CrossRef]
62. Heffernan, J.K.; Valgepea, K.; de Souza Pinto Lemgruber, R.; Casini, I.; Plan, M.; Tappel, R.; Simpson, S.D.; Köpke, M.; Nielsen, L.K.; Marcellin, E. Enhancing CO_2-Valorization Using Clostridium autoethanogenum for Sustainable Fuel and Chemicals Production. *Front. Bioeng. Biotechnol.* **2020**, *8*, 204. [CrossRef]
63. Nangle, S.N.; Ziesack, M.; Buckley, S.; Trivedi, D.; Loh, D.M.; Nocera, D.G.; Silver, P.A. Valorization of CO_2 through Lithoautotrophic Production of Sustainable Chemicals in Cupriavidus necator. *Metab. Eng.* **2020**, *62*, 207–220. [CrossRef] [PubMed]
64. Kiriukhin, M.; Tyurin, M.; Gak, E. UVC-Mutagenesis in Acetogens: Resistance to Methanol, Ethanol, Acetone, or n-Butanol in Recombinants with Tailored Genomes as the Step in Engineering of Commercial Biocatalysts for Continuous CO_2/H_2 Blend Fermentations. *World J. Microbiol. Biotechnol.* **2014**, *30*, 1559–1574. [CrossRef] [PubMed]
65. de Souza Pinto Lemgruber, R.; Valgepea, K.; Tappel, R.; Behrendorff, J.B.; Palfreyman, R.W.; Plan, M.; Hodson, M.P.; Simpson, S.D.; Nielsen, L.K.; Köpke, M.; et al. Systems-Level Engineering and Characterisation of Clostridium autoethanogenum through Heterologous Production of Poly-3-Hydroxybutyrate (PHB). *Metab. Eng.* **2019**, *53*, 14–23. [CrossRef] [PubMed]
66. Fernández-Blanco, C.; Veiga, M.C.; Kennes, C. Efficient Production of N-Caproate from Syngas by a Co-Culture of Clostridium aceticum and Clostridium kluyveri. *J. Environ. Manag.* **2022**, *302*, 113992. [CrossRef]
67. González-Tenorio, D.; Muñoz-Páez, K.M.; Valdez-Vazquez, I. Butanol Production Coupled with Acidogenesis and CO_2 Conversion for Improved Carbon Utilization. *Biomass Convers. Biorefinery* **2022**, *12*, 2121–2131. [CrossRef]
68. He, Y.; Lens, P.N.L.; Veiga, M.C.; Kennes, C. Selective Butanol Production from Carbon Monoxide by an Enriched Anaerobic Culture. *Sci. Total Environ.* **2022**, *806*, 150579. [CrossRef]
69. He, Y.; Lens, P.N.L.; Veiga, M.C.; Kennes, C. Effect of Endogenous and Exogenous Butyric Acid on Butanol Production From CO by Enriched Clostridia. *Front. Bioeng. Biotechnol.* **2022**, *10*, 828316. [CrossRef] [PubMed]
70. Dürre, P. Biobutanol: An Attractive Biofuel. *Biotechnol. J.* **2007**, *2*, 1525–1534. [CrossRef]
71. Kumar, M.; Chong, C.T.; Karmakar, S. Combustion Characteristics of Butanol- Jet A-1 Fuel Blends in a Swirl-stabilized Combustor under the Influence of Preheated Swirling Air. *Int. J. Energy Res.* **2022**, *46*, 2601–2616. [CrossRef]
72. Sun, X.; Atiyeh, H.K.; Zhang, H.; Tanner, R.S.; Huhnke, R.L. Enhanced Ethanol Production from Syngas by Clostridium ragsdalei in Continuous Stirred Tank Reactor Using Medium with Poultry Litter Biochar. *Appl. Energy* **2019**, *236*, 1269–1279. [CrossRef]

73. Klask, C.-M.; Kliem-Kuster, N.; Molitor, B.; Angenent, L.T. Nitrate Feed Improves Growth and Ethanol Production of Clostridium ljungdahlii With CO_2 and H_2, but Results in Stochastic Inhibition Events. *Front. Microbiol.* **2020**, *11*, 724. [CrossRef] [PubMed]
74. Oliveira, L.; Röhrenbach, S.; Holzmüller, V.; Weuster-Botz, D. Continuous Sulfide Supply Enhanced Autotrophic Production of Alcohols with Clostridium ragsdalei. *Bioresour. Bioprocess.* **2022**, *9*, 15. [CrossRef]
75. Pardo-Planas, O.; Atiyeh, H.K.; Phillips, J.R.; Aichele, C.P.; Mohammad, S. Process Simulation of Ethanol Production from Biomass Gasification and Syngas Fermentation. *Bioresour. Technol.* **2017**, *245*, 925–932. [CrossRef] [PubMed]
76. European Commission Biopolymers from Syngas Fermentation—SYNPOL Project—FP7. Available online: https://cordis.europa.eu/project/id/311815 (accessed on 21 December 2022).
77. European Commission Biological Production of Butanol from Syngas—SYNTOBU Project—FP7. Available online: https://cordis.europa.eu/project/id/618593 (accessed on 21 December 2022).
78. European Commission Advanced Biofuel Production with Energy System Integration—AMBITION Project—H2020. Available online: https://cordis.europa.eu/project/id/731263 (accessed on 21 December 2022).
79. European Commission BIOtechnological Processes Based on Microbial Platforms for the CONversion of CO2 from Ironsteel Industry into Commodities for Chemicals and Plastics—BIOCONCO2—H2020. Available online: https://cordis.europa.eu/project/id/761042 (accessed on 21 December 2022).
80. Ministry of Science and ICT (Republic of Korea) C1 Gas Refinery. Available online: https://cgrc.sogang.ac.kr/ (accessed on 21 December 2022).
81. LanzaTech, with the Support of Danone, Discovers Method to Produce Sustainable PET Bottles from Captured Carbon—LanzaTech. Available online: https://lanzatech.com/lanzatech-with-the-support-of-danone-discovers-method-to-produce-sustainable-pet-bottles-from-captured-carbon/ (accessed on 21 December 2022).
82. Papageorgiadis, N.; Sofka, W. Patent Enforcement across 51 Countries—Patent Enforcement Index 1998–2017. *J. World Bus.* **2020**, *55*, 101092. [CrossRef]
83. Technology-Synata Bio. Available online: https://synatabio.com/technology/ (accessed on 21 December 2022).
84. Seaboard Energy Opens New Biodiesel Plant in Kansas—Feed & Grain News. Available online: https://www.feedandgrain.com/news/seaboard-energy-opens-new-biodiesel-plant-in-kansas (accessed on 21 December 2022).
85. History-JupengBio. Available online: http://www.jupengbio.com/history (accessed on 21 December 2022).
86. Rückel, A.; Hannemann, J.; Maierhofer, C.; Fuchs, A.; Weuster-Botz, D. Studies on Syngas Fermentation with Clostridium carboxidivorans in Stirred-Tank Reactors with Defined Gas Impurities. *Front. Microbiol.* **2021**, *12*, 655390. [CrossRef]
87. Oliveira, L.; Rückel, A.; Nordgauer, L.; Schlumprecht, P.; Hutter, E.; Weuster-Botz, D. Comparison of Syngas-Fermenting Clostridia in Stirred-Tank Bioreactors and the Effects of Varying Syngas Impurities. *Microorganisms* **2022**, *10*, 681. [CrossRef]
88. On the Mend: Why INEOS Bio Isn't Producing Ethanol in Florida: Biofuels Digest. Available online: https://www.biofuelsdigest.com/bdigest/2014/09/05/on-the-mend-why-ineos-bio-isnt-reporting-much-ethanol-production/ (accessed on 21 December 2022).
89. News-LanzaTech. Available online: https://lanzatech.com/news/ (accessed on 21 December 2022).
90. European Commission Carbon Capture, Storage and Utilisation. Available online: https://energy.ec.europa.eu/topics/oil-gas-and-coal/carbon-capture-storage-and-utilisation_en (accessed on 27 February 2023).
91. European Commission Demonstrating Sustainable Value Creation from Industrial CO2 by Its Thermophilic Microbial Conversion into Acetone—PYROCO2—H2020. Available online: https://cordis.europa.eu/project/id/101037009 (accessed on 27 February 2023).
92. Handler, R.M.; Shonnard, D.R.; Griffing, E.M.; Lai, A.; Palou-Rivera, I. Life Cycle Assessments of Ethanol Production via Gas Fermentation: Anticipated Greenhouse Gas Emissions for Cellulosic and Waste Gas Feedstocks. *Ind. Eng. Chem. Res.* **2016**, *55*, 3253–3261. [CrossRef]

Disclaimer/Publisher's Note: The statements, opinions and data contained in all publications are solely those of the individual author(s) and contributor(s) and not of MDPI and/or the editor(s). MDPI and/or the editor(s) disclaim responsibility for any injury to people or property resulting from any ideas, methods, instructions or products referred to in the content.

Article

Carbon Tax or Low-Carbon Subsidy? Carbon Reduction Policy Options under CCUS Investment

Qian Zhang [1], Yunjia Wang [2] and Lu Liu [3],*

[1] Haier Group, Qingdao Hainayun Technology Holding Co., Ltd., Qingdao 266101, China
[2] Bathurst Future Agri-Tech Institute, Qingdao Agricultural University, Qingdao 266109, China
[3] College of Economics and Management, Shandong University of Science and Technology, Qingdao 266590, China
* Correspondence: magic_liu@sdust.edu.cn

Abstract: Great expectations are placed in carbon capture, utilization, and storage (CCUS) technology to achieve the goal of carbon neutrality. Governments adopt carbon tax policies to discourage manufacturing that is not eco-friendly, and subsidies to encourage low-carbon production methods. This research investigates which carbon reduction incentive policy is more viable for the supply chain under CCUS application. The most significant finding is that carbon tax and low-carbon subsidy policies are applicable to high-pollution and low-pollution supply chains with the goal of maximizing social welfare. Both policies play a significant role in reducing carbon emissions. However, it is very important for the government to set reasonable policy parameters. Specifically, carbon tax and low-carbon subsidy values should be set in the intermediate level rather than being too large or too small to achieve higher social welfare. We also find that the higher the value of carbon dioxide (CO_2) in CCUS projects, the higher the economic performance and social welfare, but the lower the environmental efficiency. Governments should properly regulate the value of CO_2 after weighing economic performance, environmental efficiency and social welfare. The findings yield useful insights into the industry-wise design of carbon emission reduction policies for CCUS and similar projects.

Keywords: CCUS; carbon tax; low-carbon subsidy; policy comparison

---

Citation: Zhang, Q.; Wang, Y.; Liu, L. Carbon Tax or Low-Carbon Subsidy? Carbon Reduction Policy Options under CCUS Investment. *Sustainability* **2023**, *15*, 5301. https://doi.org/10.3390/su15065301

Academic Editors: Zilong Liu, Meixia Shan and Yakang Jin

Received: 17 February 2023
Revised: 8 March 2023
Accepted: 15 March 2023
Published: 16 March 2023

Copyright: © 2023 by the authors. Licensee MDPI, Basel, Switzerland. This article is an open access article distributed under the terms and conditions of the Creative Commons Attribution (CC BY) license (https://creativecommons.org/licenses/by/4.0/).

1. Introduction

Climate change is accelerating the global ecology to dangerous levels. Greenhouse gases have been widely recognized as the critical cause of extreme climate events [1]. The issue related to carbon neutrality has received a large amount of attention and concern over the last several years [2]. Carbon capture, utilization, and storage (CCUS) has been accepted as a critical technology to reduce CO_2 emissions and mitigate climate change hazards [3]. CCUS technology captures, purifies, and stores the CO_2 emitted in the production process. Then, the captured CO_2 is sold to specific buyers, and utilized in enhancing oil recovery, gas recovery, and coal bed methane [4]. CCUS has been identified as an important technology for promoting carbon emission reduction [5]. According to the global status of CCUS 2021 released by the Global CCUS Institute, there are 135 CCUS projects in operation worldwide, and the number is growing rapidly. Qilu Petrochemical—Shengli Oilfield Project is a representative CCUS project in China [6]. In this project, Qilu Petrochemical first captures CO_2 from the tail gas of gasification equipment using liquefaction purification technology. Then, the captured CO_2 is sold and transported to Shengli Oil Field through land or pipeline. Ultimately, the CO_2 is used for oil displacement in Shengli Oil Field. Data show that the efficiency of CO_2 oil displacement is 40% higher than that of water. The CCUS project is expected to inject more than 1000 tons of carbon dioxide (CO_2) into 73 wells over 15 years. In addition, nearly 3 million tons of oil production and more than 12% of the oil extraction rate will be increased. Most importantly, the project can reduce CO_2 emissions by 1 million tons per year. Figure 1 clearly illustrates the CCUS supply chain system. The Quest Carbon Capture and Storage (QCCS) project, funded and operated by Royal Dutch Shell in Western Canada, is one of the world's first large-scale CCUS projects. The project captured and stored one million tons of CO_2 ahead of schedule in its first year of operation [7]. To date, Quest has captured and stored over 6 million tons of CO_2. In totally, the CCUS project plays a critical role in addressing the global greenhouse effect and climate change [8]. The International Energy Agency (IEA) predicted that CCUS will account for nearly 15% of total global emission reductions by 2070 [6].

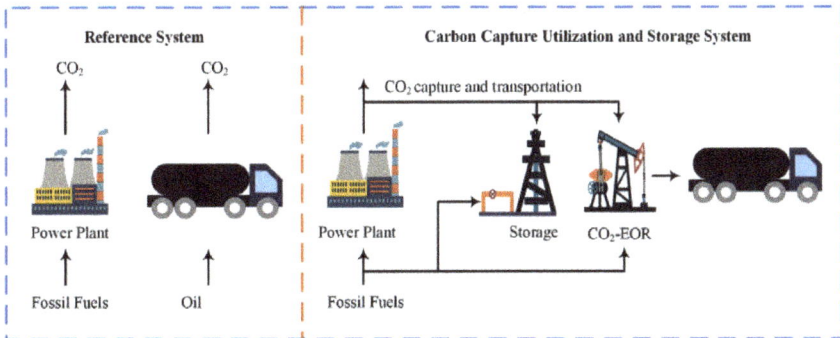

Figure 1. Illustration of CCUS supply chain system (source: figure is used with permission from Zhang et al. [4]).

As the role of CCUS projects in carbon emission reduction continues to be highlighted, the world is accelerating the investment in and construction of CCUS projects. The data from the International Energy Agency (IEA) predicts that total investment in CCUS projects will total USD 67.09 billion in the period 2026–2030, which could increase to USD 776.61 billion in the period 2056–2060 [9]. The carbon emission reduction incentive policy initiated by the government is an important way to promote enterprises to invest in CCUS projects and implement carbon emission reduction [10]. The incentive policies mainly include mandatory policies and voluntary policies [11]. Mandatory policy means that the government adopts a mandatory carbon control mechanism to regulate the high emission behavior of enterprises. Carbon cap and carbon tax are two common types of mandatory policies [12]. With a carbon cap policy, the government issues fixed carbon emission quotas to enterprises and prohibits them from exceeding carbon caps [13]. Cap-and-trade policy is a deformation of the carbon cap policy. It allows enterprises with insufficient and surplus allowances to buy and sell carbon emission rights in the carbon trading market [12]. With a carbon tax policy, the government imposes a carbon tax on enterprises that emit CO_2 [14]. To sum up, the mandatory policy is regarded as a penalty policy. In contrast, the voluntary policy is a reward policy. Typical voluntary policies include pure carbon trading policy and low-carbon subsidy policies. With a carbon trading policy, the CO_2 emission reductions are quantified as CERs. Enterprises can obtain a profit from selling CERs. With a pure carbon trading policy, manufacturers are not penalized for not implementing carbon reductions [11]. With the low-carbon subsidy policy, the government provides subsidies to the enterprises that implement carbon emission reduction [15].

Based the above discussion, we know that academia has achieved a lot in the research on carbon emission reduction incentive policies. However, since the CCUS project is in the early stage of industrial application, the research on the combination of CCUS and carbon emission reduction incentive policies is still in the initial stage. Very few studies have analyzed how to design a policy to facilitate carbon emission reduction in supply chains with CCUS applications. Our work aims to fill this gap by addressing the following research questions.

(i) How does the carbon tax policy and the low-carbon subsidy policy affect the optimal operation decision of the supply chain?
(ii) How does the government choose between carbon tax and low-carbon subsidy policy to simultaneously improve economic efficiency, environmental performance and social welfare?
(iii) In addition to policy options, what strategies can the government formulate to further enhance social welfare?

To answer the above questions, we first construct model A and model S to characterize the CCUS supply chain using carbon tax and low-carbon subsidy policies, respectively. Then, we conduct decision optimization and sensitivity analysis to optimize the CCUS supply chain. Most importantly, we compare the differences between model A and model S in terms of operational decision-making, supply chain profits, carbon emission reductions, and social welfare. Many significant conclusions are obtained in this research. The results provide a policy basis for the government to reasonably select and optimize carbon emission reduction incentive policies to achieve sustainable development of the supply chain with CCUS applications.

The novelty and significance of the study are summarized as follows. First, we are one of the few studies to characterize the supply chain decision models with CCUS technology and carbon control policy applications. The optimal operational decisions and environmental policies are deeply analyzed to guide decision makers in rational joint operational and environmental decisions. The economic efficiency, environmental performance, and social welfare of the CCUS supply chain and their influencing factors are also examined. Second, as far as we know, ours is one of the first studies in the field of operations management to explore the impacts of carbon reduction incentive policies such as carbon tax and low-carbon subsidy on the performance of CCUS supply chain. More importantly, we answer the question of which strategy is more feasible for the CCUS supply chain (achieving greater joint economic–environment–social performance) by quantitatively comparing carbon tax and low-carbon subsidy policies. The findings yield useful insights into the industry-wise design of carbon emission reduction policies for CCUS and similar projects.

2. Literature Review

We review the literature most related to our research from two aspects: (i) the carbon emission reduction incentive policies, including mandatory policy and voluntary policy; (ii) the comparison of carbon reduction incentive policies, especially the comparison of mandatory and voluntary policies. They are reviewed accordingly as follows.

2.1. Carbon Emission Reduction Incentive Policies

We first review the literature on carbon emission reduction incentive policies. The policies include mandatory policies such as cap-and-trade and carbon tax, voluntary policies such as low-carbon subsidy and reduction-and-trading.

Cap-and-trade policy is a typical mandatory emission reduction policy. Zhang et al. [16] examined the impact of power structures on the governmental cap regulation and manufacturer's low carbon strategy. Tang and Yang [17] analyzed the effects of power structure and financing mode on the capital-constrained low-carbon supply chain. Liu et al. [18] compared three carbon emission reduction modes: manufacturer emission reduction, retailer emission reduction, and joint emission reduction, and found that joint emission reduction mode has the highest social welfare. Xu et al. [19] investigated the government's optimal region-cap setting strategy by comparing grandfather-based allocation and benchmark-based allocation rules.

In addition to the carbon cap policy, the carbon tax policy is another mandatory policy (carbon penalty policy). Krass et al. [20] analyzed the impacts of regulatory policies of environmental tax, cost subsidy and consumer rebate on the choice of green technology and social welfare. They found that the combined policy can incentivize firms to adopt greener technology, thereby improving social welfare. Pal and Saha [21] explored the best application conditions of privatization policy when privatization and carbon tax can be used simultaneously for environmental improvement. Zhou et al. [22] examined the impacts of carbon tax policy and consumer environmental awareness on social welfare. Yu et al. [23] investigated the impacts of carbon emissions tax on the supply chain channel structure, and provided the applicable conditions of decentralized structure and centralized structure, respectively. Chen et al. [14] analyzed the optimal carbon tax design with respect to different power structures and green technology investment efficiencies. Zhou et al. [24] provided a comprehensive review of combined research on carbon taxes and low-carbon supply chain operations. Gopalakrishnan et al. [25] designed a footprint-balanced scheme to rationally reallocate carbon emissions and carbon tax costs using cooperative game theory methodology.

Different from the mandatory policy, the voluntary policy, i.e., low-carbon subsidy policy, has proved to be another effective way to motivate enterprises to implement carbon emission reduction. In this regard, Xu et al. [15] compared four kinds of governmental subsidy strategies, and found that subsidizing to both manufacturer and retailer is more profitable for the supply chain and the government. Bao et al. [26] uncovered that the government subsidy scheme contributes to the carbon reduction of the new energy vehicle supply chain. Zhang and Huang [27] compared consumer subsidy (CS) and the R&D subsidy (RS) programs, and found that both CS and RS programs might pose negative impacts on the environment. Ma et al. [28] summarized the role of government subsidies in promoting carbon reduction and information sharing of the supply chain. Liu et al. [11] discussed the conditions under which manufacturers choose to introduce voluntary carbon emission reduction policies and the impacts of supply chain competition. Xu et al. [29] examined the effects of horizontal integration on social welfare under the interaction of carbon tax and green subsidy, and provided the optimal subsidy and carbon tax levels.

While the above-mentioned studies are rich in research on carbon emission reduction policies and policies, studies rarely involve CCUS. CCUS emerges in the supply chain operations research field until recent years. Among the few studies, Wang and Qie [30] built an analytical real options model to explore when the supply chain should invest in CCUS. Zhang et al. [4] used a multi-objective mixed integer linear programming to optimize the CCUS supply chain with economic and environmental concerns. Ostovari et al. [8] quantified the large-scale potential of CO_2 mineralization in Europe by designing a climate-optimal supply chain for CO_2 CCUS. To sum up, the research related to CCUS supply chain operations is very limited. We enrich CCUS supply chain research by pioneering the design and comparison of carbon emission reduction incentives such as a carbon tax and a low-carbon subsidy.

2.2. The Comparison of Carbon Reduction Incentive Policies

More recently, a stream of research has emerged exploring the comparison of cap-and-trade and carbon tax policies in low-carbon supply chain. Miao et al. [31] showed that the implementation of carbon regulations such as carbon tax and cap-and-trade hurts the manufacturer. However, a well-designed subsidy scheme can improve the manufacturer's profit, and simultaneously reduce the carbon emissions. Anand and Giraud-Carrier [32] analyzed and compared cap-and-trade and carbon tax regulations, and proved that well-chosen regulation can simultaneously improve firms' profits, environmental performance, and social welfare. Hu et al. [33] compared carbon tax and cap-and-trade policies and answered which policy is more viable for Chinese remanufacturing industry. Sun and Yang [34] pointed out that a cap-and-trade policy is more effective than a carbon tax policy in improving social welfare. Chen et al. [35] also proved that a cap-and-trade policy is more efficient than a carbon tax in reducing carbon emissions and promoting clean innovation. Hasan et al. [12] analyzed the optimal inventory level and technology investment joint decisions under different carbon regulations such as cap-and-trade, carbon tax, and strict carbon limit. Zhou et al. [36] summarized the respective advantages of a carbon tax and cap-and-trade, for example, the former has higher cost efficiency and the latter has lower sector-level impacts. Fan et al. [1] indicated that when the correlation between the sales market and the permit trading market is moderate, the carbon tax policy shows an advantage in terms of technology investment compared with cap-and-trade. Yu et al. [37] investigated an online platform's decision between reselling and marketplace modes, and a government's selection between cap-and-trade and carbon tax regulations.

The comparison of carrot and stick (i.e., penalty and subsidy) policies has also drawn significant attention from a large body of research. Yin et al. [38] gave the optimal carbon emission policy to maximize social welfare by comparing the carbon tax and low-carbon subsidy policies. Huang et al. [39] compared border tax (BT) and output-based allocation (OB), where the former and latter adopt "stick" and "carrot" approaches. They found that BT shows higher superiority than OB. Ma et al. [40] analyzed the optimal government intervention strategy in the automotive low-carbon supply chain by comparing a carbon tax and a consumer subsidy. Zhu et al. [41] found that the hybrid model shows potential in outperforming the other two policies in terms of the optimal technology R&D by comparing cash subsidy, carbon regulation, and hybrid model. Guo et al. [42] found that the incentive-compatible combination policy composed of carbon tax and low-carbon subsidy is better than the pure carbon tax policy. He et al. [2] explored the impacts of the penalty and subsidy regulations on the straw-based bioenergy supply chain, and presented the respective applicable conditions of the two regulations. Dou and Choi [43] formulated optimal old product collection strategies, and designed a "carrot-and-stick" policy consisting of a carbon tax and a subsidy to motivate both the supply chain and consumers to accept the advocated strategies. Wang and Zhang [44] examined the interplay between subsidies and regulation under competition. The results showed that the effects of regulation and subsidies may offset or reinforce depending on the situation. Wu et al. [45] compared three carbon regulatory policies: carbon tax, low-low-carbon subsidy, and a mixed policy, and indicated that a mixed policy is a supply chain equilibrium strategy.

The above studies have carried out a significant amount of research on the comparison of carbon emission reduction incentive policies without considering specific carbon emission reduction technologies such as CCUS. However, virtually none of the current research has considered the impact of CCUS technology on the choice of carbon emission reduction policies. We are almost the first to conduct a comparative study of carbon emission reduction policies around the CCUS supply chain. We then selected some classical studies that focus on low-carbon supply chain to highlight the contributions of this research (see Table 1).

Table 1. Positioning of this research in the literature.

Research	Carbon Tax Policy	Low-Carbon Subsidy Policy	CCUS	Policy Comparison
Yu et al. [23]; Gopalakrishnan et al. [25]	√	×	×	×
Bao et al. [26]; Ma et al. [28]	×	√	×	×
Zhang et al. [4]; Ostovari et al. [8]	×	×	√	×
Yin et al. [38]; Wu et al. [45]	√	√	×	√
This study	√	√	√	√

3. Model Description

We consider a supply chain consisting of a manufacturer M and a retailer R. The manufacturer (she) produces products at unit cost c. The products are sold to the retailer at wholesale price w. Then, the retailer sells the products to consumers at retail price p. The demand function of products is $q = a - bp$ (Zhou et al. [22]), where a is the initial market potential demand, and b measures the price sensitivity (Yu et al. [23]).

The manufacturing process generates a large amount of carbon emissions. We assume that the initial carbon emission per unit of product is e_0. Obviously, parameter e_0 quantifies the pollution level of the manufacturer/product (Yu et al., [23]). In order to alleviate the product pollution to the environment, the manufacturer initiates the CCUS project. In this project, the manufacturer invests in research and development (i.e., R&D) cost to achieve carbon capture and carbon storage in the production process. The R&D cost of achieving carbon capture and carbon storage is $0.5he^2$, which is a quadratic function of carbon emission reduction per unit product e (Dou and Choi [43]; Deng and Liu [46]; Liu et al. [47]). In this cost function, h represents the carbon emission reduction cost coefficient. It is worth noting that in the traditional cap-and-trade policy, the remaining carbon emission quota of enterprises can be sold in the carbon trading market (Liu et al. [11]). However, in the CCUS project, what can be traded is the stored CO_2 (Wang and Qie [30]). This is the main difference between the cap-and-trade policy and the CCUS mechanism. We assume that the CO_2 selling price is u.

The government designs two kinds of carbon reduction incentive policies, namely carbon tax policy and low-carbon subsidy policy (Wu et al. [45]). We use the symbols A and S ($j = A, S$) to identify these two policies, respectively. In model A, the government imposes a carbon tax of t per unit of carbon emission. In model S, the government issues a low-carbon subsidy of s per unit of carbon emission reduction. Hence, policy A and policy S can be regarded as a punishment policy and a reward policy, respectively. Figure 2 is an illustration of models A and S.

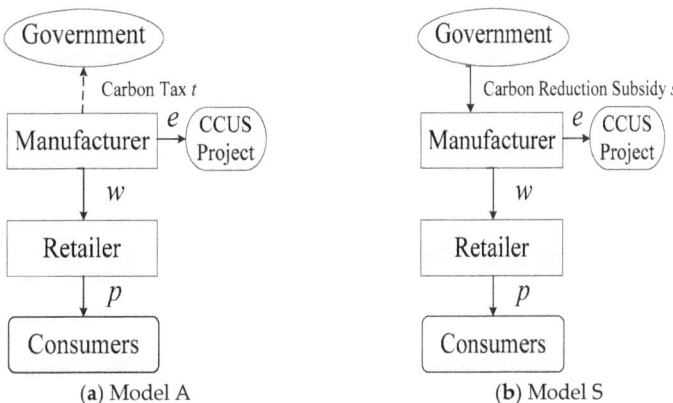

Figure 2. Illustration of supply chain structures and carbon policies.

We use a Stackelberg game to describe the relationship among the government, manufacturer and retailer, where the government, the manufacturer, and the retailer are the game leader, the sub-leader and the follower, respectively (Xu et al. [19]). The manufacturer and the retailer aim to maximize their profits, while the government's goal is to maximize social welfare (Zhang et al. [16]). To explore our research questions, we establish three game models as follows. In model A, the manufacturer earns wholesale revenue $(w^A - c)q^A$ and carbon sales revenue $ue^A q^A$ while paying for carbon capture and storage R&D cost $\frac{1}{2}h(e^A)^2$ and carbon tax $t(e_0 - e^A)q^A$ (Hu et al. [33]). The retailer earns retail revenue $(p^A - w^A)q^A$. The social welfare function consists of four parts. The first is the sum of the profits of manufacturer and retailer $\pi_M^A + \pi_R^A$. The second is the total carbon tax value of the government $t(e_0 - e^A)q^A$. The third is the cost of environmental damage caused by carbon emissions $\frac{1}{2}g((e_0 - e^A)q^A)^2$ (Sun and Yang [34]). The fourth is the consumer surplus $\frac{1}{2}d(q^A)^2$ (Liu et al. [18]). In summary, the profits of the manufacturer (π_M^A) and the retailer (π_R^A), and social welfare (SW^A) in model A are:

$$\pi_M^A = (w^A - c)q^A - \frac{1}{2}h(e^A)^2 + ue^A q^A - t(e_0 - e^A)q^A, \tag{1}$$

$$\pi_R^A = (p^A - w^A)q^A, \tag{2}$$

$$SW^A = \pi_M^A + \pi_R^A + t(e_0 - e^A)q^A - \frac{1}{2}g((e_0 - e^A)q^A)^2 + \frac{1}{2}d(q^A)^2. \tag{3}$$

In model S, the government no longer levies carbon tax on the manufacturer, but issues a carbon emission reduction subsidy $se^S q^S$ to her (Xu et al. [19]). The profits of the supply chain members and the total social welfare are characterized as follows:

$$\pi_M^S = (w^S - c)q^S - \frac{1}{2}h(e^S)^2 + (u + s)e^S q^S, \tag{4}$$

$$\pi_R^S = (p^S - w^S)q^S, \tag{5}$$

$$SW^S = \pi_M^S + \pi_R^S - se^S q^S - \frac{1}{2}g((e_0 - e^S)q^S)^2 + \frac{1}{2}d(q^S)^2. \tag{6}$$

We have characterized the supply chain decision models under carbon tax and low-carbon subsidy policies. In the following sections, we first analyze the decision-making behavior of supply chain members. Then, we conduct policy comparisons to help the government make reasonable choices between carbon tax and low-carbon subsidy policies. All proofs of propositions are contained in Appendix A.

4. Equilibrium Analysis

The supply chain decision sequence is as follows. Given carbon tax t and low-carbon subsidy s, the manufacturer first decides the optimal wholesale price w^{j*} and carbon emission reduction per unit product e^{j*}. Then, the retailer charges the optimal retail price p^{j*}. Finally, we can obtain the total carbon emissions of the supply chain E^{j*}, the profits of the supply chain members π_i^{j*} ($i = M, R$), and the overall social welfare SW^{j*}. The backward induction is used for equilibrium analysis (Zhang et al. [16]). The equilibrium results in model A are summarized as Proposition 1.

Proposition 1. *The optimal decisions, profits and carbon emissions of A model are following:*

$$w^{A*} = \frac{2be_0 ht + 2ah + 2bch - abu^2 - 2abut - abt^2}{b(4h - bu^2 - 2but - bt^2)}, \tag{7}$$

$$e^{A*} = \frac{(u+t)(a - bc - bte_0)}{4h - bu^2 - 2but - bt^2}, \tag{8}$$

$$p^{A*} = \frac{be_0 ht + 3ah + bch - abu^2 - 2abut - abt^2}{b(4h - bu^2 - 2but - bt^2)}, \tag{9}$$

$$\pi_M^{A*} = \frac{h(bc - a + bte_0)^2}{2b(4h - bu^2 - 2but - bt^2)}, \tag{10}$$

$$\pi_R^{A*} = \frac{h^2(bc - a + bte_0)^2}{b(bu^2 + 2but + bt^2 - 4h)^2}, \tag{11}$$

$$E^{A*} = \frac{h(bc - a + bte_0)(au + at - 4e_0 h - bcu - bct + bu^2 e_0 + bute_0)}{(bu^2 + 2but + bt^2 - 4h)^2}, \tag{12}$$

To investigate the influences of main parameters on the optimal decisions, profits, and carbon emissions, we perform a parametric sensitivity analysis below.

Proposition 2. *The changes of optimal decisions, profits, and carbon emissions with respect to e_0 in model A have the following features: w^{A*} and p^{A*} are increasing in e_0. e^{A*}, π_M^{A*}, and π_R^{A*} are decreasing in e_0. If $e_0 < \tilde{e}_1$, then E^{A*} is increasing in e_0; otherwise, E^{A*} is decreasing in e_0.*

Proposition 2 illustrates the effects of initial carbon emissions (i.e., product pollution level) on the decisions, profits, and carbon emissions of the supply chain. We can understand Proposition 2 from the following aspects. The total carbon tax paid by the manufacturer increases with the level of product pollution. In order to dilute the carbon tax cost, the manufacturer chooses to raise the wholesale price and reduce R&D investment in carbon emission reduction. The manufacturer's reactions to carbon tax result in two outcomes: lower sales (i.e., lower production) and lower carbon emission reductions per unit of product. The above results imply that as product pollution level increases, the manufacturer opts to reduce production rather than reduce carbon emissions per unit of product in response to the carbon tax penalty. Then, we explore the impacts of carbon selling price on the supply chain and carbon emissions under model A. The following proposition is obtained through sensitivity analysis.

Proposition 3. *The changes of optimal decisions, profits, and carbon emissions with respect to u in model A have the following features: w^{A*} and p^{A*} are decreasing in u. e^{A*}, π_M^{A*}, and π_R^{A*} are increasing in u. If $e_0 < \tilde{e}_2$, then E^{A*} is decreasing in u; otherwise, E^{A*} is increasing in u.*

Proposition 3 shows the analysis results for parameter u. As u increases, the operating cost of the supply chain decreases. As a result, the manufacturer lowers the wholesale price and increases the carbon emission reductions. The retailer lowers the retail price. The consumers increase the purchases. On the whole, the increase of u stimulates two effects: a sales-increase effect and a unit-product-carbon-emission-reduction effect. These two effects promote and inhibit the increase in total carbon emissions, respectively. Recalling Proposition 2, that when e_0 is low, the unit-product-carbon-emission-reduction effect outperforms the sales-increase effect. Thus, E^{A*} is decreasing in u. Otherwise, E^{A*} is increasing in u. Through the above analysis, we obtain the following managerial insights. When the manufacturer produces low-polluting products, increasing the CO_2 selling price is an effective means to incentivize the manufacturer to reduce carbon emissions. Otherwise, when the manufacturer produces high-polluting products, raising the CO_2 selling price increases CO_2 emissions of the supply chain. Then, we explore the impacts of carbon tax on the supply chain.

Proposition 4. *The changes of optimal decisions, profits, and carbon emissions with respect to t in model A have the following features: If $e_0 < \tilde{e}_3$, then w^{A*} is decreasing in t; otherwise, w^{A*} is increasing in t. If $e_0 < \tilde{e}_4$, then p^{A*} is decreasing in t; otherwise, p^{A*} is increasing in t. If $e_0 < \tilde{e}_5$, then e^{A*} is increasing in t; otherwise, e^{A*} is decreasing in t. If $e_0 < \tilde{e}_6$, then π_M^{A*} is increasing in t; otherwise, π_M^{A*} is decreasing in t. If $e_0 < \tilde{e}_7$, then π_R^{A*} is increasing in t; otherwise, π_R^{A*} is decreasing in t. E^{A*} is decreasing in t.*

We obtain the following findings by observing Proposition 4. First, the carbon tax policy can effectively reduce the total carbon emissions of the supply chain, in which the high-polluting manufacturer chooses to reduce carbon emissions by limiting production. The low-pollution manufacturer controls total emissions by reducing carbon emissions per unit of product. The findings can be well-explained by Proposition 2. Second, and interestingly, the carbon tax policy can simultaneously increase the economic efficiency and environmental performance of the supply chain that produces low-polluting products. The reason is that the carbon tax policy motivates the manufacturer to reduce carbon emissions, thereby increasing her carbon sales revenue. Summarizing the above results, the carbon tax policy can improve the environmental performance of the supply chain, but it may reduce the economic efficiency of the supply chain, especially the supply chain that produces high-polluting products. This result is consistent with those of Wu et al. [45]. They also found that while a carbon tax is beneficial to the ecological environment, it would also increase the economic burden on supply chains.

We already fully understand the impact of carbon tax policy on the supply chain implementing CCUS project. In the following part, we analyze the low-carbon subsidy policy. We characterize the equilibrium under the low-carbon subsidy scheme in the following proposition:

Proposition 5. *The optimal decisions and profits of S model are following:*

$$w^{S*} = \frac{abs^2 + 2absu + abu^2 - 2ah - 2bch}{b(bs^2 + 2bsu + bu^2 - 4h)}, \tag{13}$$

$$e^{S*} = \frac{(s+u)(a-bc)}{4h - bs^2 - 2bsu - bu^2}, \tag{14}$$

$$p^{S*} = \frac{3ah + bch - abs^2 - 2absu - abu^2}{b(4h - bs^2 - 2bsu - bu^2)}, \tag{15}$$

$$\pi_M^{S*} = \frac{h(a-bc)^2}{2b(4h - bs^2 - 2bsu - bu^2)}, \tag{16}$$

$$\pi_R^{S*} = \frac{h^2(a-bc)^2}{b(bs^2 + 2bsu + bu^2 - 4h)^2}, \tag{17}$$

$$E^{S*} = \frac{h(bc-a)(as + au - 4e_0h - bcs - bcu + bs^2e_0 + bu^2e_0 + 2bsue_0)}{(bs^2 + 2bsu + bu^2 - 4h)^2}. \tag{18}$$

Based on the equilibrium results, we examine the influence of e_0 on the supply chain and obtain the following proposition.

Proposition 6. *The changes of optimal decisions, profits, and carbon emissions with respect to e_0 in model S have the following features: w^{S*}, p^{S*}, e^{S*}, π_M^{S*}, and π_R^{S*} are unchanging in e_0. E^{S*} is increasing in e_0.*

Proposition 6 shows that the optimal decisions and profits under model S are not affected by parameter e_0. The reason is that the government subsidizes the manufacturer based on her carbon reductions rather than the carbon emissions. However, carbon reductions are independent of e_0. We also find from Proposition 6 that the higher the pollution level of the product, the higher the total carbon emissions of the supply chain. This conclusion is significantly different from Proposition 2.

To investigate the effects of CO_2 selling price on the low-carbon subsidy policy, we conduct a sensitivity analysis as Proposition 7.

Proposition 7. *The changes of optimal decisions, profits, and carbon emissions with respect to u in model S have the following features: The w^{S*} and p^{S*} are decreasing in u. e^{S*}, π_M^{S*}, and π_R^{S*} are increasing in u. If $a < \tilde{a}_1$, then E^{S*} is increasing in u; otherwise, E^{S*} is decreasing in u.*

The conclusion of Proposition 7 is similar to that of Proposition 3; therefore, we omit its explanation. However, we obtain the following management implications from Proposition 7. Raising the CO_2 selling price always improves the economic performance of the supply chain; however, it may detrimental to the environmental efficiency. The above results provide management implications for the government to reasonably adjust the CO_2 trading price of CCUS projects.

Proposition 8. *The changes of optimal decisions, profits, and carbon emissions with respect to s in model S have the following features: w^{S*} and p^{S*} are decreasing in s. e^{S*}, π_M^{S*}, and π_R^{S*} are increasing in s. If $a < \tilde{a}_2$, then E^{S*} is increasing in s; If $a > \tilde{a}_2$, then E^{S*} is decreasing in s.*

Proposition 8 indicates that the low-carbon subsidy policy always improves the economic efficiency of the supply chain; however, it may harm the environmental performance. This view is identical to that of Zhang et al. [48] and Yao et al. [49]. We can explain this result as follows. Low-carbon subsidy reduces operating cost and generates additional revenue for the manufacturer. Therefore, in order to obtain more low-carbon subsidy, the manufacturer increases production by reducing wholesale price; on the other hand, more carbon emission reductions are obtained by increasing carbon emission reductions per unit of product. However, a low-carbon subsidy motivates an increase in production that could lead to an increase in total carbon emissions. As is illustrated in Proposition 8, only when the basic market demand of the product is high can increasing low-carbon subsidy reduce the total carbon emissions of the supply chain.

5. Model Comparison

Different carbon policies have different impacts on economic performance, environmental efficiency and social welfare (Yu [50]). We further investigate the relationship between carbon policies and supply chain performance under CCUS applications. By comparing the optimal decisions and outcomes among different policies, we gain some managerial insights to find which policy choice is beneficial to the economy and environment. The optimal decisions are first compared as Proposition 9.

Proposition 9. *(i) When $s \geq t$, the wholesale price, retail price, and carbon emission reduction per unit product in different models have the following orders: $w^A > w^S$, $p^A > p^S$, and $e^A < e^S$. (ii) When $s < t$, the wholesale price, retail price, and carbon emission reduction per unit product in different models have the following orders: If $e_0 < \tilde{e}_8$, then $w^A < w^S$; If $e_0 > \tilde{e}_8$, then $w^A > w^S$. If $e_0 < \tilde{e}_9$, then $p^A < p^S$; If $e_0 > \tilde{e}_9$, then $p^A > p^S$. If $e_0 < \tilde{e}_{10}$, then $e^A > e^S$; If $e_0 > \tilde{e}_{10}$, then $e^A < e^S$.*

The analysis of Proposition 9 is as follows. When the low-carbon subsidy value is higher than the carbon tax value, the production volume and carbon emission reductions per unit product under low-carbon subsidy policy are higher than those under carbon tax policy. In contrast, when the low-carbon subsidy value is lower than the carbon tax value, the above comparison result is still valid only if product pollution degree is high. The comparison results are easy to understand. Model S has a significant cost advantage over model A. The cost advantage is eliminated only when $s < t$ and e_0 is low.

Next, we compare the optimal profits and carbon emissions of the two carbon emission reduction incentive policies to explore the impact of policy on the optimal supply chain performance and discuss which policy is the better one. Through theoretical analysis, the following proposition is derived.

Proposition 10. *(i) When $s \geq t$, the profits and carbon emissions in different models have the following orders: $\pi_M^A < \pi_M^S$ and $\pi_R^A < \pi_R^S$ always hold. If $e_0 < \tilde{e}_{12}$, then $E^A > E^S$; Otherwise, $E^A < E^S$. (ii) When $s < t$, the profits and carbon emissions in different models have the following orders: If $e_0 < \tilde{e}_{11}$, then $\pi_M^A > \pi_M^S$; If $e_0 > \tilde{e}_{11}$, then $\pi_M^A < \pi_M^S$. If $e_0 < \tilde{e}_{12}$, then $\pi_R^A > \pi_R^S$; If $e_0 > \tilde{e}_{12}$, then $\pi_R^A < \pi_R^S$. $E^A < E^S$ always holds.*

Proposition 10 implies the following conclusion. When $s > t$, the economic efficiency of the supply chain in model S is always better than that in model A. However, only when e_0 is low, the environmental performance of the supply chain in model S is better than that in model A. When $s > t$, the environmental performance of the supply chain in model A is always better than that in model S. Moreover, only when e_0 is low is the economic efficiency of the supply chain in model A better than that in model S. Based on the results of Proposition 9, we can make an appropriate interpretation of Proposition 10. When $s > t$, because the S model has a cost advantage over the A model, the economic efficiency of the supply chain in the S model is obviously higher than that in the A model. If $e_0 > \widetilde{e}_{12}$, i.e., the pollution degree of products is high, the manufacturer is penalized with a high carbon tax under model A. Hence, the manufacturer is more aggressive in reducing carbon emissions in the A model than in the S model, i.e., $E^A < E^S$. When $s < t$, affected by the carbon tax penalty, the manufacturer has a higher incentive to reduce carbon emissions. Thus, $E^A < E^S$ always holds. If e_0 is low enough, the carbon selling revenue of the manufacturer under the A model is higher than that of the S model. Thus, we have $\pi_M^A > \pi_M^S$ and $\pi_R^A > \pi_R^S$. Otherwise, if e_0 is high enough, the carbon tax penalty effect exceeds the CO_2 sales effect. Hence, the economic performance of the supply chain under the A model is lower than that under the S model, i.e., $\pi_M^A < \pi_M^S$ and $\pi_R^A < \pi_R^S$. The above research conclusions are consistent with the research of Chen and Hu [51], Li and Peng [52]. They also found that a carbon tax policy and a low-carbon subsidy policy have advantages in promoting economic performance and environmental efficiency growth, respectively. As the degree of pollution and environmental damage of products increases, the government's carbon control policy should be changed from low-carbon subsidy to carbon tax (Guo and Huang [53]).

We obtain the following managerial insights: First, for low-pollution supply chain, when the low-carbon subsidy value is higher and lower than the carbon tax value, the government should adopt the low-carbon subsidy policy and the carbon tax policy respectively. Second, for high-pollution supply chain, the government that prefers economic efficiency and environmental performance should choose the low-carbon subsidy policy and the carbon tax policy, respectively. The above conclusions provide policy suggestions for the government to reasonably choose the carbon emission reduction incentive policy to improve social welfare.

6. Numerical Study

In this section, we verify the above theoretical conclusions utilizing numerical examples. We also gain more conclusions and managerial insights that are difficult to obtain directly from theoretical analysis. A survey of some manufacturing enterprises and CCUS projects in China is first conducted. Based on survey data processing and statistical analysis, we set the parameters as follows:

$$a = 1000, b = 0.5, c = 30, h = 30, g = 0.0005, d = 1, t = 2, s = 2, u = 2, e_0 = 200.$$

Then, we conduct the numerical study from three aspects. First, we compare the two models to answer which model has higher economic efficiency, environmental performance, and social welfare. Second, we analyze the effects of main parameters on social welfare and address the managerial insights. The results of our experiments are shown in Figures 3–12 (The * in the figures depicts the optimal results).

6.1. Model Comparison and Optimal Mode Selection

In this part, we explore which model has lower carbon emissions, and higher economic efficiency and social welfare by comparing models A and S. Figures 3–6 are illustrations of the comparison results. As shown in these figures, the profits of the supply chain members and social welfare are decreasing or unchanging in the product initial unit carbon emissions. The total carbon emissions increase as the initial unit carbon emissions increase. The results verify Propositions 2 and 6.

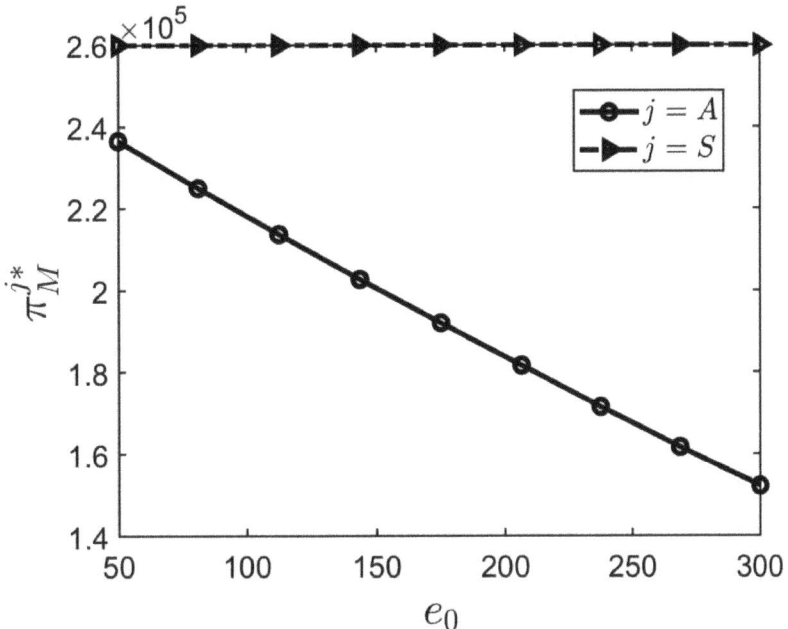

Figure 3. The impact of e_0 on manufacturers' profits under different models.

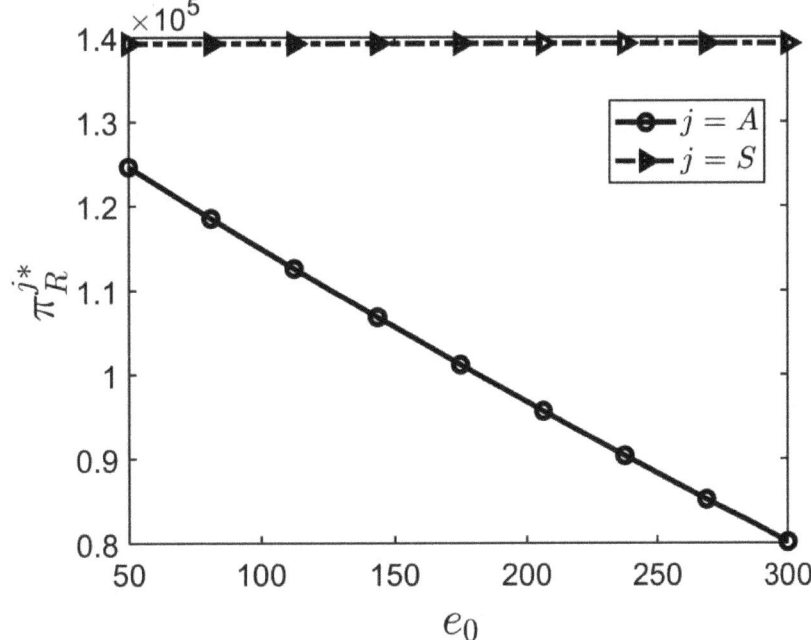

Figure 4. The impact of e_0 on retailers' profits under different models.

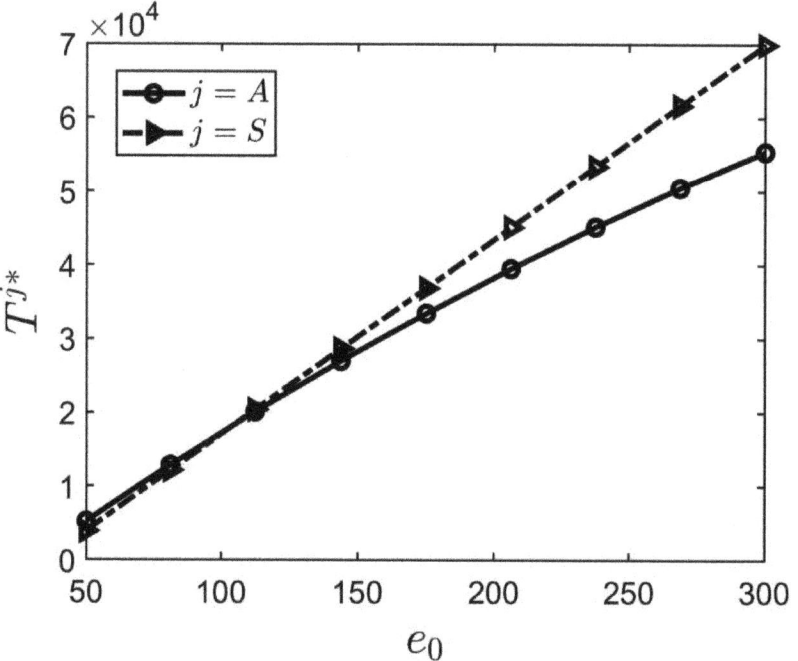

Figure 5. The impact of e_0 on carbon emissions under different models.

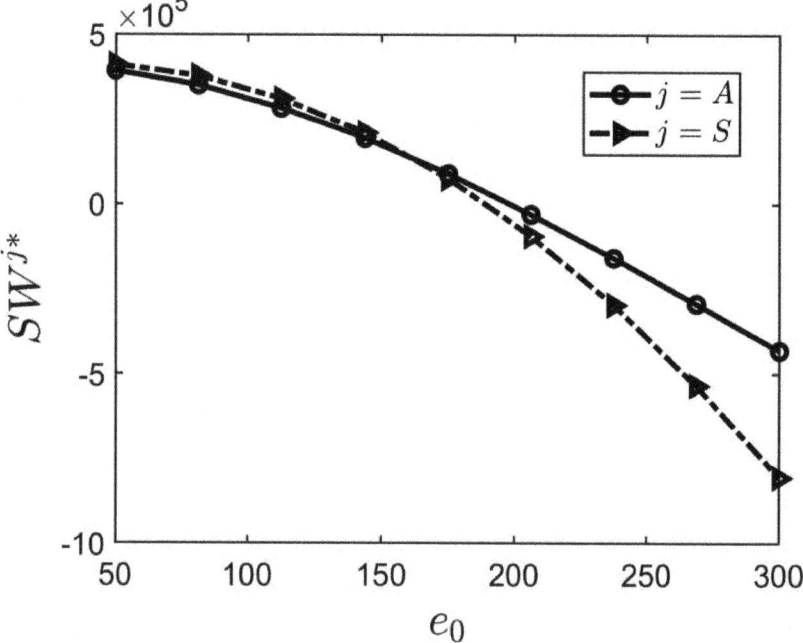

Figure 6. The impact of e_0 on social welfare under different models.

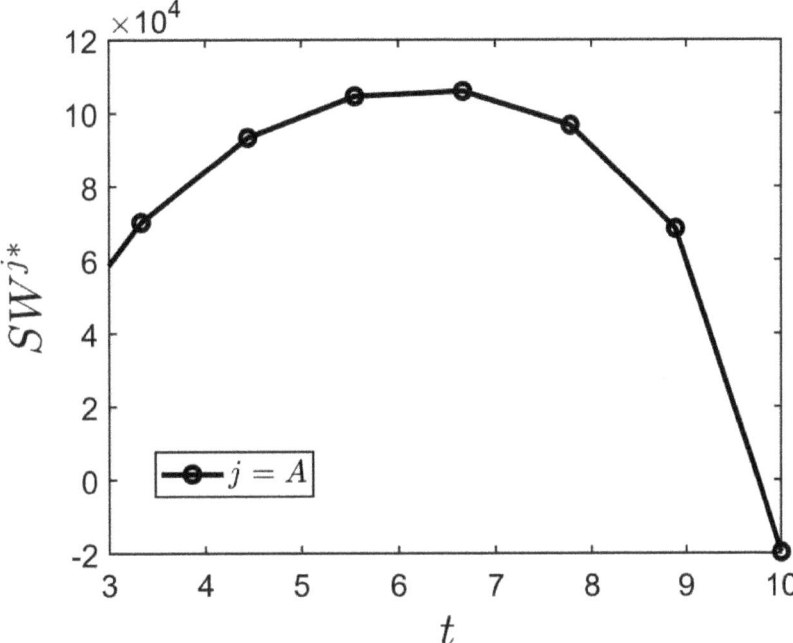

Figure 7. The impact of t on social welfare under model A.

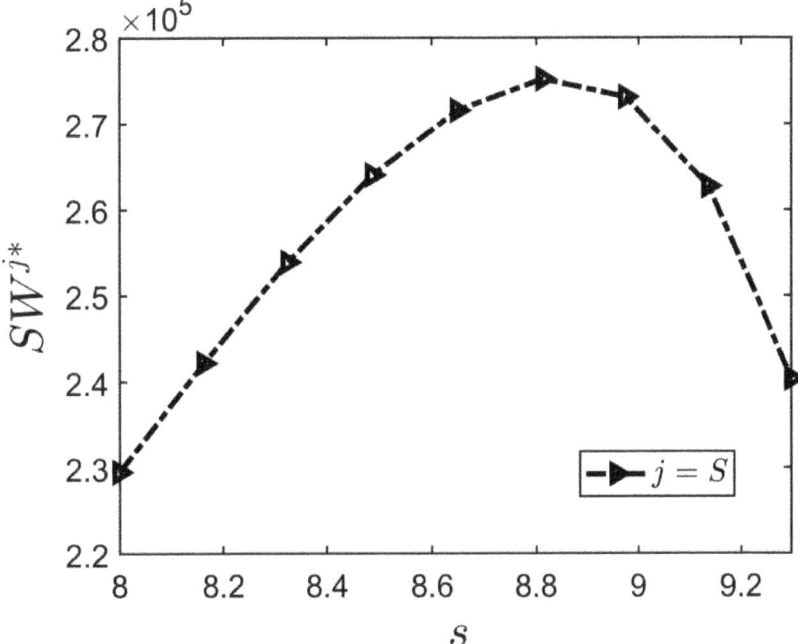

Figure 8. The impact of s on social welfare under model S.

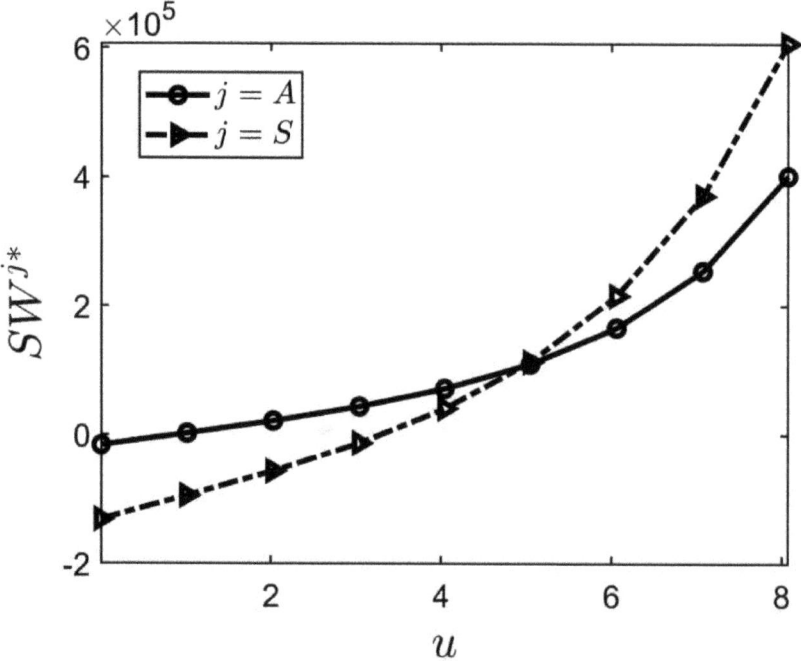

Figure 9. The impact of u on social welfare under different models.

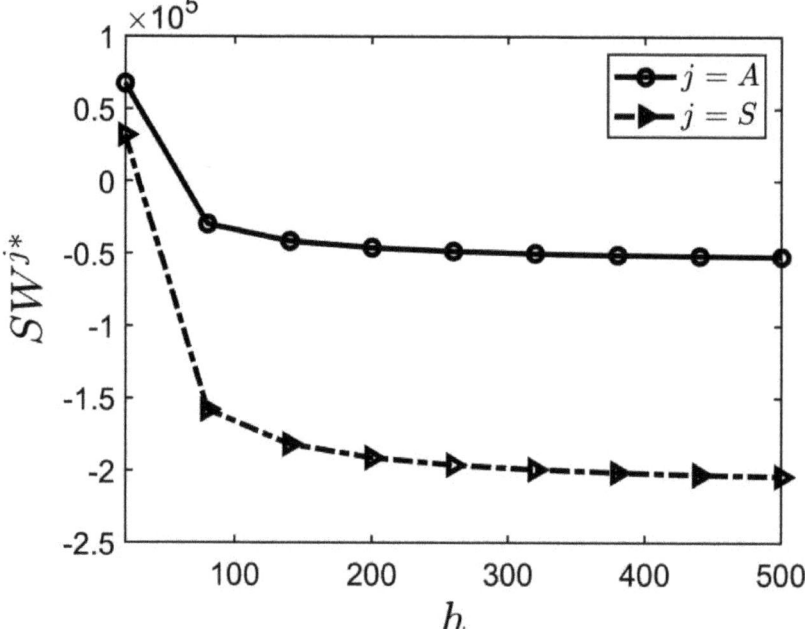

Figure 10. The impact of h on social welfare under different models.

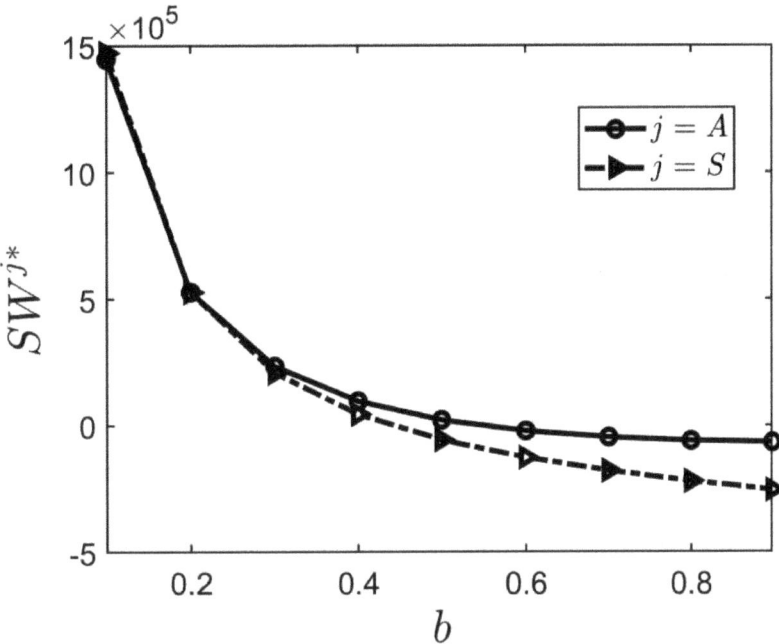

Figure 11. The impact of b on social welfare under different models.

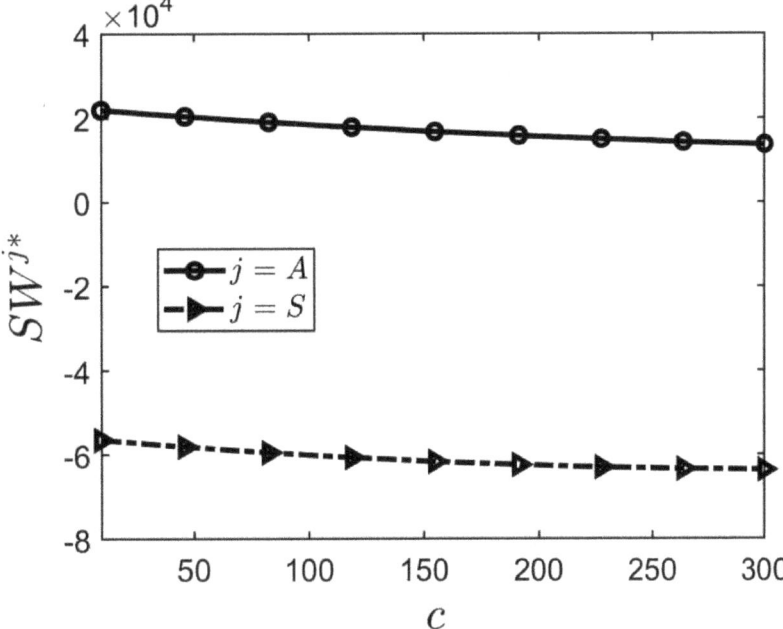

Figure 12. The impact of c on social welfare under different models.

Through model comparison, we find the following results. First, the supply chain profit in model S is higher than that in model A, and the profit gap increases as e_0 increases

(See Figures 3 and 4). Second, when parameter e_0 is relatively low, the carbon emissions (social welfare) in the S model are (is) lower (higher) than that in the A model.

The above results can be explained as follows. In models S and A, the supply chain is subsidized and penalized, respectively. That means that the operating cost of the supply chain in the S model is lower than that of the A model. Hence, the supply chain profit in model S is higher than that in model A. Meanwhile, the carbon emission reduction per unit product and the total order quantity of the supply chain under the S model are higher than those of the A model. Therefore, the total carbon emissions in model S is higher than that in model A only when parameter e_0 is low enough. The above research conclusion is consistent with the study of Shu et al. [54]. They also found that although carbon tax policy is helpful to achieve carbon emission reduction target, it may lose supply chain profit. The government should introduce low-carbon subsidy policy to achieve both environmental and economic win–win situations.

Through the above analysis, we obtain the following managerial insights. From the perspective of reducing carbon emissions and increasing total social welfare, when the manufacturer implements CCUS project, the government should use the mandatory policy (i.e., carbon tax policy) and the voluntary policy (i.e., low-carbon subsidy policy) to achieve carbon reduction incentives for high-polluting and low-polluting enterprises, respectively (Guo and Huang [53]). The above research results provide policy suggestions for the government to increase social welfare through policy selection.

6.2. Impacts of Main Parameters on Social Welfare

In this section, we first investigate the impacts of carbon tax and low-carbon subsidy on social welfare, in order to enlighten the government on how to optimally set the policy parameters. Second, we explore the effects of other main parameters on social welfare to obtain more managerial insights.

As shown in Figures 7 and 8, in model A, social welfare increases first and then decreases with the carbon tax. In model S, when the low-carbon subsidy rate is at the intermediate level, social welfare reaches the maximum. The reason is as follows. In model A, the operating cost of the supply chain increases with the carbon tax. Thus, the supply chain profit is decreasing in carbon tax. Meanwhile, it is clear that supply chain carbon emissions decrease with the carbon tax. Only when the carbon tax takes the middle value does social welfare reach the maximum. The impact of low-carbon subsidy on social welfare can be similarly explained.

Through the above analysis, we can obtain the following valuable policy recommendations for the government. Taking into account economic efficiency and environment should set the carbon tax and low-carbon subsidy values in the middle rather than too low or too high when implementing the carbon tax and low-carbon subsidy policies. This view is consistent with those of Wu et al. [45] and Yao et al. [49]. Wu et al. [45] found that the supply chain economic performance increases slightly as the carbon tax coefficient increases before declining rapidly. Yao et al. [49] found that property developers are less inclined to build low-carbon houses when low-carbon subsidy rate exceeds a certain critical point.

The above analysis provides strategies for the government to reasonably design the parameters of carbon emission reduction policies. Then, we further analyze the impact of other parameters on the supply chain and social welfare. Figure 9 shows that regardless of the model, social welfare increases with the carbon sales price. Figures 10–12 illustrate that, when the CO_2 selling price is low, social welfare under model A is higher than under model S; otherwise, the conclusion is the opposite. Therefore, the government should use a carbon tax policy and a low-carbon subsidy policy to incentivize manufacturers to reduce carbon emissions when the CO_2 selling price is low and high enough, respectively. Similarly, it can be seen from Figure 11 that when the consumer price sensitivity coefficient is low and high enough, policy S and policy A should be used, respectively, to control carbon emissions and increase social welfare.

7. Conclusions

Global climate change is affecting ecosystems, and carbon reduction has become a global consensus. CCUS is one of the key technologies to deal with global climate change, which has received worldwide attention. In order to encourage the supply chain to reduce carbon emissions by investing in CCUS, the government has introduced two kinds of policies: carbon tax and low-carbon subsidy. The former penalizes carbon emissions, while the latter rewards emissions reductions. We systematically analyze and summarize the optimal carbon emission reduction incentive policies of the government from three aspects: the choice of carbon policy, the parameter setting of carbon policy and the setting of carbon price, etc. We first compare carbon tax and low-carbon subsidy from three aspects: economic performance, environmental efficiency and social welfare. We find that under the same conditions, low-carbon subsidy policy generates higher economic performance than carbon tax policy. When the degree of product pollution is high, the environmental efficiency and social welfare under carbon tax policy are higher than those under low-carbon subsidy policy; otherwise, the conclusion is opposite. Therefore, with the aim of maximizing social welfare, the government should adopt carbon tax policy and low-carbon subsidy policy for high-polluting and low-polluting supply chains, respectively. Given the optimal carbon policy choice of the government, we further study the way to set the parameter of selected carbon policy. We find that the increase of carbon tax value always improves the environmental efficiency of supply chain, but may reduce the economic performance. In contrast, the increase of carbon subsidy value always improves the economic performance of supply chain, but may reduce the environmental efficiency. A significant and interesting finding is that social welfare level of the supply chain reaches its maximum when the carbon tax value and the carbon subsidy value are set at an intermediate level. The above results provide decision support for the government to reasonably set carbon policy parameters. Specifically, the government should set the carbon tax value and the low-carbon subsidy value at a middle level, rather than too high or too low. We also provide managerial insights for the government to reasonably regulate the CO_2 trading price in CCUS projects based on the following research conclusions. Regardless of which carbon policy is adopted, the economic performance and social welfare increase with the CO_2 trading price. Hence, in order to improve the overall social welfare, the government should formulate policies to control the CO_2 trading price within a higher range, for example, by setting a government-directed price for CO_2 similar to that for natural gas by adopting the CO_2 trading access system to properly increase the scarcity of CO_2. The above research results provide systematic decision-making suggestions for the government to reasonably formulate carbon emission reduction incentive policies for CCUS projects. In the future, we can conduct evaluative studies on the input–output of CCUS projects.

Author Contributions: All authors contributed to the study conception and design. Spot investigation and data collection were performed by Q.Z. The first draft of the manuscript was written by Q.Z. Methodology, formal analysis, and experimental work were mainly contributed by Y.W. with guidance from L.L. L.L. was involved in the critical revision of the article for important intellectual content. All authors commented on previous versions of the manuscript. All authors have read and agreed to the published version of the manuscript.

Funding: This work was supported by the National Natural Science Foundation of China (Nos. 72272089 and 71902105).

Institutional Review Board Statement: Not applicable.

Informed Consent Statement: Not applicable.

Data Availability Statement: The data and materials presented in the study are included in the article.

Acknowledgments: We thank the editors and the anonymous reviewers for their very insightful and constructive comments.

Conflicts of Interest: The authors declare no conflict of interest.

Appendix A

Proof of Proposition 1. Taking the second-order partial derivatives of π_R^A with respect to p^A yields that $\frac{d^2 \pi_R^A}{d(p^A)^2} = -2b < 0$. Hence, $\pi_R^A(p^A)$ is a concave function with respect to p^A. Then solving the first-order condition $\frac{d\pi_R^A}{dp^A} = 0$, we can obtain the optimal response function $p^{A*}(w^A) = \frac{a+bw^{Ap}}{2b}$. Substituting $p^{A*}(w^A)$ into Equation (1), we obtain the Hessian matrix of (1) with respect to w^A and e^A:

$$H_M = \begin{bmatrix} -b & -\frac{b(u+t)}{2} \\ -\frac{b(u+t)}{2} & -h \end{bmatrix}.$$

The sequential principal minors of H_M are: $-b < 0$, $\frac{b(4h - bu^2 - 2but - bt^2)}{4} > 0$. Thus, $\pi_M^A(w^A, e^A)$ is a joint concave function of w^A and e^A. Solving the first order conditional equations $\frac{d\pi_M^A}{dw^A} = 0$, $\frac{d\pi_M^A}{de^A} = 0$ yields

$$w^{A*} = \frac{2be_0 ht + 2ah + 2bch - abu^2 - 2abut - abt^2}{b(4h - bu^2 - 2but - bt^2)},$$

$$e^{A*} = \frac{(u+t)(a - bc - bte_0)}{4h - bu^2 - 2but - bt^2}.$$

Then, it is easy to further obtain that p^{A*}, π_M^{A*}, π_R^{A*}, and E^{A*} satisfy Equations (9)–(12), respectively.

Proof of Proposition 2. Differentiating the optimal decisions, profits, and carbon emissions with respect to e_0 yields

$$\frac{dw^{A*}}{de_0} = \frac{2th}{4h - bu^2 - 2but - bt^2} > 0, \quad \frac{dp^{A*}}{de_0} = \frac{th}{4h - bu^2 - 2but - bt^2} > 0,$$

$$\frac{de^{A*}}{de_0} = -\frac{bt(u+t)}{4h - bu^2 - 2but - bt^2} < 0, \quad \frac{d\pi_M^{A*}}{de_0} = -\frac{th(a - bc - bte_0)}{4h - bu^2 - 2but - bt^2} < 0,$$

$$\frac{d\pi_R^{A*}}{de_0} = -\frac{2th^2(a - bc - bte_0)}{(bu^2 + 2but + bt^2 - 4h)^2} < 0.$$

The above evidences confirm the changes of w^{A*}, p^{A*}, e^{A*}, π_M^{A*}, and π_R^{A*} on e_0 in Proposition 1. Then, we prove the relationship between E^{A*} and e_0. The differential of E^{A*} with respect to e_0 is

$$\frac{dE^{A*}}{de_0} = \frac{h(2e_0 b^2 u^2 t + cb^2 u^2 + 2e_0 b^2 ut^2 - cb^2 t^2 - abu^2 + abt^2 - 8e_0 hbt - 4chb + 4ah)}{(bu^2 + 2but + bt^2 - 4h)^2}.$$

It is easy to see that $\frac{dE^{A*}}{de_0}$ is a linear monotone function with respect to e_0. Let $\bar{e}_0 = \frac{a-bc}{bt}$. Since $e_0 < \bar{e}_0$ is a necessary condition for $q^{A*} > 0$, we assume that $0 < e_0 < \bar{e}_0$. At the left and right endpoints of the valid interval of e_0, the values of $\frac{dE^{A*}}{de_0}$ are

$$\left.\frac{dE^{A*}}{de_0}\right|_{e_0=0} = \frac{h(a-bc)(4h - bu^2 + bt^2)}{(bu^2 + 2but + bt^2 - 4h)^2} > 0 \text{ and}$$

$$\left.\frac{dE^{A*}}{de_0}\right|_{e_0=\bar{e}_0} = -\frac{h(a - bc)}{4h - bu^2 - 2but - bt^2} < 0.$$

That means that there exists a threshold $\tilde{e}_1$ such that $\frac{dE^{A*}}{de_0} > 0$ when $e_0 < \tilde{e}_1$, and $\frac{dE^{A*}}{de_0} < 0$ when $e_0 > \tilde{e}_1$. Here, we obtain Proposition 2.

Proof of Proposition 3. The supply chain system is effective only when the optimal decisions and profits are greater than 0. Hence, from $\pi_M^{A*} > 0$ and $e^{A*} > 0$, we know that $4h - bu^2 - 2but - bt^2 > 0$ and $a - bc - bte_0 > 0$. Then, the following differentials are obviously true:

$$\frac{dw^{A*}}{du} = -\frac{4h(u+t)(a-bc-bte_0)}{(bu^2+2but+bt^2-4h)^2} < 0,$$

$$\frac{de^{A*}}{du} = \frac{(a-bc-bte_0)(bu^2+2but+bt^2+4h)}{(bu^2+2but+bt^2-4h)^2} > 0,$$

$$\frac{dp^{A*}}{du} = \frac{2h(u+t)(bc-a+bte_0)}{(bu^2+2but+bt^2-4h)^2} < 0,$$

$$\frac{d\pi_M^{A*}}{du} = \frac{h(u+t)(bc-a+bte_0)^2}{(bu^2+2but+bt^2-4h)^2} > 0,$$

$$\frac{d\pi_R^{A*}}{du} = \frac{4h^2(u+t)(bc-a+bte_0)^2}{(4h-bu^2-2but-bt^2)^3} > 0.$$

Part of the conclusion of Proposition 3 is proved. Then we show how E^{A*} varies with u. The first-order condition of E^{A*} on u is

$$\frac{dE^{A*}}{du} = \frac{\left[\begin{array}{l} h(a-bc-bte_0)(-2e_0b^2u^3-3e_0b^2u^2t+3cb^2u^2+6cb^2ut+e_0b^2t^3+ \\ 3cb^2t^2-3abu^2-6abut+8e_0hbu-3abt^2+12e_0hbt+4chb-4ah) \end{array}\right]}{(4h-bu^2-2but-bt^2)^3}.$$

Let

$A_1 = -2e_0b^2u^3 - 3e_0b^2u^2t + 3cb^2u^2 + 6cb^2ut + e_0b^2t^3 + 3cb^2t^2 - 3abu^2 - 6abut + 8e_0hbu - 3abt^2 + 12e_0hbt + 4chb - 4ah.$

We have shown that $a - bc - bte_0 > 0$ and $4h - bu^2 - 2but - bt^2 > 0$. That means, the sign of $\frac{dE^{A*}}{du}$ depends on the size of A_1. Obviously, A_1 is a linear function of e_0. Meanwhile, we have

$$\left.\frac{dA_1}{du}\right|_{e_0=0} = -(a-bc)(3bu^2+6but+3bt^2+4h) < 0,$$

$$\left.\frac{dA_1}{du}\right|_{e_0=\tilde{e}_0} = \frac{2(u+t)(a-bc)(4h-bu^2-2but-bt^2)}{t} > 0.$$

The above evidences indicate that there exists a threshold $\tilde{e}_2$ such that $A_1 < 0$ when $e_0 < \tilde{e}_2$, and $A_1 > 0$ when $e_0 > \tilde{e}_2$. Summarizing the above conditions, we conclude that if $e_0 < \tilde{e}_2$, then, E^{A*} is decreasing in u; otherwise, E^{A*} is increasing in u. By combining the above results, we complete the proof of Proposition 3.

Proof of Proposition 4. Differentiating w^{A*} with respect to t yields

$$\frac{dw^{A*}}{dt} = \frac{2h(4e_0h - 2at - 2au + 2bcu + 2bct - bu^2e_0 + bt^2e_0)}{(bu^2+2but+bt^2-4h)^2}.$$

It can be seen that $\frac{dw^{A*}}{dt}$ is a linear function of e_0. Further, we have $\frac{dw^{A*}}{dt}\big|_{e_0=0} = -\frac{4h(u+t)(a-bc-bte_0)}{(bu^2+2but+bt^2-4h)^2} < 0$ and $\frac{dw^{A*}}{dt}\big|_{e_0=\tilde{e}_0} = \frac{2h(a-bc)}{bt(4h-bu^2-2but-bt^2)} > 0$.

The inequalities imply that there exists a threshold $\tilde{e}_3$ such that if $e_0 < \tilde{e}_3$, then w^{A*} is decreasing in t; otherwise, w^{A*} is increasing in t.

Similarly, differentiating p^{A*} and e^{A*} with respect to t yields

$$\frac{dp^{A*}}{dt} = \frac{h(4e_0h - 2at - 2au + 2bcu + 2bct - bu^2e_0 + bt^2e_0)}{(bu^2 + 2but + bt^2 - 4h)^2},$$

$$\frac{de^{A*}}{dt} = \frac{\left[\begin{array}{c}e_0b^2u^3 + 2e_0b^2u^2t - cb^2u^2 + e_0b^2ut^2 - 2cb^2ut - cb^2t^2 + \\ abu^2 + 2abut - 4e_0hbu + abt^2 - 8e_0hbt - 4chb + 4ah\end{array}\right]}{(bu^2 + 2but + bt^2 - 4h)^2}.$$

$\frac{dp^{A*}}{dt}$ and $\frac{de^{A*}}{dt}$ are all linear functions of e_0. Meanwhile, we obtain the following inequalities:

$$\frac{dp^{A*}}{dt}\bigg|_{e_0=0} = -\frac{2h(u+t)(a-bc)}{(bu^2+2but+bt^2-4h)^2} < 0,$$

$$\frac{dp^{A*}}{dt}\bigg|_{e_0=\tilde{e}_0} = \frac{h(a-bc)}{bt(4h-bu^2-2but-bt^2)} > 0,$$

$$\frac{de^{A*}}{dt}\bigg|_{e_0=0} = \frac{(a-bc)(bu^2+2but+bt^2+4h)}{(bu^2+2but+bt^2-4h)^2} > 0,$$

$$\frac{de^{A*}}{dt}\bigg|_{e_0=\tilde{e}_0} = -\frac{(u+t)(a-bc)}{t(4h-bu^2-2but-bt^2)} < 0.$$

The above evidences confirm the relationship between p^{A*}, e^{A*}, and t in Proposition 4. Further, by taking the derivation of π_M^{A*} with respect to t, we know that $\frac{d\pi_M^{A*}}{dt} = \frac{h(a-bc-bte_0)(au+at-4e_0h-bcu-bct+bu^2e_0+bute_0)}{(bu^2+2but+bt^2-4h)^2}$ is a linear function with respect to e_0. Let $A_2 = au + at - 4e_0h - bcu - bct + bu^2e_0 + bute_0$. Since $a - bc - bte_0 > 0$ always holds, the sign of $\frac{d\pi_M^{A*}}{dt}$ depends on A_2. At the two endpoints of the valid interval of e_0, the size of $\frac{dA_2}{dt}$ satisfies the following inequalities:

$$\frac{dA_2}{dt}\bigg|_{e_0=0} = (u+t)(a-bc) > 0, \quad \frac{dA_2}{dt}\bigg|_{e_0=\tilde{e}_0} = -\frac{(a-bc)(4h-bu^2-2but-bt^2)}{bt} < 0.$$

The results indicate that $\frac{d\pi_M^{A*}}{dt} > 0$ and $\frac{d\pi_M^{A*}}{dt} < 0$ establish when e_0 is low and high, respectively. In a similar way, we can also prove that $\frac{d\pi_R^{A*}}{dt} > 0$ and $\frac{d\pi_R^{A*}}{dt} < 0$ when e_0 is less than and greater than a certain threshold. Moreover, $\frac{dE^{A*}}{dt} < 0$ always holds in the valid interval of e_0. Combining the above results, Proposition 4 is clearly proved.

Proof of Proposition 5. The proof of Proposition 5 is similar to that of Proposition 1; thus, we have omitted it.

Proof of Proposition 6. It is easy to see that the expressions of w^{S*}, p^{S*}, e^{S*}, π_M^{S*}, and π_R^{S*} do not contain parameter e_0. Thus, w^{S*}, p^{S*}, e^{S*}, π_M^{S*}, and π_R^{S*} are unchanging in e_0. Moreover, we have $\frac{dE^{S*}}{de_0} = \frac{h(a-bc)}{4h-bs^2-2bsu-bu^2} > 0$, which means that E^{S*} is increasing in e_0. The proof is complete.

Proof of Proposition 7. It can be seen from $\pi_M^{S*} > 0$ and $e^{S*} > 0$ that both $a - bc$ and $4h - bs^2 - 2bsu - bu^2$ are greater than 0. Therefore, we have:

$$\frac{dw^{S*}}{du} = -\frac{4h(s+u)(a-bc)}{(bs^2+2bsu+bu^2-4h)^2} < 0,$$

$$\frac{de^{S*}}{du} = \frac{(a-bc)(bs^2+2bsu+bu^2+4h)}{(bs^2+2bsu+bu^2-4h)^2} > 0,$$

$$\frac{dp^{S*}}{du} = -\frac{2h(s+u)(a-bc)}{(bs^2+2bsu+bu^2-4h)^2} < 0,$$

$$\frac{d\pi_M^{S*}}{du} = \frac{h(s+u)(a-bc)^2}{(bs^2+2bsu+bu^2-4h)^2} > 0,$$

$$\frac{d\pi_R^{S*}}{du} = \frac{4h^2(s+u)(a-bc)^2}{(4h-bs^2-2bsu-bu^2)^3} > 0.$$

The changes of w^{S*}, p^{S*}, e^{S*}, π_M^{S*}, and π_R^{S*} with respect to u are vindicated. The first order derivative of the E^{S*} with respect to u is

$$\frac{dE^{S*}}{du} = -\frac{\begin{bmatrix} h(a-bc)((3bs^2+6bsu+3bu^2+4h)a+2e_0b^2s^3+6e_0b^2s^2u-3cb^2s^2+ \\ 6e_0b^2su^2-6cb^2su+2e_0b^2u^3-3cb^2u^2-8e_0hbs-8e_0hbu-4chb) \end{bmatrix}}{(4h-bs^2-2bsu-bu^2)^3}.$$

Let $A_3 = (3bs^2+6bsu+3bu^2+4h)a + 2e_0b^2s^3 + 6e_0b^2s^2u - 3cb^2s^2 + 6e_0b^2su^2 - 6cb^2su + 2e_0b^2u^3 - 3cb^2u^2 - 8e_0hbs - 8e_0hbu - 4chb.$

Since $a - bc > 0$ and $4h - bs^2 - 2bsu - bu^2 > 0$, the sign of $\frac{dE^{S*}}{du}$ depends on the size of A_3. It is obvious to see that A_3 is a linearly increasing function of a. From $q^{S*} > 0$, we know that $a > bc$. Let $\underline{a} = bc$. It is clear to see that $A_3|_{a=\underline{a}} = -2be_0(s+u)(4h-bs^2-2bsu-bu^2) < 0$. Recalling that A_3 is a continuously increasing function with respect to a. Meanwhile, the valid range of a has no upper limit. Combing the above conditions, we conclude that $A_3 < 0$ and $A_3 > 0$ when a is below and above a certain threshold, respectively. The above results imply that if $a < \tilde{a}_1$, then E^{S*} is increasing in u; otherwise, E^{S*} is decreasing in u. We complete the proof of Proposition 7.

Proof of Proposition 8. We derive the following inequalities by differentiating the variables with respect to s:

$$\frac{dw^{S*}}{ds} = -\frac{4h(s+u)(a-bc)}{(bs^2+2bsu+bu^2-4h)^2} < 0,$$

$$\frac{de^{S*}}{ds} = \frac{(a-bc)(bs^2+2bsu+bu^2+4h)}{(bs^2+2bsu+bu^2-4h)^2} > 0,$$

$$\frac{dp^{S*}}{ds} = -\frac{2h(s+u)(a-bc)}{(bs^2+2bsu+bu^2-4h)^2} < 0, \quad \frac{d\pi_M^{S*}}{ds} = \frac{h(s+u)(a-bc)^2}{(bs^2+2bsu+bu^2-4h)^2} > 0,$$

$$\frac{d\pi_R^{S*}}{ds} = \frac{(4h^2(s+u)(a-bc)^2)}{(4h-bs^2-2bsu-bu^2)^3} > 0.$$

Moreover, similar to the proof of Proposition 7, we can prove that E^{S*} is increasing and decreasing in a is $a < \tilde{a}_2$ and $a > \tilde{a}_2$, respectively. Thus, we can obtain Proposition 8.

Proof of Proposition 9. The gap between w^A and w^S is

$$w^A - w^S = -\frac{\left[\begin{array}{c} 2h(at^2 - as^2 - 2asu + 2aut - 4te_0h + bcs^2 - bct^2 + bs^2te_0 + bu^2te_0 + \\ 2bcsu - 2bcut + 2bsute_0) \end{array}\right]}{(4h - bs^2 - 2bsu - bu^2)(4h - bu^2 - 2but - bt^2)}.$$

It is easy to see that $w^A - w^S$ is a linear function of e_0. Meanwhile, we can verify that

$$(w^A - w^S)\big|_{e_0=0} = \frac{2h(s-t)(a-bc)(s+2u+t)}{(4h - bs^2 - 2bsu - bu^2)(4h - bu^2 - 2but - bt^2)},$$

$$(w^A - w^S)\big|_{e_0=\bar{e}_0} = \frac{2h(a-bc)}{b(4h - bs^2 - 2bsu - bu^2)} > 0.$$

The above conditions imply that when $s > t$, $w^A > w^S$ always holds; otherwise, $w^A > w^S$ holds only when e_0 is greater than a certain threshold. The same method can prove the relationship between p^A and p^S; therefore, we omit the analysis process. Further, taking the difference between e^A and e^S yields the following formula:

$$e^A - e^S = \frac{\left[\begin{array}{c} (e_0b^2s^2ut + cb^2s^2u + e_0b^2s^2t^2 + cb^2s^2t + 2e_0b^2su^2t + cb^2su^2 + 2e_0b^2sut^2 - cb^2st^2 + \\ e_0b^2u^3t + e_0b^2u^2t^2 - cb^2u^2t - cb^2ut^2 - abs^2u - abs^2t - absu^2 + abst^2 + 4chbs + \\ abu^2t + abut^2 - 4e_0hbut - 4e_0hbt^2 - 4chbt - 4ahs + 4aht) \end{array}\right]}{(bs^2 + 2bsu + bu^2 - 4h)(bu^2 + 2but + bt^2 - 4h)}.$$

It is obvious to see that $e^A - e^S$ is a linear function of e_0. When $e_0 = 0$ and $e_0 = \bar{e}_0$, the values of $e^A - e^S$ are as follows:

$$(e^A - e^S)\big|_{e_0=0} = -\frac{(s-t)(a-bc)(4h + bu^2 + bsu + bst + but)}{(4h - bs^2 - 2bsu - bu^2)(4h - bu^2 - 2but - bt^2)},$$

$$(e^A - e^S)\big|_{e_0=\bar{e}_0} = -\frac{(s+t)(a-bc)}{4h - bs^2 - 2bsu - bu^2} < 0.$$

The relationship between e^A and e^S in Proposition 9 obviously holds. In addition, the above analysis results support the conclusions in Proposition 9.

Proof of Proposition 10. Taking the difference between π_M^A and π_M^S yields the following equation:

$$\pi_M^A - \pi_M^S = \frac{\left[\begin{array}{c} h(bt^2(4h - bs^2 - 2bsu - bu^2)e_0^2 - 2t(a-bc)(4h - bs^2 - 2bsu - bu^2)e_0 - \\ (s-t)(a-bc)^2(s+2u+t)) \end{array}\right]}{2(4h - bs^2 - 2bsu - bu^2)(4h - bu^2 - 2but - bt^2)}.$$

Let $A_4(e_0) = bt^2(4h - bs^2 - 2bsu - bu^2)e_0^2 - 2t(a-bc)(4h - bs^2 - 2bsu - bu^2)e_0 - (s-t)(a-bc)^2(s+2u+t)$. It is clear to see that $A_4(e_0)$ is a quadratic function with respect to e_0. Since both $4h - bs^2 - 2bsu - bu^2$ and $(a-bc)(4h - bs^2 - 2bsu - bu^2)$ are greater than 0, the quadratic and first-order coefficients of $A_4(e_0)$ are greater than 0 and less than 0, respectively. The abscissa of the symmetry axis of the function $A_4(e_0)$ is $\tilde{x} = \frac{a-bc}{bt} = \bar{e}_0 > 0$. When $s > t$, the constant term of $A_4(e_0)$ is lower than 0. $A_4(e_0)$ is a continuously decreasing function of e_0 in the interval $(0, \bar{e}_0]$. Meanwhile, we also have

$$A_4(e_0)\big|_{e_0=0} = -(s-t)(a-bc)^2(s+2u+t) < 0,$$

$$A_4(e_0)\Big|_{e_0=\bar{e}_0} = -\frac{(a-bc)^2(4h-bu^2-2but-bt^2)}{b} < 0.$$

That means, $A_4(e_0) < 0$ always holds when $\bar{e}_0 \in (0, \bar{e}_0]$. Since

$$\pi_M^A - \pi_M^S = \frac{hA_4(e_0)}{2(4h-bs^2-2bsu-bu^2)(4h-bu^2-2but-bt^2)},$$

$\pi_M^A - \pi_M^S < 0$ obviously holds.

When $s < t$, the constant term of $A_4(e_0)$ is greater than 0. In addition, the following inequalities establish:

$$A_4(e_0)\Big|_{e_0=0} = -(s-t)(a-bc)^2(s+2u+t) > 0,$$

$$A_4(e_0)\Big|_{e_0=\bar{e}_0} = -\frac{(a-bc)^2(4h-bu^2-2but-bt^2)}{b} < 0.$$

Therefore, $A_4(e_0) > 0$ and $A_4(e_0) < 0$ establish when e_0 is greater than and less than a certain threshold. In summary, there exists a threshold $\tilde{e}_{10}$ such that $\pi_M^A - \pi_M^S > 0$ holds if $e_0 < \tilde{e}_{11}$; otherwise, $\pi_M^A - \pi_M^S < 0$ holds.

In a similar way, we can also prove that $\pi_R^A - \pi_R^S < 0$ always holds when $s > t$. However, when $s > t$, it has the following features: there exists a threshold $\tilde{e}_{11}$ such that if $e_0 < \tilde{e}_{12}$, then $\pi_R^A - \pi_R^S > 0$; otherwise, $\pi_R^A - \pi_R^S < 0$.

Similar to the above analysis, we can prove that $E^A - E^S$ is decreasing in e_0 when $e_0 \in (0, \bar{e}_0]$. In addition, when $e_0 = 0$, we have

$$(E^A - E^S)\Big|_{e_0=0} = \frac{h(s+u)(a-bc)^2}{(bu^2+2bsu+bs^2-4h)^2} - \frac{h(t+u)(a-bc)^2}{(bu^2+2but+bt^2-4h)^2}.$$

Let $A_5(x) = \frac{h(x+u)(a-bc)^2}{(bx^2+2bxu+bu^2-4h)^2}$. The first derivative of $A_5(x)$ with respect to x is $\frac{dA_5(x)}{dx} = \frac{(h(a-bc)^2(3bx^2+6bxu+3bu^2+4h))}{(4h-bx^2-2bxu-bu^2)^3} > 0$. The above findings imply that if $s > t$, then $(E^A - E^S)\Big|_{e_0=0} > 0$; otherwise, $(E^A - E^S)\Big|_{e_0=0} < 0$. Meanwhile, it is easy to verify that $(E^A - E^S)\Big|_{e_0=\bar{e}_0} < 0$. Combining the above information, we obtain the results stated in Proposition 10.

References

1. Fan, X.; Chen, K.; Chen, Y. Is price commitment a better solution to control carbon emissions and promote technology investment? *Manag. Sci.* **2022**, *69*, 325–341. [CrossRef]
2. He, N.; Jiang, Z.; Huang, S.; Li, K. Evolutionary game analysis for government regulations in a straw-based bioenergy supply chain. *Int. J. Prod. Res.* **2022**, 1–22. [CrossRef]
3. Sun, L.; Chen, W. Impact of carbon tax on CCUS source-sink matching: Finding from the improved ChinaCCS DSS. *J. Clean. Prod.* **2022**, *333*, 130027. [CrossRef]
4. Zhang, S.; Zhuang, Y.; Tao, R.; Liu, L.; Zhang, L.; Du, J. Multi-objective optimization for the deployment of carbon capture utilization and storage supply chain considering economic and environmental performance. *J. Clean. Prod.* **2020**, *270*, 122481. [CrossRef]
5. Qureshi, F.; Yusuf, M.; Kamyab, H.; Zaidi, S.; Khalil, M.; Khan, M.; Alam, M.; Masood, F.; Bazli, L.; Chelliapan, S.; et al. Current trends in hydrogen production, storage and applications in India: A review. *Sustain. Energy Technol. Assess.* **2022**, *53*, 102677. [CrossRef]
6. Liu, B.; Liu, S.; Xue, B.; Lu, S.; Yang, Y. Formalizing an integrated decision-making model for the risk assessment of carbon capture, utilization, and storage projects: From a sustainability perspective. *Appl. Energy* **2021**, *303*, 117624. [CrossRef]
7. Qureshi, F.; Yusuf, M.; Kamyab, H.; Vo, D.; Chelliapan, S.; Joo, S.; Vasseghian, Y. Latest eco-friendly avenues on hydrogen production towards a circular bioeconomy: Currents challenges, innovative insights, and future perspectives. *Renew. Sustain. Energy Rev.* **2022**, *168*, 112916. [CrossRef]
8. Ostovari, H.; Müller, L.; Mayer, F.; Bardow, A. A climate-optimal supply chain for CO_2 capture, utilization, and storage by mineralization. *J. Clean. Prod.* **2022**, *360*, 131750. [CrossRef]

9. Chen, D.; Jiang, M. Assessing the socioeconomic effects of carbon capture, utility and storage investment from the perspective of carbon neutrality in China. *Earth's Future* **2022**, *10*, e2021EF002523. [CrossRef]
10. Yang, L.; Xu, M.; Yang, Y.; Fan, J.; Zhang, X. Comparison of subsidy schemes for carbon capture utilization and storage (CCUS) investment based on real option approach: Evidence from China. *Appl. Energy* **2019**, *255*, 113828. [CrossRef]
11. Liu, L.; Feng, L.; Jiang, T.; Zhang, Q. The impact of supply chain competition on the introduction of clean development mechanisms. *Transp. Res. Part E Logist. Transp. Rev.* **2021**, *155*, 102506. [CrossRef]
12. Hasan, M.; Roy, T.; Daryanto, Y.; Wee, H. Optimizing inventory level and technology investment under a carbon tax, cap-and-trade and strict carbon limit regulations. *Sustain. Prod. Consum.* **2020**, *25*, 604–621. [CrossRef]
13. Ding, H.; Zhao, Q.; An, Z.; Tang, O. Collaborative mechanism of a sustainable supply chain with environmental constraints and carbon caps. *Int. J. Prod. Econ.* **2016**, *181*, 191–207. [CrossRef]
14. Chen, X.; Yang, H.; Wang, X.; Choi, T. Optimal carbon tax design for achieving low carbon supply chains. *Ann. Oper. Res.* **2020**, 1–28. [CrossRef]
15. Xu, L.; Wang, C.; Miao, Z.; Chen, J. Governmental subsidy policies and supply chain decisions with carbon emission limit and consumer's environmental awareness. *RAIRO Oper. Res.* **2019**, *53*, 1675–1689. [CrossRef]
16. Zhang, S.; Wang, C.; Yu, C.; Ren, Y. Governmental cap regulation and manufacturer's low carbon strategy in a supply chain with different power structures. *Comput. Ind. Eng.* **2019**, *134*, 27–36. [CrossRef]
17. Tang, R.; Yang, L. Impacts of financing mechanism and power structure on supply chains under cap-and-trade regulation. *Transp. Res. Part E Logist. Transp. Rev.* **2020**, *139*, 101957. [CrossRef]
18. Liu, H.; Kou, X.; Xu, G.; Qiu, X.; Liu, H. Which emission reduction mode is the best under the carbon cap-and-trade mechanism? *J. Clean. Prod.* **2021**, *314*, 128053. [CrossRef]
19. Xu, X.; Zhang, M.; Chen, L.; Yu, Y. The region-cap allocation and delivery time decision in the marketplace mode under the cap-and-trade regulation. *Int. J. Prod. Econ.* **2022**, *247*, 108407. [CrossRef]
20. Krass, D.; Nedorezov, T.; Ovchinnikov, A. Environmental taxes and the choice of green technology. *Prod. Oper. Manag.* **2013**, *22*, 1035–1055. [CrossRef]
21. Pal, R.; Saha, B. Pollution tax, partial privatization and environment. *Resour. Energy Econ.* **2015**, *40*, 19–35. [CrossRef]
22. Zhou, Y.; Hu, F.; Zhou, Z. Pricing decisions and social welfare in a supply chain with multiple competing retailers and carbon tax policy. *J. Clean. Prod.* **2018**, *190*, 752–777. [CrossRef]
23. Yu, W.; Shang, H.; Han, R. The impact of carbon emissions tax on vertical centralized supply chain channel structure. *Comput. Ind. Eng.* **2020**, *141*, 106303. [CrossRef]
24. Zhou, X.; Wei, X.; Lin, J.; Tian, X.; Lev, B.; Wang, S. Supply chain management under carbon taxes: A review and bibliometric analysis. *Omega* **2021**, *98*, 102295. [CrossRef]
25. Gopalakrishnan, S.; Granot, D.; Granot, F.; Sosic, G.; Cui, H. Incentives and emission responsibility allocation in supply chains. *Manag. Sci.* **2021**, *67*, 4172–4190. [CrossRef]
26. Bao, B.; Ma, J.; Goh, M. Short- and long-term repeated game behaviours of two parallel supply chains based on government subsidy in the vehicle market. *Int. J. Prod. Res.* **2020**, *58*, 7507–7530. [CrossRef]
27. Zhang, J.; Huang, J. Vehicle product-line strategy under government subsidy programs for electric/hybrid vehicles. *Transp. Res. Part E Logist. Transp. Rev.* **2021**, *146*, 102221. [CrossRef]
28. Ma, C.; Yang, H.; Zhang, W.; Huang, S. Low-carbon consumption with government subsidy under asymmetric carbon emission information. *J. Clean. Prod.* **2021**, *318*, 128423. [CrossRef]
29. Xu, C.; Wang, C.; Huang, R. Impacts of horizontal integration on social welfare under the interaction of carbon tax and green subsidies. *Int. J. Prod. Econ.* **2019**, *222*, 107506. [CrossRef]
30. Wang, X.; Qie, S. When to invest in carbon capture and storage: A perspective of supply chain. *Comput. Ind. Eng.* **2018**, *123*, 26–32. [CrossRef]
31. Miao, Z.; Mao, H.; Fu, K.; Wang, Y. Remanufacturing with trade-ins under carbon regulations. *Comput. Oper. Res.* **2016**, *89*, 253–268. [CrossRef]
32. Anand, K.; Giraud-Carrier, F. Pollution regulation of competitive markets. *Manag. Sci.* **2020**, *66*, 4193–4206. [CrossRef]
33. Hu, X.; Yang, Z.; Sun, J.; Zhang, Y. Carbon tax or cap-and-trade: Which is more viable for chinese remanufacturing industry? *J. Clean. Prod.* **2019**, *243*, 118606. [CrossRef]
34. Sun, H.; Yang, J. Optimal decisions for competitive manufacturers under carbon tax and cap-and-trade policies. *Comput. Ind. Eng.* **2021**, *156*, 107244. [CrossRef]
35. Chen, Y.; Wang, C.; Nie, P.; Chen, Z. A clean innovation comparison between carbon tax and cap-and-trade system. *Energy Strategy Rev.* **2020**, *29*, 100483. [CrossRef]
36. Zhou, J.; Wu, D.; Chen, W. Cap and trade versus carbon tax: An analysis based on a CGE model. *Comput. Econ.* **2021**, *59*, 853–885. [CrossRef]
37. Yu, Y.; Li, X.; Xu, X. Reselling or marketplace mode for an online platform: The choice between cap-and-trade and carbon tax regulation. *Ann. Oper. Res.* **2022**, *310*, 293–329. [CrossRef]
38. Yin, X.; Chen, X.; Xu, X.; Zhang, L. Tax or subsidy? Optimal carbon emission policy: A supply chain perspective. *Sustainability* **2020**, *12*, 1548. [CrossRef]

39. Huang, X.; Tan, T.; Toktay, L. Carbon leakage: The impact of asymmetric regulation on carbon-emitting production. *Prod. Oper. Manag.* **2020**, *30*, 1886–1903. [CrossRef]
40. Ma, J.; Hou, Y.; Wang, Z.; Yang, W. Pricing strategy and coordination of automobile manufacturers based on government intervention and carbon emission reduction. *Energy Policy* **2021**, *148*, 111919. [CrossRef]
41. Zhu, X.; Chiong, R.; Wang, M.; Liu, K.; Ren, M. Is carbon regulation better than cash subsidy? The case of new energy vehicles. *Transp. Res. Part A Policy Pract.* **2021**, *146*, 170–192. [CrossRef]
42. Guo, J.; Wang, G.; Wang, Z.; Liang, C.; Gen, M. Research on remanufacturing closed loop supply chain based on incentive-compatibility theory under uncertainty. *Ann. Oper. Res.* **2022**, 1–22. [CrossRef]
43. Dou, G.; Choi, T. Does implementing trade-in and green technology together benefit the environment? *Eur. J. Oper. Res.* **2021**, *295*, 517–533. [CrossRef]
44. Wang, X.; Zhang, S. The interplay between subsidy and regulation under competition. *IEEE Trans. Syst. Man Cybern. Syst.* **2022**, *53*, 1038–1050. [CrossRef]
45. Wu, H.; Sun, Y.; Su, Y. Which is the best supply chain policy: Carbon tax, or a low-carbon subsidy? *Sustainability* **2022**, *14*, 6312. [CrossRef]
46. Li, P.; Xu, S.; Liu, L. Channel structure and greening in an omni-channel tourism supply chain. *J. Clean. Prod.* **2022**, *375*, 134136. [CrossRef]
47. Liu, L.; Feng, L.; Xu, B.; Deng, W. Operation strategies for an omni-channel supply chain: Who is better off taking on the online channel and offline service? *Electron. Commer. Res. Appl.* **2020**, *39*, 100918. [CrossRef]
48. Zhang, S.; Yu, Y.; Zhu, Q.; Qiu, C.; Tian, A. Green innovation mode under carbon tax and innovation subsidy: An evolutionary game analysis for portfolio policies. *Sustainability* **2020**, *12*, 1385. [CrossRef]
49. Yao, Q.; Shao, L.; Yin, Z.; Wang, J.; Lan, Y. Strategies of property developers and governments under carbon tax and subsidies. *Front. Environ. Sci.* **2022**, *10*, 916352. [CrossRef]
50. Yu, P. Carbon tax/subsidy policy choice and its effects in the presence of interest groups. *Energy Policy* **2020**, *147*, 111886. [CrossRef]
51. Chen, W.; Hu, Z. Using evolutionary game theory to study governments and manufacturers' behavioral strategies under various carbon taxes and subsidies. *J. Clean. Prod.* **2018**, *201*, 123–141. [CrossRef]
52. Li, H.; Peng, W. Carbon tax, subsidy, and emission reduction: Analysis based on DSGE model. *Complexity* **2020**, *2020*, 6683482. [CrossRef]
53. Guo, J.; Huang, R. A carbon tax or a subsidy? Policy choice when a green firm competes with a high carbon emitter. *Environ. Sci. Pollut. Res.* **2021**, *29*, 12845–12852. [CrossRef] [PubMed]
54. Shu, T.; Huang, C.; Chen, S.; Wang, S.; Lai, K. Trade-old-for-remanufactured closed-loop supply chains with carbon tax and government subsidies. *Sustainability* **2018**, *10*, 3935. [CrossRef]

Disclaimer/Publisher's Note: The statements, opinions and data contained in all publications are solely those of the individual author(s) and contributor(s) and not of MDPI and/or the editor(s). MDPI and/or the editor(s) disclaim responsibility for any injury to people or property resulting from any ideas, methods, instructions or products referred to in the content.

Article

Expansion of Geological CO₂ Storage Capacity in a Closed Aquifer by Simultaneous Brine Production with CO₂ Injection

Seungpil Jung

SK Earthon Co., Ltd., Seoul 03188, Republic of Korea; phil.jung@sk.com; Tel.: +82-2-2121-5114

Abstract: Structural trapping is the primary mechanism for intensive CO_2 sequestration in saline aquifers. This is the foundation for increasing global CO_2 storage; gradual switch to preferable trapping mechanisms, such as residual saturation, dissolution, and mineral trapping, will require a long-time scale. The major constraints limiting the storage capacity of structural trapping are formation pressure and structure size. Over-pressure owing to CO_2 injection causes a disruption of seal integrity indicating a failure in geological sequestration. The other constraint on storage capacity is a spill point determining geological storage volume. Overflowing CO_2, after filling the storage volume, migrates upward along the aquifer geometry with buoyancy. This study proposes a methodology to maximize CO_2 storage capacity of a geological site with a substructure created by an interbedded calcareous layer below spill point. This study provides various conceptual schemes, i.e., no brine production, simultaneous brine production and pre-injection brine production, for geological CO_2 storage. By the comparative analysis, location of brine producer, production rate, and distance between injector and producer are optimized. Therefore, the proposed scheme can enhance CO_2 storage capacity by 68% beyond the pressure and migration limits by steering CO_2 plume and managing formation pressure.

Keywords: pressure-limited capacity; migration-limited capacity; brine production; interbedded impermeable layer; spill point

Citation: Jung, S. Expansion of Geological CO₂ Storage Capacity in a Closed Aquifer by Simultaneous Brine Production with CO₂ Injection. *Sustainability* **2023**, *15*, 3499. https://doi.org/10.3390/su15043499

Academic Editors: Zilong Liu, Meixia Shan and Yakang Jin

Received: 19 December 2022
Revised: 12 January 2023
Accepted: 9 February 2023
Published: 14 February 2023

Copyright: © 2023 by the author. Licensee MDPI, Basel, Switzerland. This article is an open access article distributed under the terms and conditions of the Creative Commons Attribution (CC BY) license (https:// creativecommons.org/licenses/by/ 4.0/).

1. Introduction

Global climate change caused by rapidly increasing greenhouse gas content in the atmosphere poses a great risk to humans. To mitigate this risk, the Paris Climate Change Accord defines the limit of temperature increase to avoid irreversible changes and encourages each member country to meet its own CO_2-reduction targets [1]. For mitigation measures, energy transition from fossil fuels to renewable energy has progressed, but it can only reduce the emission amount of greenhouse gases. Reducing the emission amount alone is not sufficient, and hence, a measure to reduce the overall greenhouse gas content is required. Carbon capture, utilization, and storage (CCUS) is recommended as a key technique that enables the removal of significant amounts of CO_2 from a project perspective.

Aquifers are considered prospective sites for geological sequestration of greenhouse gases. Aquifers containing brine can dissolve CO_2. Particularly, aquifers can sequester CO_2 with high storage efficiency by maintaining the injected CO_2 in a liquid or supercritical phase owing to the hydrostatic pressure and geothermal temperature. However, it has a technical problem of salt precipitation [2], which may clog pore throat resulting in decreasing CO_2 injectivity, and a limitation on the injection volume. The limitation is that the injection volume should not exceed the fracturing pressure of the formation, particularly in a closed aquifer. Injected fluid can induce fractures in formations [3], micro-seismic events [4], and earthquakes [5]. Additionally, a formation pressure higher than the cap rock capillary entry pressure can break the stability of the CO_2 storage system. The fracturing pressure against regional pressurization due to CO_2 injection is a key constraint on the CO_2 storage capacity.

The size of the geological structure is another constraint on the CO_2 storage. The CO_2 storage resource management system (SRMS) highlights a high confidence in the commercial storage of geological formations, as supported by confinement [6]. According to the principle of hydrocarbon fill-and-spill, the remaining CO_2 after displacing the native formation water of the structure migrates upward beyond the spill point. It may be untraceable during the migration along the aquifer; i.e., whether it reaches the biosphere may not be clear. During migration, it can be trapped by residual saturation, mineralization, dissolution, or other structures. However, it can also leak to the surface. The geological structure is the most reliable confinement for geological storage projects. The structure size is generally regarded as pre-determined, but it can be increased if the geological conditions meet specific requirements. Interbedded impermeable layer below the spill point is a potential opportunity for storage capacity expansion. If the injected CO_2 plume is steered to the substructure generated by the cap rock with the interbedded impermeable layer, the storage capacity can be increased without an additional injector.

A traditional strategy is to inject CO_2 and let it flow upwards [7]. Once an injector is drilled in the lower part of the aquifer, CO_2 tends to migrate upward because of its lower density than that of brine. If the upper seal rock functions effectively, injected CO_2 can be trapped in the structure. This concept is applied to depleted oil or gas reservoirs [8], which have a proven sealing structure and reduced formation pressure due to the long period of hydrocarbon production. However, it is challenging to apply this concept to aquifers. The increased aquifer pressure due to the injected CO_2 may result in the mechanical instability of the formation. To mitigate this risk, several concepts of pressure management were suggested in earlier studies [9–13].

Here, an injection scheme with an additional substructure below the spill point is investigated for the structural trap in the aquifer. It applies the strategy of pressure management to prevent the formation of fractures and to steer CO_2 plumes as long as the brine producer does not produce injected CO_2. This study investigates the potential candidate schemes, which are a combination of various types of injectors and producers, from the perspective of quantities of trapped CO_2 in the main structure and substructure, and evolution of the CO_2 plume.

2. Background
2.1. Storage Capacity

Many CO_2 resource management systems were developed for storage estimation, classification, and categorization by various institutes. Most existing storage-resource methodologies approach a volumetric basis [14–17]. It consists of calculating the geological volume in which CO_2 can potentially be stored, and the storage efficiency factor is applied to the calculated volume. The geological storage volume can be calculated numerically using input parameters, despite the inherent uncertainty. The problem is that the storage efficiency factor comprises complicated dynamic factors such as trapping mechanisms, boundary condition, number and type of well, water extraction, and other in situ parameters, which are all related. The volumetric method can only be utilized if sufficient data from regional storage projects are available. Although storage efficiency factors can be revised to reflect pilot tests or actual injection projects, it is technically difficult to evaluate the actual efficiency factor of an individual project. Furthermore, the CCUS industry is still in its initial stage [18–21], and hence, a database for the efficiency factor was never constructed.

The dynamic approach can consider site-specific dynamic factors [22–24]. It simulates the dynamic behavior, which is the result of site operation based on the static model from geological data. The benefits of the dynamic approach are as follows:

- Field specific storage capacity;
- Optimization tool for the development design and operation plan;
- Sensitivity analysis on the uncertainties;
- Fate of CO_2 plume over time.

As with all the simulation methods, it still requires a high cost of computation time and resources and qualified input data for accurate results. However, it is preferable to avoid determining an ambiguous storage efficiency factor.

2.2. Injection Strategy and Pressure Management

The traditional injection scheme involves injecting CO_2 from the lower injector of the aquifer, then let CO_2 upward and trapped by the seal rock [7]. The injected CO_2 is driven to rise through the aquifer due to its lower density than that of the aquifer brine. This results in a larger contact area between the CO_2 and brine with a longer exposure time. Thus, the injected CO_2 plume can be trapped by a stable mechanism such as residual saturation, dissolution, or mineralization. However, the CO_2 quantities trapped by these mechanisms are difficult to measure. Additionally, the remaining mobile CO_2 not trapped by the above mechanisms may migrate outward from the storage site. The structural trap is a reliable storage site to apply the injection scheme of injecting CO_2 from the lower part of the aquifer without migration risk.

However, a structural trap is likely to be a closed system with boundaries by fault, pinch-out, or reservoir heterogeneity. If industrial-scale CO_2 is injected into the compartmentalized formation, it may cause a significant pressure build-up in the aquifer, which can trigger micro-seismic rupture, fault reactivation, and seal fracturing. This can limit the storage capacity of the project. Pressure management by brine production was proposed as a mitigation method. The common advantage of pressure management is that it increases the storage capacity by impeding the pressure build-up to the pressure limit. Additionally, it provides a margin between the actual pressure and estimated pressure limit from a conservative perspective. The methodology is categorized in detail as follows:

- Passive extraction;
- Simultaneous brine production;
- Pre-injection brine production.

The principle of passive extraction is to produce brine from the formation equaling the pressure build-up by CO_2 injection [12]. It exploits the pressure buildup to generate a pressure drawdown of the producer, which is drilled at a location close to the injector. This prevents overpressure near the producer, although its impact on pressure management is limited. The pressure propagates into the entire formation, and therefore, it is not sufficient to relieve the formation pressure by only limiting the pressure drawdown of the producer. Although passive extraction restricts CO_2 breakthrough in the producer, CO_2 breakthrough eventually occurs in the long cycle of the injection operation. This is much worse for pressure management.

Simultaneous brine production [9–11] is an alternative method to passive extraction. The methodology was verified through the industrial-scale project, Gorgon CCS in Australia [25]. As long as the producer is drilled sufficiently far away from the injector, it can produce more brine in a decisive manner with a lower risk of CO_2 breakthrough. However, the brine producer should be drilled in this position for effective pressure propagation for pressure management purposes. However, the efficiency of pressure management has a trade-off with the CO_2 breakthrough. The location of the brine producer should be optimized for each project.

Furthermore, the brine production scheme advances pre-injection brine production [13]. This creates the CO_2 storage site low-pressured zone which provides the same environment as the depleted gas or oil reservoir for the CO_2 storage site. Although it delays the CO_2 injection time in the project life cycle, it is an effective method of pressure management with additional benefits. First, it can reduce drilling costs because the injector can be utilized as a producer prior to injection. It also provides operational benefits such as a smaller area of review (AoR) and less post-injection monitoring. Additionally, through prior brine production as an extended pressure drawdown test, pressure drawdown data can be acquired to estimate the injectivity for actual CO_2 injection. Injection design using injectivity analysis can mitigate the risks of CO_2 injection projects.

In this study, a pressure management scheme for the target storage site is proposed considering the geological and engineering conditions. Several cases representing each injection scheme were applied to the target storage site. By analyzing the pressure behavior and evolution of the CO_2 plume, operational conditions, such as the location of the brine producer and brine production method, were optimized to enhance the CO_2 storage capacity.

3. Evaluation Method

3.1. Geological Description of the Target Aquifer

The target of this study is a hypothetical aquifer, the geological concept of which is based on actual field data. The aquifer was formed between the ancient delta system facies and the uplift carbonate platform facies. The aquifer formation is dominated by marine delta sediments, although there is also a carbonate platform with a stable lateral distribution. As the transgression continued to a late age, carbonate deposition did not occur. The underwater channel and river mouth bar within the delta front facies mainly developed until the seawater reached the carbonate compensation depth. The shore–marine delta deposition system generated thick layers that were stably developed with good lateral continuity. With this geological background, an interbedded calcareous layer can be deposited and a substructure can develop within the main structure. The aquifer structure, including the substructure considered in this study, is illustrated in Figure 1.

3.2. Storage Resources of This Research

This study focuses on the technical storage volume for comparison among storage scheme candidates regardless of project maturity, commerciality, regulations, and other social conditions. The technical storage volume is calculated using dynamic simulations, which can reflect the geological properties and operational conditions of the potential CO_2 sequestration project. To calculate the technical storage volume, the following assumptions on the specific criteria are applied.

3.2.1. Pressure Threshold

A cap rock is the weakest part of geological storage because it is generally more fragile than aquifer formations in terms of lithology [26]. The role of the cap rock is to provide sealing storage so that the injected CO_2 cannot migrate out of the geological storage. Thus, the mechanical stability of the cap is an important aspect of CO_2 geological storage projects. In this study, the upper part of the aquifer formation immediately below the cap rock was considered as a reference point. The fracturing of the cap rock is assumed to occur when the pressure at the point is higher than the fracturing pressure of the cap rock. The fracturing pressure of the cap rock can be estimated by formation leak-off test, destructive testing of the sampled core, mud loss while drilling, or empirical correlations [27]. For this research, the fracturing pressure with depth was estimated by information from offset wells, the seismic velocity, and the empirical correlation [28], which indicated that the fracturing pressure of the cap rock was 27.6 MPa.

For a more reasonable analysis, a geo-mechanical simulation integrated with flow simulation is required. Although the coupled model using the multiphase flow and the geo-mechanical process has been applied to the CO_2 sequestration study [29], several limitations, i.e., instability in low permeability zone and computational bottleneck in industry scaled model were issued [30]. This integrated model cannot guarantee the accuracy of the estimation despite its high computational cost. Considering the estimation uncertainty, 80% of the fracturing pressure of the cap rock was applied as the pressure threshold from a practical perspective. The CO_2 injection duration was from the initial injection to the shut-down point as soon as the local pressure of the reference point reached the pressure threshold for fracturing the cap rock.

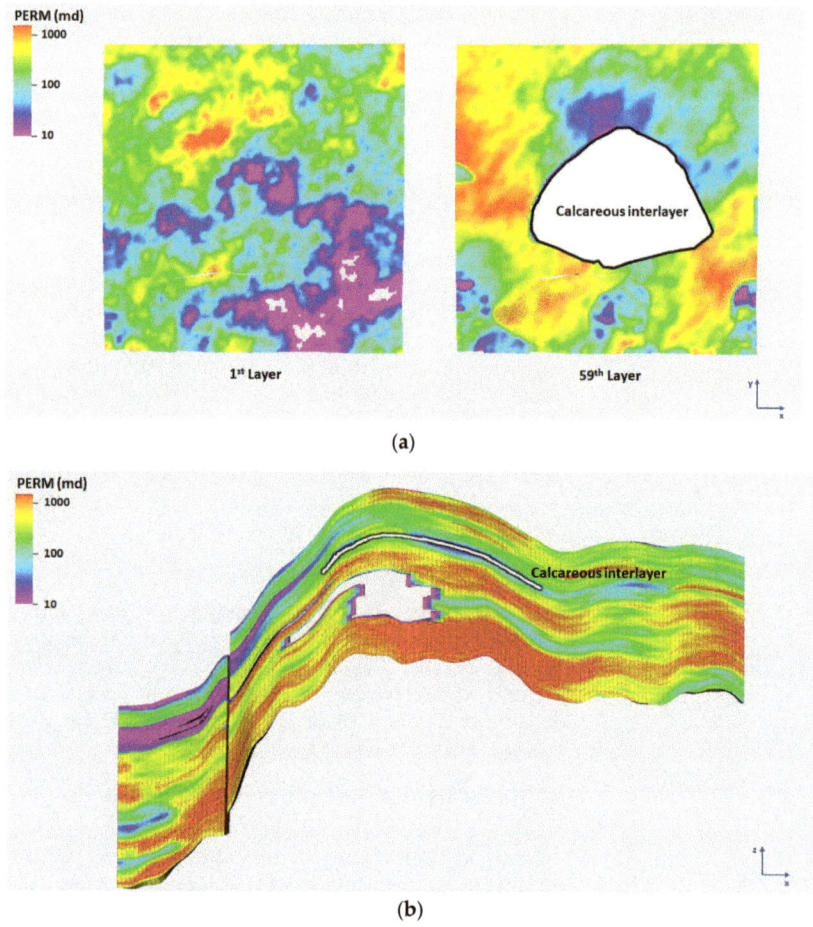

Figure 1. Permeability maps depicting the calcareous interlayer which is a cap rock of substructure. (**a**) Planar view (**Left**: top layer, **Right**: 59th layer), (**b**) cross sectional view.

3.2.2. Sequestration Mechanism

The injected CO_2 can be stored in the aquifer through the following four mechanisms:

- Structural and stratigraphic trap;
- Residual saturation trap;
- Dissolution trap;
- Geochemical trap.

Estimates obtained by numerical simulation have a wide range of uncertainty in CO_2 storage capacity, especially for residual traps, solution traps, and mineralization, which are caused by highly uncertain input data for the simulation. These trapping mechanisms categorized into the secondary mechanisms depend on the highly site-specific coefficients and take hundreds of years for the process [20]. The quantity of sequestrated CO_2 is related to the actual cash flow enabled by the government or the trade of carbon credit. From a conservative point of view, the CO_2 quantity in a structure is certified by the storage capacity of the aquifer. The mobile phase and the residual trapped inside the structure were applied as the criteria of a CO_2 storage capacity. For this purpose, the aquifer was categorized into three regions, as shown in Figure 2. Regions 1 and 3 were the main structures above the spill point and beyond the structure diverging outward, respectively. Region 2 was the other region, including the substructure below the main structure.

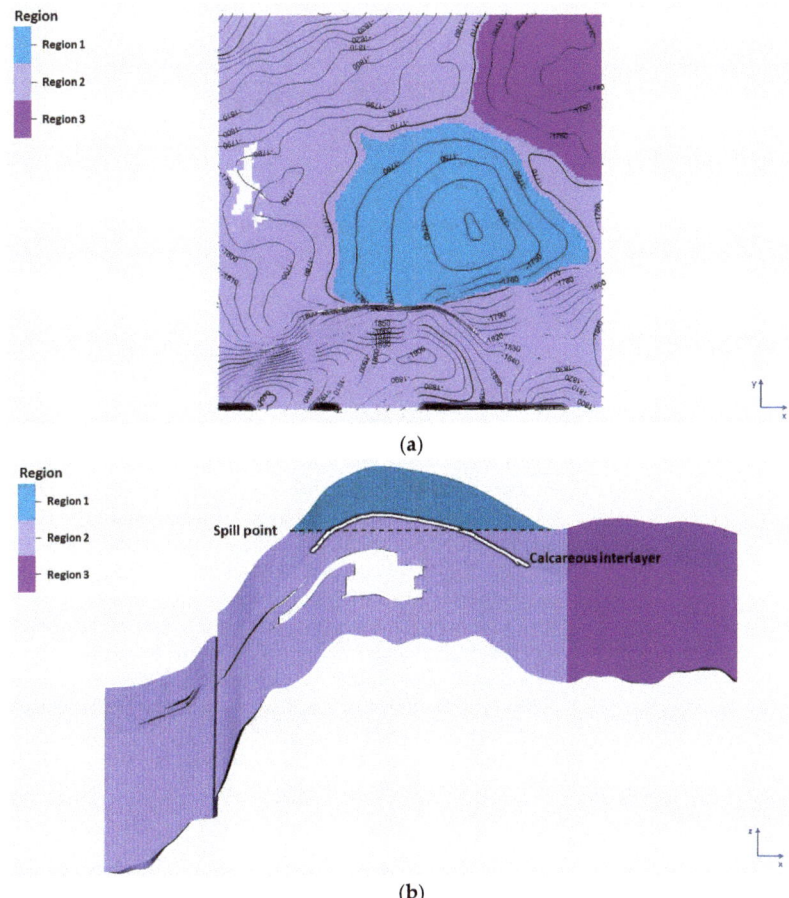

Figure 2. Gradient images illustrating the region classification of the model. (**a**) Planar view, (**b**) cross sectional view.

3.2.3. Timescale

There are two important points in a CO_2 sequestration project: the time of injection cessation and the time of immobilization of free-phase CO_2 [31]. The former is already described in the pressure-threshold section. The injection was stopped as soon as the local formation pressure reached the threshold pressure defined above. The sequestration type of injected CO_2 can be transformed into the above sequestration mechanisms. Generally, while the mobile phase of CO_2 migrates, it is trapped in pore volumes up to the critical saturation or dissolves into the aquifer brine. It can also be mineralized by chemical interactions with specific components of the formation rock. It may have transformed for tens of hundreds of years, but the quantity of transformation decreased after a stabilized CO_2 plume. Through test simulations for the target site, it was confirmed that the transition rate from the mobile phase gradually decreased and the mobile CO_2 plume stabilized approximately 200 years later, as shown in Figure 3. In this research, the storage capacity of CO_2 was determined by the quantity in the structures at 200 years later from commencing injection as the time of immobilization of CO_2.

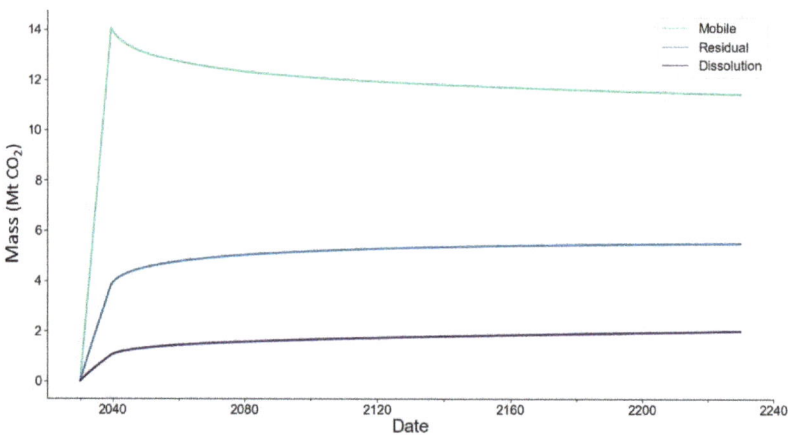

Figure 3. Line graph depicting the temporal evolution of injected CO_2 plume.

Summarily, an accurate quantity of CO_2 sequestration on a reasonable time scale established by the project requires an industrial perspective. It is not only the operational information of a company for ESG management, but also financial information directly related to the commercial activity of the company. If the quantity of CO_2 sequestration is certified as Certificated Emissions Reduction, then it is traded in the carbon credit market, which can be additional earnings for a company. Therefore, the quantity of sequestrated CO_2 should be certified by a strict evaluation method, including monitoring for a sufficiently long period of time. The trapped CO_2 in a structure is more convenient for surveillance than trapped by other mechanisms of residual trapping, dissolution, and mineralization outside of a structure. While these mechanisms relying on parameters with high uncertainty have difficulties in tracking for a sufficiently long time, CO_2 trapped in a structure can be monitored within a limited region with a more confident method. Hence, the quantity of structural and residual saturation trap in a structure is exploited as the main CO_2 storage capacity in this research.

3.3. Model Description

For CO_2 sequestration simulation in a saline aquifer, an isothermal compositional model was utilized [32]. The detailed parameters of the simulation are summarized in Table 1. It is a CO_2-H_2O system in which each component dissolves other components. Sodium chloride, which is the only solid component, determines the salinity of the aquifer. This is a key factor that affects the dissolution of CO_2 in the water phase. As the solid phase is not considered in this study, sodium chloride is assumed to remain in the aqueous phase only.

Table 1. Parameters of CO_2 sequestration model.

Phase	CO_2 rich, H_2O rich
Fluid component	CO_2, H_2O, NaCl
Mutual solubility	Phase-partitioning [33]
Molecular diffusion factor	(Unit: m^2/day)
• Water phase	H_2O: 0.0005/CO_2: 0.001/NaCl: 0.005
• Gas phase	H_2O: 0.001/CO_2: 0.001

A geological model of the potential sequestration site was constructed using the properties summarized in Table 2. The model has 3,232,584 grid blocks covering a 7.5 km by 7.5 km area and 100 m thickness. The relative permeability of CO_2–H_2O system [32] applied to a dynamic simulation is shown in Figure 4.

Table 2. Geological property of aquifer model.

Model dimension (i, j, k)	(152, 153, 139)
Porosity (max, mean, min)	(34%, 24%, 12%)
Permeability (max, mean, min)	(7794 md, 506 md, 3 md)
Vertical/horizontal ratio	0.31
Initial pressure	17.4 MPa @1746 m
Initial temperature	89 °C
Gas residual saturation	30%
Salinity	8%

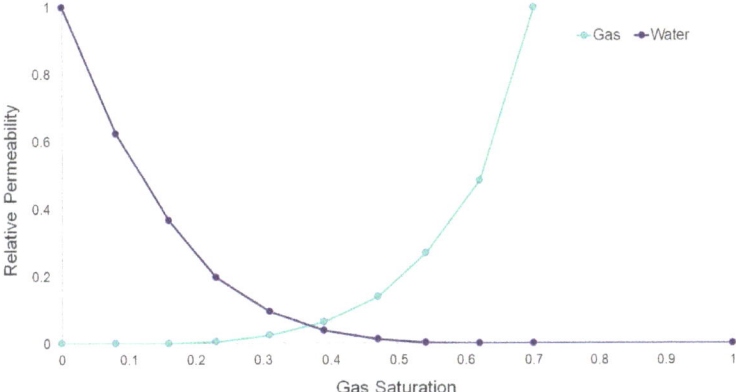

Figure 4. Line graph of the relative permeability of CO_2-H_2O system.

For computational efficiency, the aquifer model can be extended by attaching an analytic aquifer, which is a widely applied concept in reservoir engineering. The model can be constructed by focusing on the area of review where the injected CO_2 plume migrates or remains. The outside of the model interacts with the numerical model, communicating the pressure by flowing aquifer brine only. The properties of the analytical aquifer are listed in Table 3. The Carter–Tracy aquifer model [34] was adopted as an analytic model, and the influence function assumed that the radius of the analytic aquifer was twice that of the numerical aquifer.

Table 3. Property of analytic aquifer.

Analytic model	Carter–Tracy
Radius numerical aquifer	3800 m
Influence function	2
Permeability	300 md
Porosity	29%

3.4. CO_2 Injection Scenarios

For a comparative analysis of the CO_2 capacity, six scenarios in Table 4 were defined in combination with CO_2 injection and brine production. The first two cases were designed to investigate the effects of the injector location on CO_2 storage. The location of the CO_2 injector is illustrated in Figure 5. The location of the lower injector is below the substructure, which can be filled first with CO_2. The upper injector is located within the main structure immediately below the cap rock. The following four cases were designed to optimize the injection schemes with brine production. CASE 3 applies pre-injection brine production using a single dual-mode well. Brine production and CO_2 injection were performed sequentially in the same well. The last three cases were designed to analyze the effects of the producer location as shown in Figure 6. The schemes involve CO_2 simultaneous injection with brine production. CASE 4 has a brine producer in opposite direction of the migration pathway. CASE 5 is a revised case of CASE 4 by moving the location of the brine producer close to the migration pathway. The location of the brine producer in CASE 6 is under the crest of the substructure which is sufficient distance apart from the migration pathway. Additionally, the permeability of the completion intervals of CASE 5 and 6 was much higher than that of CASE 4.

The injection condition was to inject CO_2 at a constant rate of 2 Mt/year. When the reference pressure reaches the fracture pressure, CO_2 injection ceases by shut-in. The production condition is a constant rate of production with a lower limit of bottom-hole pressure. The producer was set to produce water at 5000 m^3/day to maintain a bottom-hole pressure higher than 5 MPa. For the cases of simultaneous brine production, brine production stops just before CO_2 reaches the brine producers.

Table 4. Scenario description in combination with CO_2 injection and brine production.

Case	Injector	Producer	Description
CASE 1	Lower	N/A	Injector below substructure
CASE 2	Upper	N/A	Injector right below cap rock of the main structure
CASE 3	Upper	Injector	Pre-injection brine production for 3 years
CASE 4	Upper	Lower	Producer in opposite to the migration direction
CASE 5	Upper	Lower	Producer in the migration pathway
CASE 6	Upper	Lower	Producer away from the migration pathway

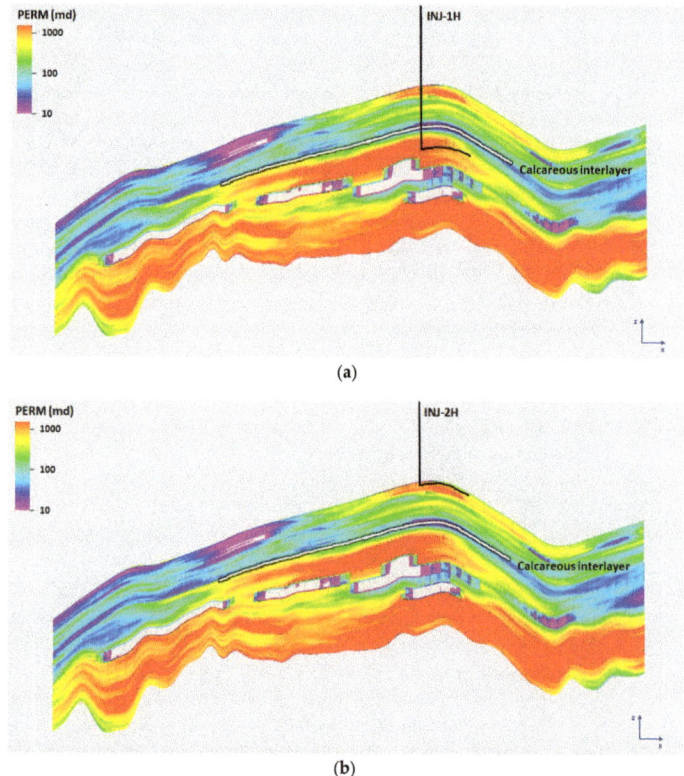

(a)

(b)

Figure 5. Permeability maps depicting the location of the injector. (**a**) Lower injector (CASE 1), (**b**) upper injector (CASE 2–CASE 6).

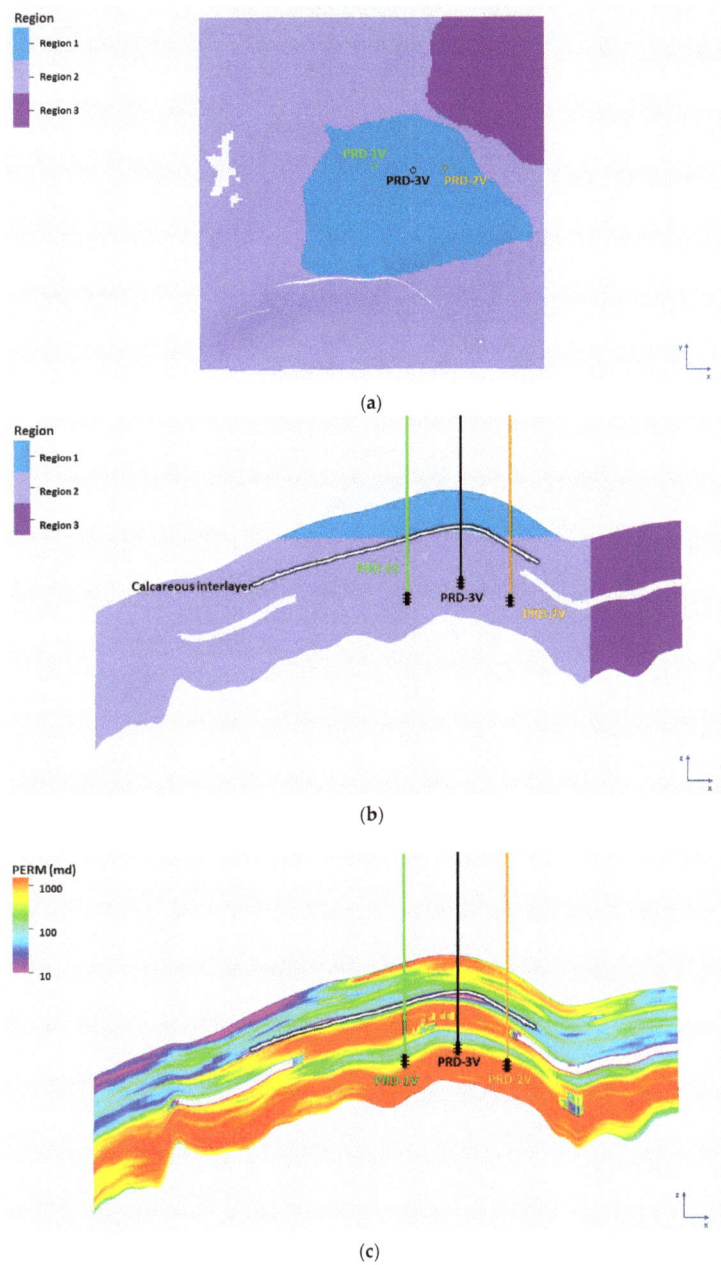

Figure 6. Gradient images showing the location of producer; Producer opposite direction of the migration (PRO-1V for CASE 4), producer close to the migration pathway (PRO-2V for CASE 5), and producer away from the migration pathway (PRO-3V for CASE 6). (**a**) Well location on a planar region map, (**b**) well location on a cross sectional region map, (**c**) well location on a cross sectional permeability map.

4. Results and Discussion

The pressure management strategy was optimized by comparing cases with various combinations of injectors and producers to meet the following requirements. The formation pressure does not exceed the threshold pressure for the fracturing pressure of the cap rock. The injected CO_2 was not produced by brine production. The injected CO_2 can be trapped in the structure as much as possible, particularly in the substructure, by steering the CO_2 plume.

Table 5 summarizes the injection time, total injection mass, and CO_2 capacity, and shows the pressure constraints and effects of pressure management. The lower injector of CASE 1 has a limited effect on the pressure constraint because the injector is located far away from the reference point for the formation fracture, as shown in Figure 5. Additionally, it takes a long time to propagate the pressure buildup from the aquifer bottom due to the calcareous barrier ceiling of the substructure. Additional CO_2 can be injected during the extended time over CASE 2 until the formation pressure reaches the threshold pressure of formation fracture. However, it is the highest limit because the scheme has an obstacle for brine production due to the CO_2 production associated with brine production. The migration direction due to buoyancy is the main reason for not locating the brine producer, even in the upper part of the aquifer. More injection volume without brine production causes the over-pressure around the weak formation, which can be a potential risk on the structure stability.

Table 5. Summary of injection period, injected mass, and storage capacity of CO_2.

Case	Injection Time (year)	Injected Mass (Mt CO_2)	CO_2 Capacity (Mt CO_2)
CASE 1	8.22	16.45	10.13
CASE 2	6.42	12.83	9.79
CASE 3	8.39	16.79	11.87
CASE 4	9.47	18.93	12.92
CASE 5	9.38	18.76	12.77
CASE 6	13.22	26.43	16.44

Pressure management has a significant effect on the CO_2 injection mass summarized in Table 5. It enables at most 6.8 years of additional injection time compared to the cases without brine production. From the formation pressure behaviors shown in Figure 7, the formation pressure of CASE 3, pre-injection brine production, was reduced linearly below the initial formation pressure before ceasing brine production. Then, the formation pressure increased discretely for 3 months, i.e., a shut-in period of 3 months to convert the brine producer into a CO_2 injector. Once CO_2 was injected into the aquifer, the formation pressure increased rapidly until it reached the pressure threshold of the fracturing pressure of the cap rock. The simultaneous brine production case of CASE 4 exhibits a trend of formation pressure increase with a gradual slope. The target production rate is set to 5000 m^3/d, but it cannot meet the target due to the lower limit of bottomhole pressure, as shown in Figure 8. The simultaneous injection with higher rate production in CASE 5 has a dual slope, and its intersection is the point at which brine production stops due to CO_2 breakthrough. Despite the relatively short period of brine production (4.4 years) indicated in Figure 8, a low formation pressure can be maintained due to the higher production rate. Well completion in the interval with higher permeability enables the brine production as much as CASE 4 in a shorter period. CASE 6 shows the longest CO_2 injection time owing to the higher production rate and delayed CO_2 breakthrough. It drives CO_2 injection for 13.2 years, which is longer than that of any other pressure management case.

The injected CO_2 masses categorized by sequestration mechanism and region are summarized in Table 6. The combination of sequestration mechanisms and sequestrated regions can provide a basis for prioritizing the sequestration concept. The sequestration mass by structural traps and residual saturation traps in the main- and the sub-structure was determined as the CO_2 capacity, as discussed in Section 3.2. The mass beyond the spill point cannot be certified as a proven storage even if it is trapped by saturation residual or dissolution in the simulation results. Although the sequestrated mass by residual and dissolution is known to be stable, the actual quantity is highly uncertain depending on the parameters or coefficients.

Table 6. Classification of sequestrated CO_2 by mechanism and region at year 2230.

	Region	Mobile (Mt CO_2)	Residual (Mt CO_2)	Dissolution (Mt CO_2)
CASE 1	Main structure	3.38	1.51	0.49
	Substructure	5.25	4.14	1.65
	Beyond spill	0.01	0.02	0.01
CASE 2	Main structure	7.37	2.24	0.60
	Substructure	0.18	1.12	0.55
	Beyond spill	0.25	0.36	0.17
CASE 3	Main structure	9.02	2.39	0.60
	Substructure	0.52	1.77	0.83
	Beyond spill	0.65	0.70	0.32
CASE 4	Main structure	9.57	2.37	0.57
	Substructure	0.98	2.27	1.03
	Beyond spill	0.90	0.86	0.40
CASE 5	Main structure	9.47	2.37	0.57
	Substructure	0.93	2.21	1.01
	Beyond spill	0.92	0.88	0.41
CASE 6	Main structure	10.98	2.42	0.54
	Substructure	3.04	3.65	1.52
	Beyond spill	2.17	1.45	0.65

The effects of the injector location were analyzed in two cases, as described in Section 3.3. Only CASE 1 had an injector below the substructure. It fills the substructure first, and then the injected CO_2 migrates to the main structure. Because of the buoyancy of CO_2, its dominant migration direction is vertical. This resulted in a negligibly small mass of migration beyond the spill. While vertically migrating long distances from the aquifer bottom to the top, it sequestrates the large amount of residual trapped CO_2. CASE 1, which sequestrates the most mass in the substructure, contains CO_2 in the main structure as well, as shown in Figure 9a. However, the amount in the main structure is relatively small compared to the whole size of the main structure, and hence, CASE 1 cannot be an efficient sequestration concept. Contrarily, CASE 2 cannot fill the substructure as shown in Figure 9b. The calcareous layer, which is a cap rock of the substructure, acts as a bottom barrier, and buoyancy causes the CO_2 plume to migrate out of the structure.

For the pre-injection brine production in CASE 3, a single well was drilled and utilized as a brine producer and a CO_2 injector sequentially. It is less effective from a pressure management perspective compared to that of CASE 4, 5, or 6 because of the rapid build-up of formation pressure. When shut-in for well conversion, the formation pressure of the reference location recovers immediately. Once the CO_2 injection commences, the recovered formation pressure, which is still below the initial pressure, starts to build up rapidly and finally reaches the pressure threshold of the fracturing pressure, as shown in Figure 7b. As CO_2 is injected into the under-pressured zone by brine production, the scheme displaces the zone with CO_2 rather than overflowing beyond the spill, as shown in Figure 9c. Although it has the disadvantages of project delay and less efficiency, it can be selected for reasons of less CAPEX (capital expenditure) and confident operation based on the test analysis through prior drawdown.

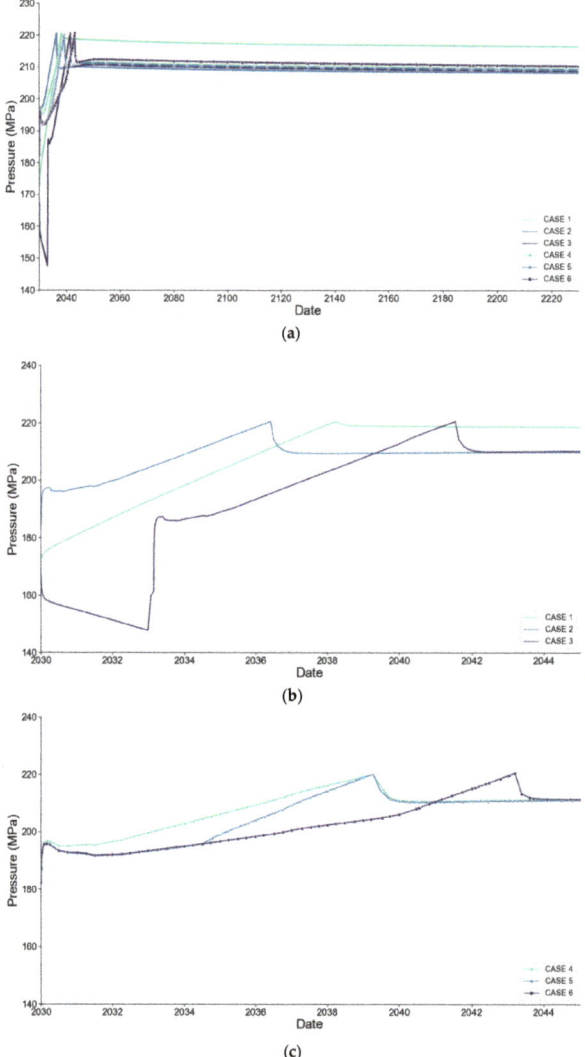

Figure 7. Line graphs showing the comparison of reference pressure below cap rock. (**a**) Pressure behavior for whole 200 years, (**b**) pressure behavior for initial 15 years (CASE1–CASE3), (**c**) pressure behavior for initial 15 years (CASE4–CASE6).

CASE 4 implements pressure management by simultaneous brine production using an additional well. The results show that this is a measure to extend the sequestration capacity by overcoming pressure constraints. It enables an additional 3.1 years of CO_2 injection compared to that of the control case (CASE 2). The additional injection mass was almost 6.1 Mt, although the migration mass beyond the spill increased proportionally to the total injection mass. It still cannot make better use of the substructure for sequestration site considering only 0.98 Mt sequestration mass in the substructure. The brine producer was drilled below the substructure, away from the CO_2 injector, and perforated in the interval with poor permeability. The effects of pressure management were not directly propagated to the reference point of the cap rock. Additionally, the producer is located in the opposite direction to the migration pathway; therefore, there is no drive for the CO_2

plume to dig beneath the cap rock of the substructure. The CO_2 plume from the main structure developed to be distributed both in the substructure and beyond the spill, as shown in Figure 9d.

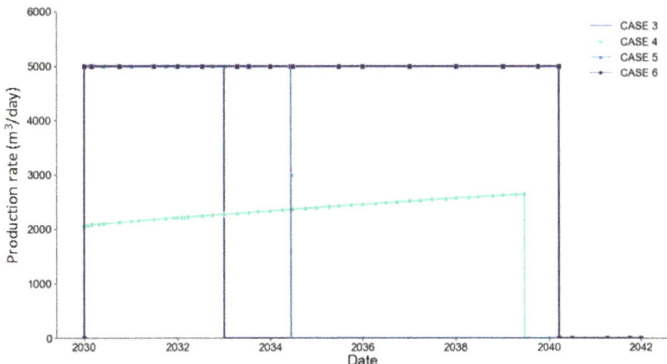

Figure 8. Line graph depicting the summary of brine production (CASE3–CASE6).

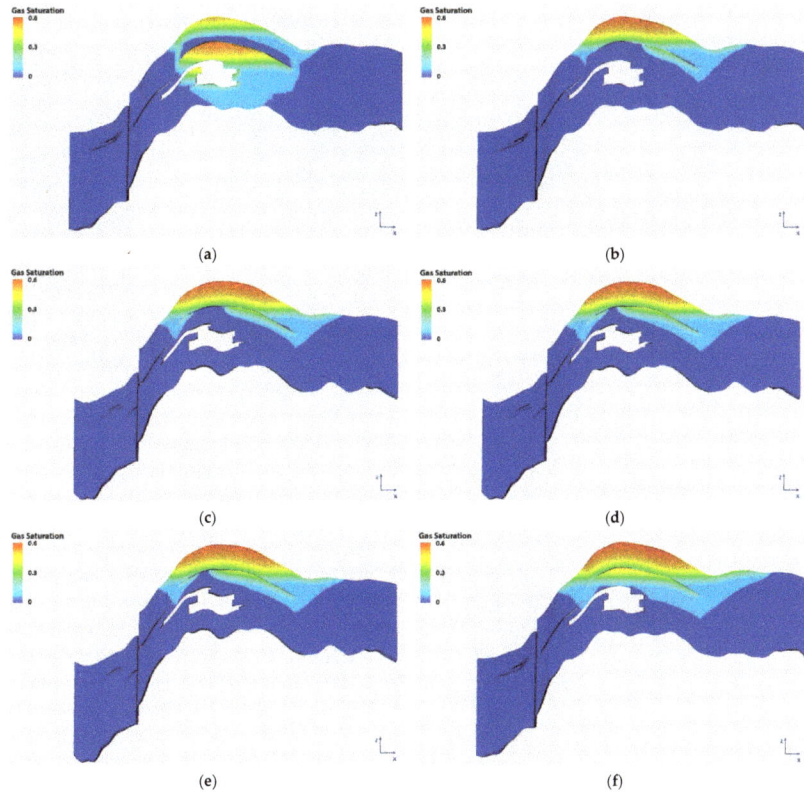

Figure 9. Gradient images depicting the comparison of the distributions of CO_2 at 200 years later from the commencement of CO_2 sequestration project. (**a**) CASE 1, (**b**) CASE 2, (**c**) CASE 3, (**d**) CASE 4, (**e**) CASE 5, (**f**) CASE 6.

CASE 5 has a producer inside the substructure along the migration pathway. The formation pressure was reduced around the brine producer, which drove the CO_2 plume to migrate beneath the top of the substructure. If the brine producer is close to the migration pathway, the steering effects to the substructure can be enhanced, rather than migrating beyond the spill as illustrated in Figure 10. However, early CO_2 breakthrough eventually leads to cessation of CO_2 injection. It acts as another constraint of the CO_2 storage capacity judging from less CO_2 capacity than that of CASE 4.

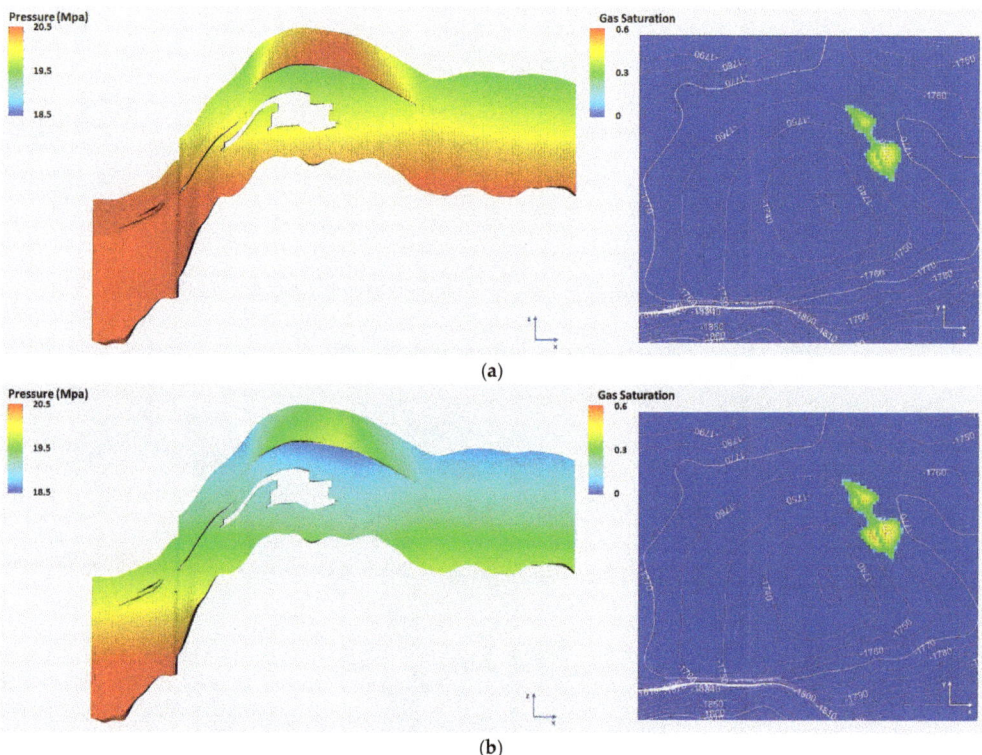

Figure 10. Gradient images showing the behavior of CO_2 plume around the intersection point of the structure outside and the substructure inside. (**a**) Pressure map (cross sectional) and saturation map (planar) of CASE 4 at year 2034, (**b**) pressure map (cross sectional) and saturation map (planar) of CASE 5 at year 2034.

The trade-off between CO_2 steering and pressure management is confirmed through CASE 4 and CASE 5. The key optimization point for maximizing storage capacity is the producer location. Thus, the location of the producer in CASE 6 is moved to the center of the substructure away from the migration pathway. The requirement of the brine producer is to make under-pressured zone near the migration pathway, but sufficient distance away from the migration pathway to delay CO_2 breakthrough. Through the application of this scheme, the substructure accounted for 18% of the total storage capacity. The contribution of the substructure secures an additional 2.1 Mt of CO_2 storage capacity compared to that of CASE 4. The distribution of sequestrated CO_2 plume in the substructure is shown in Figure 9f. The residual trap is depicted in light green color according to the footprint of the CO_2 plume movement. The remaining mobile CO_2 after being trapped by residual saturation is distributed in the top of the substructure with a high concentration of CO_2.

5. Conclusions

Several injection strategies were investigated to overcome the constraints of formation fracture pressure and structure size, which are major constraints that limit the storage capacity in a closed aquifer. The lower injection scheme below substructure showed a good performance only among the cases without pressure management. However, it has no upside potential for the injection period extension because of the systematic limits of CO_2 production associated with brine production. The pressure management with brine extraction was verified as an effective technology from a pressure constraint perspective. It enabled the additional CO_2 injection range from 31% to 106% compared to the cases without pressure management. Focusing on the CO_2 storage capacity, sequestration mass was increased by 6.65 Mt, which is an improvement by 68%. The brine extraction scheme was optimized to maximize the storage capacity by overcoming the constraint of the structure size in the target site, which is part of a closed aquifer with a substructure inside the main structure. Through the optimization, the location of the brine producer was concluded a key parameter for steering the injected CO_2 plume toward the inside of the substructure. For a brine producer, a sufficient distance from the migration pathway and perforation in the high permeability interval have a significant impact on the direction of CO_2 plume movement and evolution.

Conclusively, a CO_2 sequestration strategy to overcome the constraints limiting the CO_2 storage capacity is suggested. Brine extraction extends the CO_2 injection period and evolves the injected CO_2 plume inside the substructure. Through verification of the applicability of brine production, the proposed CO_2 injection scheme can provide key insight into the optimization of a target aquifer considering its geological characteristics. Moreover, the methodology of this study is expected as a practical tool for tacking injected CO_2 plume when combined with the post-injection monitoring data.

Funding: The author received financial support for this article from SK earthon Co., Ltd. (http://www.skearthon.com/).

Institutional Review Board Statement: Not applicable.

Informed Consent Statement: Not applicable.

Data Availability Statement: Not applicable.

Acknowledgments: This publication has been produced with support and permission from the China business division, SK earthon. The author acknowledges the VP of the China Business division, Dooyun Park. The review comments on the geological description of the target aquifer from Sr. Geologist, Peter Bahn, are much appreciated. The author would like to thank to Rev. Fr. Daniel M. Kim who motivated the author for this study.

Conflicts of Interest: The author declares no conflict of interest.

References

1. UNFCCC. Paris Agreement. In Proceedings of the Conference of the Parties COP 21, Paris, France, 30 November–12 December 2015.
2. Cui, G.; Hu, Z.; Ning, F.; Jiang, S.; Wang, R. A Review of Salt Precipitation during CO_2 Injection into Saline Aquifers and Its Potential Impact on Carbon Sequestration Projects in China. *Fuel* **2023**, *334*, 126615. [CrossRef]
3. Espinoza, D.N.; Santamarina, J.C. CO_2 breakthrough—Caprock Sealing Efficiency and Integrity for Carbon Geological Storage. *Int. J. Greenh. Gas Control* **2017**, *66*, 218–229. [CrossRef]
4. Goertz-Allmann, B.P.; Kühn, D.; Oye, V.; Bohloli, B.; Aker, E. Combining Microseismic and Geomechanical Observations to Interpret Storage Integrity at the In Salah CCS Site. *Geophys. J. Int.* **2014**, *198*, 447–461. [CrossRef]
5. Ellsworth, W.L.; Giardini, D.; Townend, J.; Ge, S.; Shimamoto, T. Triggering of the Pohang, Korea, Earthquake (M w 5.5) by Enhanced Geothermal System Stimulation. *Seismol. Res. Lett.* **2019**, *90*, 1844–1858. [CrossRef]
6. Frailey, S.; Koperna, G.; Tucker, O. The CO_2 Storage Resources Management System (SRMS): Toward a Common Approach to Classifying, Categorizing, and Quantifying Storage Resources. In Proceedings of the 14th Greenhouse Gas Control Technologies Conference, Melbourne, Australia, 21–26 October 2018.
7. Kumar, A.; Ozah, R.; Noh, M.; Pope, G.A.; Bryant, S.; Sepehrnoori, K.; Lake, L.W. Reservoir Simulation of CO_2 Storage in Deep Saline Aquifers. *SPE J.* **2005**, *10*, 336–348. [CrossRef]

8. Bouquet, S.; Gendrin, A.; Labregere, D.; Le Nir, I.; Dance, T.; Xu, J.; Cinar, Y. CO2CRC Otway Project, Australia: Parameters Influencing Dynamic Modeling of CO2 Injection into a Depleted Gas Reservoir. In Proceedings of the SPE Offshore Europe Oil and Gas Conference and Exhibition, Aberdeen, UK, 8–11 September 2009.
9. Bergmo, P.E.S.; Grimstad, A.-A.; Lindeberg, E. Simultaneous CO_2 Injection and Water Production to Optimise Aquifer Storage Capacity. *Int. J. Greenh. Gas Control* **2011**, *5*, 555–564. [CrossRef]
10. Buscheck, C.A.; Sun, Y.; Chen, M.; Hao, Y.; Wolery, T.J.; Bourcier, W.L.; Court, B.; Celia, M.A.; Friedmann, S.J.; Aines, R.D. Active CO_2 Reservoir Management for Carbon Storage: Analysis of Operational Strategies to Relieve Pressure Buildup and Improve Injectivity. *Int. J. Greenh. Gas Control* **2012**, *6*, 230–245. [CrossRef]
11. Court, B.; Bandilla, K.W.; Celia, M.A.; Buscheck, T.A.; Nordbotten, J.M.; Dobossy, M.; Janzen, A. Initial Evaluation of Advantageous Synergies Associated with Simultaneous Brine Production and CO_2 Geological Sequestration. *Int. J. Greenh. Gas Control* **2012**, *8*, 90–100. [CrossRef]
12. Dempsey, D.; Kelkar, S.; Pawar, R. Passive Injection: A Strategy for Mitigating Reservoir Pressurization, Induced Seismicity and Brine Migration in Geologic CO_2 Storage. *Int. J. Greenh. Gas Control* **2014**, *28*, 96–113. [CrossRef]
13. Buscheck, T.A.; Bielicki, J.M.; White, J.A.; Sun, Y.; Hao, Y.; Bourcier, W.L.; Carroll, S.A.; Aines, R.D. Managing Geologic CO_2 Storage with Pre-injection Brine Production in Tandem Reservoirs. *Energy Procedia* **2017**, *114*, 4757–4764. [CrossRef]
14. Bachu, S.; Bonijoly, D.; Bradshaw, J.; Burress, R.; Christensen, N.; Holloway, S.; Mathiassen, O. *Estimation of CO_2 Storage Capacity in Geological Media*; Carbon Sequestration Leadership Forum (CSLF): Washington, DC, USA, 2007.
15. National Energy Technology Laboratory. *Carbon Sequestration Atlas of the United States and Canada*; US Department of Energy: Washington, DC, USA, 2008.
16. Gorecki, C.; Sorensen, J.; Bremer, J.; Ayash, S.; Knudsen, D.; Holubnyak, Y.; Smith, S.; Steadman, E.; Harju, J. *Development of Storage Coefficients for CO_2 Storage in Deep Saline Formations*; IEA Greenhouse Gas R&D Programme (IEA GHG): Cheltenham, UK, 2009.
17. Brennan, S.T.; Merrill, M.D.; Buursink, M.L.; Warwick, P.D.; Cahan, S.M.; Cook, T.A.; Corum, M.D.; Craddock, W.H.; DeVera, C.A. *National Assessment of Geologic Carbon Dioxide Storage Resources: Methodology Implementation*; Blondes, M.S., Ed.; US Department of the Interior, US Geological Survey: Washington, DC, USA, 2013.
18. Michael, K.; Golab, A.; Shulakova, V.; Ennis-King, J.; Allinson, G.; Sharma, S.; Aiken, T. Geological Storage of CO_2 in Saline Aquifers—A Review of the Experience from Existing Storage Operations. *Int. J. Greenh. Gas Control* **2010**, *4*, 659–667. [CrossRef]
19. Hosa, A.; Esentia, M.; Stewart, J.; Haszeldine, S. Injection of CO_2 into Saline Formations: Benchmarking Worldwide Projects. *Chem. Eng. Res. and Design* **2011**, *89*, 1855–1864. [CrossRef]
20. Kelemen, P.; Benson, S.M.; Pilorgé, H.; Psarras, P.; Wilcox, J. An Overview of the Status and Challenges of CO_2 Storage in Minerals and Geological Formations. *Front. Clim.* **2019**, *1*, 9. [CrossRef]
21. Global CCS Institute. *The Global Status of CCS: 2021*; Global CCS Institute: Melbourne, Australia, 2021.
22. Le Guenan, T.; Rohmer, J. Corrective Measures Based on Pressure Control Strategies for CO_2 Geological Storage in Deep Aquifers. *Int. J. Greenh. Gas Control* **2011**, *5*, 571–578. [CrossRef]
23. Cameron, D.A.; Durlofsky, L.J. Optimization of Well Placement, CO_2 Injection Rates, and Brine Cycling for Geological Carbon Sequestration. *Int. J. Greenh. Gas Control* **2012**, *10*, 100–112. [CrossRef]
24. Gorecki, C.D.; Ayash, S.C.; Liu, G.; Braunberger, J.R.; Dotzenrod, N.W. A Comparison of Volumetric and Dynamic CO_2 Storage Resource and Efficiency in Deep Saline Formations. *Int. J. Greenh. Gas Control* **2015**, *42*, 213–225. [CrossRef]
25. Trupp, M.; Ryan, S.; Mendoza, I.B.; Leon, D.; Scoby-Smith, L. Developing the World's Largest CO_2 Injection System—A History of the Gorgon Carbon Dioxide Injection System. In Proceedings of the 15th Greenhouse Gas Control Technologies Conference, Abu Dhabi, United Arab Emirates, 16 March 2021.
26. Song, J.; Zhang, D. Comprehensive Review of Caprock-sealing Mechanisms for Geologic Carbon Sequestration. *Environ. Sci. Technol.* **2013**, *47*, 9–22. [CrossRef] [PubMed]
27. Zhang, J.; Yin, S.-X. Fracture Gradient Prediction: An Overview and an Improved Method. *Pet. Sci.* **2017**, *14*, 720–730. [CrossRef]
28. Matthews, W.R.; Kelly, J. How To Predict Formation Pressure and Fracture Gradient. *Oil Gas J.* **1967**, *60*, 92–98.
29. Masoudi, R.; Abd Jalil, M.A.; Tan, C.P.; Press, D.; Keller, J.; Anis, L.; Darman, N.; Othman, M. Simulation of Chemical Interaction of Injected CO_2 and Carbonic Acid Based on Laboratory Tests in 3D Coupled Geomechanical Modelling. In Proceedings of the International Petroleum Technology Conference, Beijing, China, 26 March 2013.
30. Gai, X.; Dean, R.H.; Wheeler, M.F.; Liu, R. Coupled Geomechanical and Reservoir Modeling on Parallel Computers. In Proceedings of the SPE Reservoir Simulation Symposium, Houston, TX, USA, 3 February 2003.
31. Bachu, S. Review of CO_2 Storage Efficiency in Deep Saline Aquifers. *Int. J. Greenh. Gas Control* **2015**, *40*, 188–202. [CrossRef]
32. Schlumberger Information Solutions. *Eclipse Technical Description*; Schlumberger: Houston, TX, USA, 2018.
33. Spycher, N.; Pruess, K. A Phase-partitioning Model for CO_2-brine Mixtures at Elevated Temperature and Pressures. *Transp. Porous Media* **2009**, *82*, 173–196. [CrossRef]
34. Carter, R.; Tracy, G. An Improved Method for Calculating Water Influx. *Trans. AIME* **1960**, *219*, 415–417. [CrossRef]

Disclaimer/Publisher's Note: The statements, opinions and data contained in all publications are solely those of the individual author(s) and contributor(s) and not of MDPI and/or the editor(s). MDPI and/or the editor(s) disclaim responsibility for any injury to people or property resulting from any ideas, methods, instructions or products referred to in the content.

Editorial

An Overview of Advances in CO$_2$ Capture Technologies

José Ramón Fernández

Instituto de Ciencia y Tecnología del Carbono, Spanish Research Council (INCAR-CSIC), Francisco Pintado Fe, 26, 33011 Oviedo, Spain; jramon@incar.csic.es; Tel.: +34-985119090

CO$_2$ emissions generated by human activities reached the highest ever annual level of 36.3 Gt in 2021, due to the extremely rapid growth of the energy demand observed after the COVID-19 crisis [1]. There is a consensus to consider CO$_2$ (the main greenhouse gas emitted into the atmosphere) a great contributor to climate change [2]. In a scenario where the demand for energy is expected to increase by 15% by the middle of this century [3], the predominant use of fossil fuels will continue in the coming decades, especially in certain industrial sectors, in order to avoid excessive disruption to the existing energy supply chain that could negatively affect the global economy [4].

A substantial CO$_2$ emission abatement is urgent to meet the global climate targets agreed in the Paris Agreement aimed at limiting the global temperature increase to only 1.5–2 °C above pre-industrial levels [2]. Apart from the development of renewable energy sources, switching to lower carbon alternatives, or the improvement of energy efficiency in existing processes—such as CO$_2$ Capture and Storage (CCS) or Utilization (CCU)—should play a key role in the successful transition towards deep decarbonization of the global production system [5]. It has been estimated that CCS/CCU should contribute to reducing about one third of overall CO$_2$ emissions by 2050 [2], but CO$_2$ capture technologies are being developed slower than desired due to technological, infrastructural and policy barriers. As a result, most of these technologies are still relatively far from being implemented at a commercial scale at present [6,7].

Basically, in all CO$_2$ capture technologies the objective is to separate and concentrate CO$_2$ generated in stationary emissions sources, such as power generation plants or industrial processes (e.g., steel mills, refineries, cement plants, etc.). Despite the great progress made over the last decade to reduce the energy penalty and capital cost in these technologies [8–11], CO$_2$ capture is still the most demanding step (around 70% of the total cost) of the complete chain of processes required to permanently store CO$_2$ or to use it as a feedstock for subsequent chemical transformation [7].

CO$_2$ capture technologies can be classified into three groups: post-combustion, pre-combustion and oxy-combustion processes. In post-combustion systems, the CO$_2$ is removed from flue gases generated in previous fuel combustion. These flue gases typically contain a relatively low concentration of CO$_2$ (between 5 and 15% vol.), which makes it necessary to operate with a great volume of gases, leading to a large equipment cost. The separation of CO$_2$ from highly diluted gases is typically carried out using chemical solvents that require a large amount of energy for their regeneration [12]. In pre-combustion systems, the carbonaceous fuel is converted into syngas through steam reforming, gasification or partial oxidation, which is followed by a water-gas-shift reaction to obtain a mixture of H$_2$ and CO$_2$ at high pressure (i.e., between 20 and 50 bar). Then, CO$_2$ is separated, and the resulting H$_2$ can be used as carbon-free fuel or as clean feedstock for the production of ammonia, methanol or synthetic fuels. The concentration of CO$_2$ in the gases before its separation is significantly higher (i.e., 15–60% vol.), which allows for more compact CO$_2$ capture equipment. The higher starting CO$_2$ concentrations could allow the use of solvents (e.g., physical absorption) that typically demand lower energy for regeneration [13]. Finally, in the oxy-combustion systems, the fuel is burnt with almost pure oxygen rather than air, which results in virtually 100% of CO_2 and avoids costly CO_2 purification steps downstream. However, air separation to produce pure O_2 requires high energy consumption, and strict safety procedures are needed to avoid air infiltration during oxy-combustion [14].

Currently, the industrial sector accounts for around 20% of overall CO_2 emissions, and about 70–80% of these emissions come from energy intensive industries, such as steelmaking, cement manufacture, chemical sector or paper manufacture [15]. Therefore, the decarbonisation of these industries is essential to reach the climate neutral targets in the coming decades. Some developed countries are implementing climate-positive solutions in order to drastically reduce the emissions of CO_2 into the atmosphere. Nurdiawati and Urban [15] show the substantial decarbonisation efforts planned in Sweden to achieve a deep reduction in greenhouse gas emissions by 2050, by means of great financial and political support. Numerous R&D programmes are focused on promoting renewable energies, circular economy and CCUS technologies. A good example is the HYBRIT project, in which the main Swedish steel producer leads the production of H_2 from renewable sources, which is subsequently used as a reducing agent of iron ore (instead of coke) to obtain sponge iron [16]. A recent supply chain analysis reported by Karlsson et al. [17] for the Swedish building and construction sector reveals that the implementation of energy efficiency measures, promotion of biofuels usage, renewable electrification and CCS in primary steel and cement production may lead to almost zero emissions of CO_2 by 2045. Australia presents a different situation, as its power system is dominated by the use of coal, although there is great potential for the expansion of renewable energies. Aboumahboub et al. [18] developed a comprehensive multi-sectorial model to evaluate the capacity of Australia's energy system to drastically reduce its dependence of fossil fuels in the short-to-medium term. Their results indicate that the transition to a low-carbon scenario to comply with the Paris Agreement makes necessary the rapid replacement (in less than 20 years) of coal-fired power generation through the combination of solar photovoltaic and wind energies, as well as the electrification and use of hydrogen in energy-intensive industry sectors.

Amine-based chemical absorption is currently the most technically mature CO_2 capture technology. However, this process, typically proposed as a post-combustion technology, still presents serious challenges for its commercialization, such as the high energy demand (about 4 GJ/t CO_2), the tendency of solvent degradation in the presence of SO_X and/or NO_X and the high cost of high-performance amines [7,12,19]. Alternative chemical solvents are being developed to increase the CO_2 sorption capacity at a lower cost. Ethylenediamine (EDA) is a promising solvent, less corrosive, with a higher capacity for the capture of CO_2, and it consumes less energy for regeneration than conventional alkanolamines such as monoethanolamine (MEA) [20]. The combination of post-combustion with MEA absorption in biomass-fired power plants and the subsequent storage of CO_2 in geothermal systems appears as a feasible negative CO_2 emissions option, as the calculated energy penalty is limited to 6 MJ/kg CO_2, and the estimated cost for the CO_2 avoided is around 50 EUR/t CO_2 [21].

As mentioned above, another CO_2 capture pathway that has reached a significantly high technology-readiness level (TRL) is oxy-fuel combustion. Recent studies have focused on solving the existing limitations of this technology. Ahn and Kim [22] demonstrated the feasibility of introducing flue gas recirculation (FGR) in a 0.5 MW boiler, in order to stabilize the flame generated through the fuel combustion in O_2-enriched atmospheres, while the generation of NOx was considerably reduced. As a result of that, flue gases with more than 90% of CO_2 can be obtained.

Although pressure swing adsorption (PSA) is a well-known technology for the separation of CO_2, there is great interest in developing advanced materials with improved CO_2 sorption capacity and selectivity. Cheng et al. [23] use a high-performance zeolite to study the adsorption of CO_2 from a flue gas in three consecutive beds, in order to achieve separated streams of CO_2 and N_2 with gas purity above 90%. Modelling and experimental results demonstrate that the proposed PSA configuration is able to reach the targeted gas purities with a moderate energy consumption of 1.2 GJ/t CO_2. The use of modified zeolites to improve the CO_2 capture and/or reduce the cost of the sorbent is also a subject of study. Coal fly ash zeolites appear as an attractive option in order to use this typical waste of fuel combustion instead of its disposal. Laboratory tests for CO_2 adsorption onto this type of material show promising results (about 123 mg/g of sorbent) operating at temperatures around 60 °C, and subsequent regeneration at about 150 °C [24].

Among the emerging CO_2 capture technologies, calcium looping offers a competitive energy efficiency and moderate cost for the removal of CO_2 in both pre-combustion and post-combustion systems [13]. Recent studies demonstrate the beneficial effect of CaO (supported over iron oxide) in the gasification of biomass, not only to separate CO_2 from the product gas, thereby increasing the production of H_2, but also for the removal of the HCl generated during the gasification [25]. Calcium looping can also be applied for thermochemical storage thanks to the cyclic carbonation and calcination of calcium-based materials. In these systems, the energy required for the process (i.e., for the calcination of $CaCO_3$ that is highly endothermic) is supplied from intermittent renewable sources that are able to provide high-temperature heat (e.g., solar). When energy production is needed, the resulting CaO obtained from the calcination is carbonated, generating high-quality heat at temperatures between 600 and 750 °C. A recent techno-economic study revealed that this type of calcium looping system is able to produce electricity at prices ranging from 140 to 20 USD/MWh for energy inputs of between 50 and 1000 MW, while the CO_2 capture cost ranges from 45 to 27 USD/tCO_2-captured [26].

Conflicts of Interest: The author declares no conflict of interest.

References

1. International Energy Agency. *Global Energy Review: CO_2 Emissions in 2021*; IEA: Paris, France, 2022.
2. Masson-Delmotte, V.; Zhai, P.; Pörtner, H.O.; Roberts, D.; Skea, J.; Shukla, P.R.; Pirani, A.; Pean, C.; Pidcock, R.; Connors, S.; et al. *Global Warming of 1.5 °C: An IPCC Special Report on the Impacts of Global Warming of 1.5 °C above Pre-Industrial Levels and Related Global Greenhouse Gas Emission Pathways, in the Context of Strengthening the Global Response to the Threat of Climate Change, Sustainable Development, and Efforts to Eradicate Poverty*; IPCC: Geneva, Switzerland, 2018.
3. McKinsey & Company. Global Energy Perspective 2022. Executive Summary. 2022. Available online: https://www.mckinsey.com/~{}/media/McKinsey/Industries/Oil%20and%20Gas/Our%20Insights/Global%20Energy%20Perspective%202022/Global-Energy-Perspective-2022-Executive-Summary.pdf (accessed on 10 October 2022).
4. International Energy Agency. *Global Energy Review 2020*; International Energy Agency: Paris, France, 2020; Available online: https://www.iea.org/reports/global-energy-review-2020 (accessed on 10 October 2022).
5. Havercroft, I.; Consoli, C. *Is the World Ready for Carbon Capture and Storage? Global CCS Institute Report*; Global CCS Institute: Docklands, VIC, Australia, 2018.
6. Olivier, G.J.; Schure, K.M.; Peters, J.A.H.W. *Trends in Global CO_2 and Total Greenhouse Gas Emissions: 2017 Report*; PBL Netherlands Environmental Assessment Agency: The Hague, The Netherlands, 2017.
7. Regufe, M.J.; Pereira, A.; Ferreira, A.; Ribeiro, A.M.; Rodrigues, A.E. Current developments of Cabon Capture Storage and/or Utilization-Looking for net-zero emissions defined in the Paris Agreement. *Energies* **2021**, *14*, 2406. [CrossRef]
8. Global CCS Institute. *The Global Status of CCS: 2017*; Global CCS Institute Report; Global CCS Institute: Docklands, VIC, Australia, 2017.
9. Fernández, J.R.; Garcia, S.; Sanz-Pérez, E. CO_2 capture and utilization editorial. *Ind. Eng. Chem. Res.* **2020**, *15*, 6767–6772. [CrossRef]
10. Santos, M.P.S.; Hanak, D.P. Carbon capture for decarbonisation of energy-intensive industries: A comparative review of techno-economic feasibility of solid looping cycles. *Front. Chem. Sci. Eng.* **2022**, *16*, 1291–1317. [CrossRef]
11. Fernandez, J.R. Process Simulations and Experimental Studies of CO_2 Capture. *Energies* **2022**, *15*, 544. [CrossRef]
12. Liang, Z.; Rongwong, W.; Liu, H.; Fu, K.; Gao, H.; Cao, F.; Zhang, R.; Sema, T.; Henni, A.; Sumon, K.; et al. Recent progress and new developments in post-combustion carbon-capture technology with amine based solvents. *Int. J. Greenh. Gas Control* **2015**, *40*, 26–54. [CrossRef]
13. Abanades, J.C.; Arias, B.; Lyngfelt, A.; Mattisson, T.; Wiley, D.E.; Li, H.; Ho, M.T.; Mangano, E.; Brandani, S. Emerging CO_2 capture systems. *Int. J. Greenh. Gas Control* **2015**, *40*, 126–166. [CrossRef]
14. Stanger, R.; Wall, T.; Spörl, R.; Paneru, M.; Grathwohl, S.; Weidmann, M.; Scheffknecht, G.; McDonald, D.; Myöhänen, K.; Ritvanen, J.; et al. Oxyfuel combustion for CO_2 capture in power plants. *Int. J. Greenh. Gas Control* **2015**, *40*, 55–125. [CrossRef]
15. Nurdiawati, A.; Urban, F. Towards Deep Decarbonisation of Energy-Intensive Industries: A Review of Current Status, Technologies and Policies. *Energies* **2021**, *14*, 2408. [CrossRef]

16. Toktarova, A.; Karlsson, I.; Rootzen, J.; Görasson, L.; Odenberger, M.; Johnson, F. Pathways for Low-Carbon Transition of the Steel Industry—A Swedish Case Study. *Energies* **2020**, *13*, 3840. [CrossRef]
17. Karlsson, I.; Rootzen, J.; Toktarova, A.; Odenberger, M.; Johnson, F.; Görasson, L. Roadmap for Decarbonization of the Building and Construction Industry—A Supply Chain Analysis Including Primary Production of Steel and Cement. *Energies* **2020**, *13*, 4136. [CrossRef]
18. Aboumahboub, T.; Brecha, R.J.; Shrestha, H.; Fuentes, U.; Geiges, A.; Hare, W.; Schaeffer, M.; Welder, L.; Gidden, M.J. Decarbonization of Australia's Energy System: Integrated Modeling of the Transformation of Electricity, Transportation, and Industrial Sectors. *Energies* **2020**, *13*, 3805. [CrossRef]
19. Ma, C.; Zhang, W.; Zheng, Y.; An, A. Economic Model Predictive Control for Post-Combustion CO_2 Capture System Based on MEA. *Energies* **2021**, *14*, 8160. [CrossRef]
20. Villarroel, J.A.; Palma-Cando, A.; Viloria, A.; Ricaurte, M. Kinetic and Thermodynamic Analysis of High-Pressure CO_2 Capture Using Ethylenediamine: Experimental Study and Modeling. *Energies* **2021**, *14*, 6822. [CrossRef]
21. Gladysz, P.; Sowizdzal, A.; Miecznik, M.; Hacaga, M.; Pajak, L. Techno-Economic Assessment of a Combined Heat and Power Plant Integrated with Carbon Dioxide Removal Technology: A Case Study for Central Poland. *Energies* **2020**, *13*, 2841. [CrossRef]
22. Ahn, J.; Kim, H.J. Combustion Characteristics of 0.5 MW Class Oxy-Fuel FGR (Flue Gas Recirculation) Boiler for CO_2 Capture. *Energies* **2021**, *14*, 4333. [CrossRef]
23. Cheng, C.; Kuo, C.; Yang, M.; Zhuang, Z.; Lin, P.; Chen, Y.; Yang, H.; Chou, C. CO_2 Capture from Flue Gas of a Coal-Fired Power Plant Using Three-Bed PSA Process. *Energies* **2021**, *14*, 3582. [CrossRef]
24. Boycheva, S.; Marinov, I.; Zgureva-Filipova, D. Studies on the CO_2 Capture by Coal Fly Ash Zeolites: Process Design and Simulation. *Energies* **2021**, *14*, 8379. [CrossRef]
25. Dashtestani, F.; Nusheh, M.; Siriwongrungson, V.; Hongrapipat, J.; Materic, V.; Yip, A.; Pang, S. Effect of the Presence of HCl on Simultaneous CO_2 Capture and Contaminants Removal from Simulated Biomass Gasification Producer Gas by $CaO-Fe_2O_3$ Sorbent in Calcium Looping Cycles. *Energies* **2021**, *14*, 8167. [CrossRef]
26. Martinez-Castilla, G.; Guio-Perez, D.C.; Papadokonstantakis, S.; Pallares, D.; Johnson, F. Techno-Economic Assessment of Calcium Looping for Thermochemical Energy Storage with CO_2 Capture. *Energies* **2021**, *14*, 3211. [CrossRef]

Disclaimer/Publisher's Note: The statements, opinions and data contained in all publications are solely those of the individual author(s) and contributor(s) and not of MDPI and/or the editor(s). MDPI and/or the editor(s) disclaim responsibility for any injury to people or property resulting from any ideas, methods, instructions or products referred to in the content.

Article

Technological Innovation Efficiency of Listed Carbon Capture Companies in China: Based on the Dual Dimensions of Legal Policy and Technology

Xiaofeng Xu [1,2,3], Dongdong He [1,2], Tao Wang [4,5], Xiangyu Chen [1,2,*] and Yichen Zhou [6,7,*]

1. International School of Law and Finance, East China University of Political Science and Law, Shanghai 200042, China
2. Legal Research Center of Optimizing Doing Business Environment, East China University of Political Science and Law, Shanghai 200042, China
3. Center of Green Finance and Energy Development, East China University of Political Science and Law, Shanghai 200042, China
4. College of Environmental Science and Engineering, Tongji University, Shanghai 200092, China
5. UNEP-Tongji Institute of Environment for Sustainable Development, Tongji University, Shanghai 200092, China
6. Faculty of Civil Engineering, RWTH Aachen University, 52074 Aachen, Germany
7. TUV Rheinland (Shanghai) Co., Ltd., Shanghai 200072, China
* Correspondence: yichenzhou0930@126.com or yichen.zhou@rwth-aachen.de (Y.Z.); 2011150372@ecpul.edu.cn (X.C.)

Abstract: To achieve carbon neutrality and improve emission reduction efficiency, capturing carbon dioxide from the air on a large scale and promoting the application and innovation of carbon capture technology (CCUS) are the most important goals. This study undertakes an annual and comprehensive evaluation of the policy and the technological innovation efficiency (TIE) of 10 listed companies in China using the DEA model and the Malmquist index analysis method. The number of relevant laws and policies is significant, but they are not well coordinated. The static evaluation results indicate that the complete factor production rate is low, generally lower than 0.9, and the technical innovation efficiency is weak, mainly because of technological backwardness. The dynamic evaluation results indicate that the changes in total factor productivity (TFP) each year are primarily affected by changes in technological progress. This suggests that most domestic enterprises are still exploring technological innovation (TI) and operational business models. Finally, this study proposes measures to improve the TIE of carbon capture technology enterprises in China, including giving full play to the role of the government, expanding effective investment, and improving innovational ability.

Keywords: CCUS; carbon neutrality; DEA; Malmquist

Citation: Xu, X.; He, D.; Wang, T.; Chen, X.; Zhou, Y. Technological Innovation Efficiency of Listed Carbon Capture Companies in China: Based on the Dual Dimensions of Legal Policy and Technology. *Energies* **2023**, *16*, 1118. https://doi.org/10.3390/en16031118

Academic Editors: Zilong Liu, Meixia Shan and Yakang Jin

Received: 2 October 2022
Revised: 5 November 2022
Accepted: 22 November 2022
Published: 19 January 2023

Copyright: © 2023 by the authors. Licensee MDPI, Basel, Switzerland. This article is an open access article distributed under the terms and conditions of the Creative Commons Attribution (CC BY) license (https://creativecommons.org/licenses/by/4.0/).

1. Introduction

The dual-carbon target strategy is a requirement for China to promote high-quality development. Whether China's dual-carbon goal strategy can be successfully realized depends on the effect of specific policies after implementation. The government needs to formulate administrative regulations and departmental rules to restrict the carbon emission behavior of emission control enterprises [1]. The market mechanism can ensure carbon emission reduction in the most cost-effective way and, at the same time, promote the enterprise innovation of emission reduction technology to control the total carbon emission [2]. Therefore, achieving the dual-carbon goal needs the help of government-level administrative regulations, market-level price mechanisms, and TI. Government departments can promote the short-term carbon peaking goal by formulating administrative rules [3]. However, from the perspective of long-term planning, increasing emission reduction motivation will breed movement-style carbon reduction behavior. The carbon market price signal has a significant role in promoting short-term carbon emission reduction behavior. However, in the long run, TI is still the fundamental way to achieve the dual-carbon goal [4].

Carbon dioxide capture, utilization, and storage (CCUS) refers to the capture and separation of CO_2 from energy utilization, industrial processes, and other emission sources or air and transporting it to suitable sites for utilization or storage by tankers, pipelines, ships, etc., ultimately achieving CO_2 emission reduction. It is necessary for China to achieve carbon peaking and carbon neutrality. CCUS technology can not only achieve near-zero emissions from fossil energy utilization but also promote profound emission reduction in industries such as steel and cement where emission reduction is difficult. Moreover, it is vital to enhance the flexibility of the power system under carbon constraints, ensure a safe and stable power supply, offset the CO_2 and non-CO_2 greenhouse gas emissions that are difficult to reduce, and finally achieve the goal of carbon neutralization. In recent years, carbon capture, utilization, and storage (CCUS) has received extensive attention from the international community as a critical technology for mitigating climate change. Researchers have focused on the emission reduction potential, cost, and application prospects of CCUS. The technology was systematically and comprehensively evaluated. The conclusion was that CCUS technology is an indispensable combination of emission reduction technologies for the realization of global climate goals and has the potential to achieve a cumulative emission reduction effect of 100 billion tons by the middle of the 21st century. The emission reduction potential must be further increased. Considering that CCUS can effectively reduce the risk of stranded assets and offers social and environmental benefits, China should take CCUS as a strategic technology and coordinate its top-level policy based on its resource endowment and the primary national conditions of "rich coal, poor soil, and little gas". China must design and accelerate the construction of technical systems, explore market incentives, strengthen international cooperation, and promote the development of CCUS technology.

For companies that conduct research and development in carbon capture technology, there will also be rewards, including improved reputation and economic benefits. For example, Guanghui Energy announced that it plans to invest in the construction of an integrated project for three million tons of carbon dioxide capture, pipeline transportation, and oil displacement. In terms of energy acquisition, China is currently highly dependent on coal for power generation and is the world's largest steel producer. Relevant assessments indicate that natural gas and coal power generation using CCUS technology are less cost competitive and may not be cost competitive with other renewable energy sources. In addition to cost factors, large-scale deployment of CCUS faces different challenges, such as environmental risks, technical challenges, lack of funding, social opposition, and policy uncertainty. Other, more cost-competitive low- or zero-emission options may emerge, and CCUS will become less attractive. Of course, the emergence of new low-carbon technologies is speculative, and their applicability will need to be tested in practice. Overall, the current CCUS technology is still in the stage of conception to experimentation, and there is still a long way to go before large-scale commercial application. Therefore, we considered China's listed companies with carbon capture-related technologies as samples, modeled their R&D investment and financial data, and explored the TIE and influencing factors of carbon capture technology companies. The innovation of this paper is in conducting an empirical study on the profit of carbon capture technology companies using carbon capture technology innovation; if carbon capture technology can lead to an increase in corporate profits, then related companies will have more significant incentives to engage in its development and utilization.

During the past few years, relevant policies and guidelines have been successively issued, including the Outline of the National Medium- and Long-Term Science and Technology Development Plan, China's Climate Change Action Plan, and China's Special Action on Climate Change Science and Technology, etc., clearly proposing CCUS technology as an essential plan and key development technology for future national development. At the same time, the Chinese CCUS Technology Development Roadmap Research was also issued, which is a series of policies to guide and encourage large domestic energy companies in carrying out R&D demonstration projects. Currently, China clearly defines CCUS technology as a significant demonstration project to guide and support carbon peaks and carbon neutralization work, which requires the promotion of large-scale CCUS technology research and industrial application.

2. Literature Review

2.1. CCUS Technologies Review

2.1.1. Development Status of Carbon Capture Technology

Carbon capture is the first link in CCUS and the main cost source in the CCUS process. Carbon is mainly captured from industrial exhaust gases and the atmosphere, and the higher the CO_2 concentration, the lower the capture cost. According to the order of carbon capture and combustion, carbon capture technologies can be divided into pre-combustion capture, post-combustion capture, and oxy-fuel combustion capture [5]. The cost of pre-combustion capture is relatively low, and the efficiency is high, but the applicability is not high. Although post-combustion capture is widely used, it has higher relative energy consumption and cost. Oxy-fuel combustion has high requirements for the operating environment and is still in the demonstration stage. According to the separation process, carbon capture technology is mainly divided into physical absorption technology, chemical absorption technology, membrane separation technology, low-temperature separation technology, etc. [6].

Carbon utilization is key to reducing the cost of CCUS implementation. Currently, geological utilization is the primary mode in China, and chemical and biological utilization are relatively rare. Specifically, CO_2-EOR technology in geological utilization can sequester a large amount of CO_2 and increase oil production, considering the economic and environmental benefits, and has high feasibility in the short term [7]. Chemical utilization is based on the main characteristics of chemical conversion, converting CO_2 and reactants into target products to achieve resource utilization; it has low requirements for CO_2 concentration and common implementation cost, which has development value. Bioutilization is the primary means of biological conversion; when using CO_2 for biomass synthesis, the concentration of CO_2 is high, and the implementation cost is high, but the yield per ton of CO_2 is also relatively high.

However, CO_2 emissions far exceed their utilization capacity, and the CO_2 that cannot be used must be stored using storage technology. Carbon sequestration is mainly divided into technologies such as saline aquifer storage and depleted oil and gas reservoir storage [8]. Between them, the saline water layer is widely distributed and has good closure and the ideal storage effect. Depleted oil and gas reservoirs usually have a complete, closed, and stable geological environment, which can ensure the safety of storage. Still, there is a particular risk of leakage, requiring multi-directional monitoring technology.

2.1.2. Research Status of Carbon Capture Technology

M. Paweł reviewed the research on various CCUS technologies, summed up the advantages and development constraints of different technologies, and proposed a CCUS technology with good application prospects, pointing out that in the future, CCUS technology urgently needs to strengthen the research on cost and risk control. Developing new materials and TIs will boost the development of this technology. At the same time, multi-scale monitoring technology will ensure the safety of the entire project implementation and provide effective solutions for alleviating global warming [9]. P. Wienchol focused on three typical thermal power and cement industries and summarized the overall development of CCUS in the three specific initiatives. It was pointed out that the application of CCUS technology in specific industries in China is still in the early stage of development, and the synergistic effects of CCUS technology (such as increased energy consumption in thermal power plants, reaction with products in cement plants, etc.), installation costs, environmental impacts, and risks are all limits to the large-scale application of CCUS. It was recommended to refine the emission reduction technology plans for thermal power, steel, and cement industries, provide exceptional funding support for CCUS technology R&D and demonstration projects in specific industries, and further promote the integrated demonstration and commercial application of the CCUS chain [10]. Chenggang Wang conducted research and found that improving carbon capture technology will drive enterprises' TIE development [11]. Gao found an inverted U-shaped relationship between intellectual property protection and manufacturing technical efficiency. Strengthening intellectual property protection at this stage can significantly improve the technical efficiency of China's manufacturing industry [12]. Yang used CiteSpace software to quantitatively analyze the CCUS technology patents in the Derwent patent database and pointed out that the United States and China are the major CCUS technology research countries, and most of China's patent holders are universities and institutions [13]. Waste gas treatment and CO_2 flooding are the hot technologies for CCUS patents; in carbon capture, the hot technology is desorption tower technology, and converting CO_2 into inorganic salts using calcium-containing compounds is an emerging field. In carbon utilization technology, CO_2 is used to generate carbon fiber. As a hot technology, mixed utilization is an innovative research direction; in carbon sequestration, the technical patents for tanks and release devices are hot technologies, and packaging technologies in different environments and device inventions for transportation are new research directions.

2.2. Review of TI

Through a study of the literature, we found that TI research fields have mainly concentrated on three aspects. One is the influence of internal and external factors on innovation activities. The second is the analysis of the TI capability of companies. The third is researching methods of measuring the efficiency of TI.

2.2.1. Research on the Influencing Factors of TI

An in-depth analysis of the factors that affect the TI of enterprises can give us a deeper understanding of the development level of the TI field of enterprises. Therefore, scholars have been focused on "factors affecting TI of enterprises" and divided the influencing factors into two aspects: external and internal. Regarding the research on external influencing factors, Yana Rubashkina found that rules and regulations promulgated by the government play an essential role in TI through a study of European manufacturing enterprises [14]. Wu found in his study that the national rules and regulations have a specific promotional effect on the innovation performance of enterprises [15]. Hong analyzed innovation from the perspective of efficiency. He found that the efficiency of enterprise TI is negatively correlated with government support, and the increase in subsidies will reduce the efficiency of TI [16]. Bigliardi et al. performed the same research and proved a significant relationship between government subsidies and TIE [17]. From the perspective of internal influencing factors, Dabic found that the degree of internationalization positively affects the innovation of enterprises. As the degree of internationalization increases, the knowledge and technology absorbed will encourage the enterprise to better carry out innovation activities [18]. Through research on pharmaceutical companies, Laermann-Nguyen discovered a specific relationship between independent innovation ability and TI ability [19]. Shearmur took export enterprises as the object of investigation and analysis and found that, compared with non-export enterprises, export enterprises have a higher level of innovation that is proportional to their export scale [20]. Through the above research, we gained a good understanding of the various factors that affect the efficiency of enterprise TI, but more in-depth research is needed.

2.2.2. Research on the Evaluation of TI Capability

Sumrit and Detcharat researched management capabilities, established a scientific comprehensive evaluation model, and concluded that management capabilities can promote

TI capability improvement [21]. Hai conducted research on the automobile industry as an example, established a TIE evaluation model, and concluded that the TIE of the automobile industry is not ideal. The scale of innovation input cannot obtain good returns for output and should be increased. Investing in large-scale innovative talents also requires increased government subsidies and the emphasis on output capacity, as well as relevant strategies to promote the rapid development of the industry [22]. C. Feniser used the classic theory of TRIZ to analyze enterprises regarding the two aspects of risk management and TI and constructed a reasonable and complete innovation evaluation system [23]. K. Lee studied and evaluated the TI of SMEs using the hidden Markov model and Viterbi algorithm and determined in-depth related indicators [24]. Through the analysis and summary of the above literature, we found that scholars cannot test TI's ability without the selection and analysis of influencing factors. We discovered that research on TI and related influencing factors is essential.

2.2.3. Related Research on TI Evaluation Methods

There are many methods to study the efficiency of TI, such as DEA, stochastic frontier analysis, Tobit method, etc. By reading the relevant literature, we summarized the methods for studying the efficiency of TI of enterprises as follows. Wang used the Malmquist index analysis method to evaluate the TIE of manufacturing enterprises and found the main influencing factor that promotes TIE [25]. Lanoie's research determined that government environmental policies positively impact environmental efficiency, and different policy intensities have heterogeneous effects [26]. Wong's research indicated that the positive impact of green process innovation on green innovation efficiency and revenue is evident, while green product innovation is negative. Wong proposed that green innovation includes product and process innovation [27]. Ghisetti and Rennings evaluated green innovation efficiency in energy consumption and environmental pollution [28]. Zhang used the input–output method to evaluate the input–output ratio of TI of industrial enterprises and conducted a comparative study [29]. Guo and Yang evaluated the efficiency of green innovation in each region [30]. Liu et al. researched the sustainability of the coal industry, taking environmental, production, and other factors into account [31]. Wang pointed out that the lag in environmental efficiency has a hindering effect on TIE [32]. Using a case study approach, Rumanti devised a new model for green innovation. For other assessments related to green innovation, see [33]. Govindan et al. reviewed multi-criteria decision-making methods for evaluating and selecting suppliers using green technology [34]. Sun et al. used the TOPSIS method to establish a model to evaluate the impact of TIE on the ecology and economic benefits [35]. Guo evaluated the level of green technology innovation from the perspective of green development and discussed the role of government environmental regulation [36]. Lin used the DEA method to assess green technology's innovation efficiency in 28 manufacturing industries in China from 2006 to 2014 [37]. The research of Lee and Choi suggested that the innovation effect leads to environmental productivity in the Korean manufacturing industry. Not only should each sector strive to improve performance, but the government must also formulate specific measures to improve overall competitiveness [38].

2.3. Current Status of Laws and Policies on Carbon Capture Technology in China

Currently, China has relatively comprehensive rules and regulations in energy conservation and emission reduction, clean and renewable energy. However, CCUS technology is still a blank page [39]; laws such as the Environmental Protection Law, Administrative Penalty Measures for Environmental Protection, the Environmental Impact Assessment Law, and the Water Pollution Prevention and Control Law do not include CCUS technology-related content. There is no CCUS legal framework for enterprises to refer to in their policies and measures. The entire CCUS chain involves different industries and departments, including national, local, enterprise, petroleum, coal, electric, chemical industry, etc. A corresponding regulatory system, overall coordination mechanism, and industrialization layout guidance policy have not been issued, especially regarding CCUS technology [40]. After realizing that the economic incentive measures for carbon emission reduction, the long-term high-cost investment and low-profit return will inevitably lead to a long-term profit and loss imbalance after enterprises carry out large-scale CCUS projects; enterprises have chosen to carry out small or no CCUS projects, which has greatly hindered the advancement of CCUS technology in China.

3. Model and Index Selection

3.1. Model

This study mainly adopted the data envelopment analysis (DEA) model with the non-parametric method, which integrates mathematics, operation research, and other contents and is a standard method for evaluating relative effectiveness in economics. Compared with traditional data analysis, DEA greatly reduces the influence of human factors and improves the objectivity of the evaluation results; thus, the relative efficiency obtained is more practical. According to the factors of production rate, scale efficiency, pure technical efficiency, and redundancy rate of DEA output, optimization directions and approaches can be proposed in a targeted manner. This paper mainly used the input-oriented variable returns to scale (VRS) DEA-BCC analysis model (which was put forth by Banker, Charness, Cooper) and the DEA-Malmquist index model to evaluate the TIE of China's carbon capture-listed companies. In this way, we could explore whether enterprises obtained better income by engaging in carbon capture technology innovation. The typical and commonly used data envelopment analysis (DEA) methods are CCR (which was put forth by Charnes, Cooper, and Rhodes) and BCC models:

$$\min[\theta - \varepsilon(e^- S^- + e^+ S^+)] \text{ s.t.} \begin{cases} \sum_{j=1}^{n} x_j \lambda_j + S^- = \theta x_k \\ \sum_{j=1}^{n} y_j \lambda_j - S^+ = y_k \\ \lambda \gg 0, j = 1, 2, \ldots, n \\ S^+ = \left(S_1^+ S_2^+, S_3^+, \ldots S_q^+\right)^T \gg 0 \\ S^- = \left(S_1^- S_2^-, S_3^-, \ldots, S_p^-\right)^T \gg 0 \end{cases}$$

In the formula: θ is the technical efficiency value, $0 \leq \theta \leq 1$; ε is the non-Archimedes infinitesimal; S^-, $S^+ \geq 0$ are the input and output slack variables, respectively; e T1 is the m-dimensional unit vector; e T2 is an n-dimensional unit vector; $\lambda_i \geq 0$ is a weight variable; x is the j input of the j decision-making unit; y_j is the m output of the j decision-making unit.

Banker, Charnes, and Cooper considered that the decision-making unit could not produce at the optimal production scale because of factors such as imperfect competition and capital constraints in practice, so they improved the CCR model. They introduced controls into the model and proposed a BCC model with variable returns to scale. In addition, DEA requires that the data must be cross-sectional data at the same time as evaluating the relative efficiency of the unit, and the time dimension cannot be introduced for analysis. To solve this problem, the Malmquist index was introduced to analyze the cross-sectional data in the time dimension, that is, panel data. Therefore, the dynamic change law of the technical efficiency of the evaluation unit can be obtained.

The Malmquist index is called the TFP index (TFP), which Sten first proposed in 1953 to analyze the data changes in different periods more vividly. Later, through the continuous improvement by scholars such as Charnes based on DEA, the DEA-Malmquist index model was further decomposed, including four parts: technological progress change, technical efficiency change, pure technical efficiency changes, and scale efficiency change.

$$M(x_t, y_t, x_{t+1}, y_{t+1}) = \frac{D^{t+1}(x_{t+1}, y_{t+1})}{D^t(x_t, y_t)} \times \left[\frac{D^t(x_{t+1}, y_{t+1})}{D^{t+1}(x_{t+1}, y_{t+1})} \times \frac{D^t(x_t, y_t)}{D^{t+1}(x_t, y_t)}\right]^{\frac{1}{2}} = \text{EFFCH} \times \text{TECH}$$

The results determine the dynamic changes in efficiency between different periods, which can not only indicate the changes in the overall year but also help to analyze the differences of each decision-making unit between years. The DEA-Malmquist index model and the DEA model can form complementary effects. The DEA model measures the efficiency value of each decision-making unit in a period but cannot observe the change of each decision-making unit in a period. The DEA-Malmquist index model can make up for this shortcoming and keep the evolution of the efficiency value of each enterprise in the time series. Therefore, this paper first constructed a DEA model to conduct a static analysis of the measurement results of each enterprise and, second, created a DEA-Malmquist index model to perform a dynamic analysis of the measurement results of each enterprise from 2018 to 2021. The TIE value of China's carbon capture listed companies can be measured from two aspects, dynamic and static, to reveal their TI capabilities more comprehensively.

3.2. Variable Selection

3.2.1. Study Area and Data Source

As shown in Table 1, we selected the data of 10 listed companies from 2018 to 2021 as the research area. At the same time, when collecting and arranging relevant data, the primary data sources were the official websites of the SHSE and SZSE and the WIND and Cathay Pacific databases.

Table 1. Panel data sample of carbon capture technology listed companies.

Company Name	Securities Code
Moon Environment Technology Co., Ltd.	000811
Hangzhou Oxygen Plant Group Co., Ltd.	002430
Xizi Clean Energy Equipment Manufacturing Co., Ltd.	002534
Hunan Kaimeite Gases Co., Ltd.	002549
Sunresin New Materials Co., Ltd.	300487
SPIC Yuanda Environmental-Protection Co., Ltd.	600292
Haohua Chemical Science and Technology Co., Ltd.	600378
Wuxi Huaguang Environment and Energy Group Co., Ltd.	600475
Shuangliang Eco-energy Systems Co., Ltd.	600481
Guanghui Energy Co., Ltd.	600256

3.2.2. Selection of Input–Output Indicators

As shown in Table 2, the selection of indicators is an indispensable step in empirical research, and the scientific degree of their selection has an important influence on the realization of the research purpose [25]. The carbon capture technology industry has the characteristics of strong dependence on environmental protection, and it is an industry that is highly dependent on policies and has more concentrated R&D resources. The domestic market demand continues to expand with the continuous improvement of carbon neutrality target policy requirements. In the face of a broad market, a product or TI is regarded as a decisive factor in the competition among carbon capture technology companies. The earlier companies can develop advanced carbon capture technologies, the more likely they are to occupy a larger market share. Concerning the technical innovation efficiency articles for the relevant listed companies, the specific indicators constructed by the relevant research were sorted out, and the characteristics of the carbon capture technology industry itself were considered. The input indicators screened in this paper included the number of R&D employees and R&D expenses; output indicators included patent licensing volume and operating income. The specific list is as follows:

Table 2. Input–output indicators of the TI of carbon capture technology enterprises.

	Indicator Name	Symbol
Input indicators	R&D expenses	X1
Input indicators	Number of R&D employees	X2
Output indicators	Main business income	Y1
Output indicators	Number of patents granted	Y2

4. Results and Analysis

4.1. Static Analysis

4.1.1. Data Analysis in 2018

As shown in Table 3, in 2018 the average technical efficiency (ATE) was 0.2342, the average pure technical efficiency (APTE) was 0.4809, and the average scale efficiency (ASE) was 0.5229. The low innovation efficiency was the result of the combination of pure technical efficiency and scale efficiency, and APTE was the main reason, with the upside close to 52%. According to the results, only Guanghui Energy had a technical efficiency of 1, which was at the frontier of efficiency. There was no enterprise with a technical efficiency between 0.9 and 1, indicating a trend of polarization among the listed companies. The TIE of other enterprises was relatively weak. Among them, Sunresin New Materials had the lowest score, and the technical efficiency was only 0.0053. Compared with the traditional lithium extraction technology, the DLE technology with independent intellectual property rights of Sunresin New Materials was more efficient and low-carbon, did not use any solvent, and would not affect the aquifer in the salt-lake area. Still, the technical level needed further improvement. The technical efficiency of most enterprises was even less than 0.5.

Table 3. 2018 Enterprise Data.

Firm	Crste	Vrste	Scale	
Moon Environment Technology Co., Ltd.	0.116	0.16	0.723	drs
Hangzhou Oxygen Plant Group Co., Ltd.	0.126	0.809	0.156	drs
Xizi Clean Energy Equipment Manufacturing Co., Ltd.	0.073	0.106	0.69	drs
Hunan Kaimeite Gases Co., Ltd.	0.084	0.492	0.171	irs
Sunresin New Materials Co., Ltd.	0.053	0.356	0.149	irs
SPIC Yuanda Environmental-Protection Co., Ltd.	0.298	0.347	0.858	irs
Haohua Chemical Science and Technology Co., Ltd.	0.122	0.357	0.34	drs
Wuxi Huaguang Environment and Energy Group Co., Ltd.	0.321	1	0.321	drs
Shuangliang Eco-energy Systems Co., Ltd.	0.149	0.182	0.821	drs
Guanghui Energy Co., Ltd.	1	1	1	-
mean	0.2342	0.4809	0.5229	

The pure technical efficiency of Wuxi Huaguang, Environment and Energy Group, reached 1, indicating that this enterprise's low technical efficiency value was mainly affected by scale efficiency. The scale efficiency of Shuangliang Eco-energy Systems and SPIC Yuanda Environmental-Protection was above 0.8, indicating that the technical inefficiency of these two enterprises was mainly due to the influence of pure technical inefficiency. SPIC Yuanda Environmental-Protection Company's business focuses on the general contracting of desulfurization, denitrification, dust removal projects, desulfurization and denitrification franchise operations, water engineering and operations, catalyst manufacturing and regeneration, ecological restoration projects, and dust collector equipment manufacturing and installation. It is at the forefront of the industry and has the first domestic 10,000-ton carbon capture device in Hechuan, Chongqing. It was prominent in scale but lacked TI.

In addition, the companies that had not reached the frontier were basically in a state of decreasing scale, and insufficient TI constrained their efficiency value.

4.1.2. Data Analysis in 2019

As shown in Table 4, the TIE in 2019 was not much different from the previous year, with an ATE of 0.1271, an APTE of 0.4261, and an ASE of 0.415. The main reason for the inefficiency of innovation was the combination of PTE and SE. According to the results, the technical efficiency of all enterprises was lower than 0.2, which was relatively low. This was not much different from the previous year, indicating the same problems in 2018 and 2019. Except for Haohua Chemical Science and Technology, whose pure technical efficiency reached 0.838, the TIE of other companies was relatively weak. Among them, Xizi Clean Energy Equipment Manufacturing was the company with the lowest score of only 0.059. Among them, the scale efficiency of Hangguo and SPIC Yuanda Environmental-Protection reached 0.9, indicating that the low technical efficiency value of these two companies was mainly affected by pure technical efficiency. Xizi Clean Energy Equipment Manufacturing mainly produces waste heat boilers that use waste heat from various industrial processes, wastes, or waste liquids and the heat generated by the combustion of combustible substances to heat water to a working quality. The business of Xizi Clean Energy is mainly equipment manufacturing. It focuses on the sun's thermal energy to generate electricity, which may be one reason its carbon capture technology is progressing more slowly. Unlike the previous year, among the companies that did not reach the frontier in 2019, there were more companies with increasing returns to scale, indicating that some companies were actively adjusting the structure of corporate resource investment, but the effect was not yet apparent.

Table 4. 2019 Enterprise Data.

Firm	Crste	Vrste	Scale	
Moon Environment Technology Co., Ltd.	0.14	0.339	0.412	drs
Hangzhou Oxygen Plant Group Co., Ltd.	0.079	0.665	0.118	drs
Xizi Clean Energy Equipment Manufacturing Co., Ltd.	0.058	0.059	0.971	irs
Hunan Kaimeite Gases Co., Ltd.	0.121	0.386	0.314	irs
Sunresin New Materials Co., Ltd.	0.045	0.187	0.239	irs
SPIC Yuanda Environmental-Protection Co., Ltd.	0.175	0.183	0.955	irs
Haohua Chemical Science and Technology Corp., Ltd.	0.184	0.838	0.219	drs
Wuxi Huaguang Environment and Energy Group Co., Ltd.	0.166	0.733	0.227	drs
Shuangliang Eco-energy Systems Co., Ltd.	0.181	0.445	0.408	drs
Guanghui Energy Co., Ltd.	0.122	0.426	0.287	drs
mean	0.1271	0.4261	0.415	

4.1.3. Data Analysis in 2020

As shown in Table 5, the TIE of listed companies in 2020 slightly improved compared to the previous year. The ATE was 0.1708, APTE was 0.5749, and ASE was 0.4259. There was large room for improvement in SE and PTE. According to the results, the technical efficiency of Guanghui Energy dropped to 0.069, but its scale efficiency was 0.990, indicating that the decline in its technical efficiency was mainly affected by the low PTE, which was only 0.069. The score of Sunresin New Materials was the lowest at only 0.046, which was not much different from the scores of the previous two years. Moon Environment Technology mastered the core technologies of the −271 to 800 °C temperature range, 0~90 MPa full pressure, from the conventional single working fluid to mixed working fluid and slight molecule special gas compression. It continues to lead the technological progress of the industry. In technical services, Haohua Chemical Science and Technology has apparent advantages in pressure swing adsorption gas separation technology (PSA), and it is one of the world's three largest PSA technology service providers. It can also be seen from the table that the PTE of Binglun Environment and Haohua Chemical Science and Technology reached 1, indicating that the low technical efficiency values of these two companies were mainly affected by scale efficiency.

Table 5. 2020 Enterprise Data.

Firm	Crste	Vrste	Scale	
Moon Environment Technology Co., Ltd.	0.229	1	0.229	drs
Hangzhou Oxygen Plant Group Co., Ltd.	0.089	0.888	0.1	drs
Xizi Clean Energy Equipment Manufacturing Co., Ltd.	0.101	0.527	0.191	drs
Hunan Kaimeite Gases Co., Ltd.	0.399	0.512	0.78	drs
Sunresin New Materials Co., Ltd.	0.046	0.213	0.218	irs
SPIC Yuanda Environmental-Protection Co., Ltd.	0.123	0.144	0.857	irs
Haohua Chemical Science and Technology Corp., Ltd.	0.172	1	0.172	drs
Wuxi Huaguang Environment and Energy Group Co., Ltd.	0.276	0.908	0.303	drs
Shuangliang Eco-energy Systems Co., Ltd.	0.204	0.488	0.419	drs
Guanghui Energy Co., Ltd.	0.069	0.069	0.99	irs
mean	0.1708	0.5749	0.4259	

4.1.4. Data Analysis in 2021

As shown in Table 6, in 2021 the TIE of listed companies dropped slightly again, with an ATE of 0.1544, an APTE of 0.6515, and an ASE of 0.2909. However, there was still room for an improvement in scale efficiency of nearly 70%. According to the results, the technical efficiency of Hunan Kaimeite Gases was relatively high. It was only 0.435, but its pure technical efficiency reached 0.925, which was high, indicating that scale efficiency was its main factor. Kaimet Gas is a listed company that uses the tail gas (waste gas) and flare gas emitted by petroleum and petrochemical enterprises as raw materials, and separates, purifies, and recycles the valuable components.

Table 6. 2021 Enterprise Data.

Firm	Crste	Vrste	Scale	
Moon Environment Technology Co., Ltd.	0.198	1	0.198	drs
Hangzhou Oxygen Plant Group Co., Ltd.	0.102	1	0.102	drs
Xizi Clean Energy Equipment Manufacturing Co., Ltd.	0.045	0.436	0.104	drs
Hunan Kaimeite Gases Co., Ltd.	0.435	0.925	0.471	drs
Sunresin New Materials Co., Ltd.	0.056	0.152	0.371	irs
SPIC Yuanda Environmental-Protection Co., Ltd.	0.139	0.214	0.653	drs
Haohua Chemical Science and Technology Corp., Ltd.	0.124	1	0.124	drs
Wuxi Huaguang Environment and Energy Group Co., Ltd.	0.257	1	0.257	drs
Shuangliang Eco-energy Systems Co., Ltd.	0.09	0.194	0.464	drs
Guanghui Energy Co., Ltd.	0.098	0.594	0.165	drs
mean	0.1544	0.6515	0.2909	

The company recovers the high-purity carbon dioxide produced by the purification of tail gas, which is applied to many fields such as food, metallurgy, tobacco, agriculture, the chemical industry, and electronics. Currently, Kaimet Gas has nine subsidiaries in Yueyang, Huizhou, and other places. The production of high-purity carbon dioxide in 2020 will be

460,000 tons. The technical level was relatively high, and it is necessary and feasible to expand the scale further. The technical efficiency scores of Hangguo and Sunresin New Materials were the lowest, at only 0.045 and 0.056, respectively. These two companies have been at the bottom in terms of technical efficiency in recent years, and the problem of low TIE is more severe and should be adjusted in time. In addition, the pure technical efficiency of Binglun Environment, Hangzhou Oxygen Plant Group, Haohua Chemical Science and Technology, and Huaguang Huanneng reached 1, indicating that these four companies' low technical efficiency value was mainly affected by scale efficiency. The common scale efficiency of Hangzhou Oxygen Plant Group was primarily due to the contraction of the equipment business in recent years because of downstream fluctuations such as those in the steel and chemical industries. This also indicated that the industry and policies greatly influence the carbon capture industry. Since the independent research and development of the first lithium bromide refrigerator, Shuangliang Eco-energy Systems has provided society with more than 30,000 energy-saving devices, which is equivalent to building 25 fewer 600-megawatt thermal power plants and reducing carbon dioxide emissions by 100 million tons per year. In addition, 27 hectares of forest will be rebuilt, saving 2.83 billion m^3/year of water. However, there is still room for further improvement in its market size. Energy-saving equipment is still one of the essential development directions in the field of carbon capture.

4.2. Dynamic Analysis

4.2.1. Year Perspective

In Table 7, we notice that the average TFP of the 10 listed companies from 2018 to 2021 was 0.930, a decrease of 7%. The average change in technical efficiency was 1.618, and the technical efficiency increased by 61.8%. Among them, the average change in pure technical efficiency was 1.302, an increase of 30.2%; the average change in scale efficiency was 1.243, an increase of 24.3%. The average technological progress was 0.575, down 42.5%. From the table, we can conclude that the decline in the TFP of the listed companies was mainly affected by technological progress, indicating that the lag in technological progress has led to the decline in TFP. From a vertical perspective, the TFPs in 2019–2020 and 2020–2021 were both greater than 1, and the growth rates were relatively large, reaching 12.2% and 2.6%, respectively, of which technological progress had the greater impact, increasing by 18%, 4%, and 24.3%, respective to each period. Through the analysis of each period, especially in 2018–2019, the technical efficiency increased by 430.93%, and the pure technical efficiency and scale efficiency increased by more than 100%. However, because of the decrease of 87.1% in technological progress, the TFP decreased by 30.3%. This suggested that the changing trend of technological progress was consistent with the changing trend of TFP, and the annual change of TFP was mainly affected by changes in technological progress.

Table 7. Corporate Malmquist Index and its Decomposition Index.

Year	Effch	Techch	Pech	Sech	Tfpch
2018–2019	5.393	0.129	2.465	2.188	0.697
2019–2020	0.952	1.18	0.97	0.981	1.122
2020–2021	0.826	1.243	0.922	0.895	1.026
Mean	1.618	0.575	1.302	1.243	0.93

4.2.2. Enterprise Perspective

As shown in Table 8, to analyze each enterprise more clearly, this paper classified 10 enterprises according to the measurement results, which were divided into solid growth type (TFP $\geq$ 1.1), weak growth type (1$\leq$ TFP < 1.1), and soft reduction type. Type (0.9 $\leq$ TFP < 1) and strong reduction type (TFP < 0.9) were four categories. There were two strong growth enterprises, Binglun Environment and Hunan Kaimeite Gases. Hunan Kaimeite Gases had the highest TFP, reaching 1.493, an increase of 49.3%. Both technical efficiency and technological progress were improved, and the increase in technological progress was 128%. The improvement of KMT gas TFP was mainly affected by technical efficiency. Likewise, Moon Environment Technology was also primarily affected by technical efficiency. The technological progress of all enterprises was relatively low, indicating that it was the key to further improvement of TFP. There were two weak growth enterprises, Haohua Chemical Science and Technology and Shuangliang Eco-energy Systems. The improvement of Haohua Chemical Science and Technology and Shuangliang Eco-energy Systems was mainly affected by the advance in pure technical efficiency. Still, their technological progress also declined significantly. Weak reduction enterprises included Hangzhou Oxygen Plant Group, Hangguo, and Sunresin New Materials. It can be seen from the analysis that the decline in TFP of these three enterprises was mainly caused by technological progress. Increasing the research and development of enterprise technology should be their focus. There were three strong reduction enterprises, SPIC Yuanda Environmental-Protection, Wuxi Huaguang Environment and Energy Group, and Guanghui Energy. Among them, Guanghui Energy saw a decline in TFP because of technological progress and technical efficiency. The technical efficiency of SPIC Yuanda Environmental-Protection and Huaguang Huaneng Energy increased significantly. The slow progress of technology leads to the decline in the production efficiency of enterprises. Therefore, promoting the research and development of new technologies should be the main development direction in the future. It is worth noting that Guanghui Energy had the most significant decline among the 10 companies, with a decrease of 63.5%, and technological progress also dropped by 56.3%. As a resource-based enterprise, Guanghui Energy mainly relies on the sales of resources for its business and has not invested much in carbon capture technology.

Table 8. Malmquist Index of Enterprise Technology Innovation and Its Decomposition Index.

Firm	Effch	Techch	Pech	Sech	Tfpch
Moon Environment Technology Co., Ltd.	2.05	0.56	1.81	1.133	1.148
Hangzhou Oxygen Plant Group Co., Ltd.	1.787	0.541	1	1.787	0.967
Xizi Clean Energy Equipment Manufacturing Co., Ltd.	1.77	0.56	1.744	1.015	0.991
Hunan Kaimeite Gases Co., Ltd.	2.28	0.655	1.267	1.8	1.493
Sunresin New Materials Co., Ltd.	1.886	0.512	1.306	1.444	0.966
SPIC Yuanda Environmental-Protection Co., Ltd.	1.498	0.571	1.424	1.052	0.855
Haohua Chemical Science and Technology Corp., Ltd.	1.486	0.708	1.409	1.054	1.053
Wuxi Huaguang Environment and Energy Group Co., Ltd.	1.457	0.597	1	1.457	0.869
Shuangliang Eco-energy Systems Co., Ltd.	1.63	0.653	1.55	1.052	1.064
Guanghui Energy Co., Ltd.	0.835	0.437	0.86	0.97	0.365
Mean	1.618	0.575	1.302	1.243	0.93

5. Discussion

First, the specific CCUS implementation path should be determined according to locally suitable geological conditions. The capture of carbon dioxide is the first stage of CCUS technology and the most critical process in the development of CCUS. The premise of developing CCUS technology is to have a sufficient "carbon source" guarantee, and carbon capture is crucial to obtaining a high-quality and abundant "carbon source." Currently, CCUS projects at home and abroad are based on carbon capture as a carrier, relying on efficient capture technology to export usable "carbon products." From the perspective of coverage technology, China's carbon dioxide capture sources currently cover various technologies such as pre-combustion, post-combustion, and oxy-fuel combustion capture in coal-fired power plants. Suitable emission sources include power plants, steel plants, cement plants, smelters, chemical fertilizers, synthetic fuel plants, and hydrogen production plants based on fossil raw materials, among which fossil fuel power plants are the most important source of carbon dioxide capture. Among the three carbon collection technologies, the two technologies of post-combustion separation and pre-combustion separation are relatively mature, and oxy-fuel combustion is still in the demonstration stage. The application of carbon capture technology is mainly concentrated in the oil and gas, coal, and power industries.

Second, various mechanisms should be established to coordinate with multiple departments to pave the way for the orderly implementation of CCUS. Accelerating the research and development of carbon emission futures products and promoting the construction of a green and low-carbon economic system with a market-oriented mechanism is of great significance for the realization of "improving the market trading system of futures and spot linkages and enhancing the price influence of carbon emission rights." Through the collective participation of many factors and means, such as TI, infrastructure construction, and the transformation of investment and financing models and behaviors, the market mechanism is used to tap the market value of the carbon emission reduction industry, and the formation of a unified carbon price in the market can help guide capital flow, to encourage more enterprises to actively participate in the development and innovation of carbon emission reduction technologies, to accelerate the emission reduction process and achieve the strategic goal of carbon neutrality.

Third, the government should accelerate the establishment of the CCUS legal framework. On the basis of improving the existing policies, policies and regulations for the trial implementation of CCUS-related industries should be formulated, such as the method of sealing and selecting land, development and utilization plan templates, ledger management system, ecological compensation methods and standards, ecological environment monitoring goals, environmental governance tasks and financial guarantee, early risk warning mechanism, emergency accident handling plan, and safety accident responsibility identification and accountability, etc. On the one hand, it provides constraints and a basis for project construction and operation, and on the other hand, it gives the basic experience that enables the legal development of CCUS technology. Through social group standards, local standards, and industry standards, enterprises can improve the CO_2 capture method, capture purity, utilization method, pipe network design, and pipeline transportation volume. There will be laws to obey regarding storage site selection and sealing techniques, and national standards for the development of CCUS technology.

6. Conclusions

This study enriches the evaluation of the policy and the TIE of listed companies in China. There are numerous studies on the importance of CCUS and firm innovation, and most of them affirm the government's positive effect on CCUS. Researchers have also analyzed government funding as an indispensable part of firms' innovation in CCUS. However, few scholars have studied the relationship between CCUS and corporate profitability. Fewer studies have investigated the extent to which the use of technology can improve the business's profitability. This study bolsters the empirical evidence that advances in CCUS are helping companies capture more market share and make more profits. It makes recommendations commensurate with the dual-carbon strategy and aims to foster innovation in CCUS, which will be a favorable reference for firms seeking ecological sustainability and economic win-win. The limitation of this paper is that it did not analyze and summarize the factors affecting the TIE of enterprises in order to carry out correlation detection on the TIE value of relevant listed companies and to deeply explore the factors influencing the TIE of enterprises. The future research direction will be to study and analyze the problems of personnel input and environmental input, TI input, profitability and total profit output, and environmental issues such as enterprise scale and capital structure.

The development of domestic CCUS projects is still in the stage of technology accumulation. For innovative companies, although domestic CCUS must be improved in terms of policies, systems, funds, and projects, the number of existing players and the competition pressure are small. For potential entrants or investment institutions on the track, the CCUS track is not crowded, there is no leading enterprise, and it is still a wide open field. Some CCUS technologies in China have been commercialized. In terms of scale, China already has the engineering capacity to capture, utilize, and store carbon dioxide on a large scale and is actively raising the CCUS industrial cluster of the whole process. CCUS is an indispensable technology option for carbon-neutral transition, which requires policy incentives, and it has joined forces with enterprises to accelerate the promotion. In China's future CCUS ecosystem, all parties should work together and continue to calibrate suitable policies for China's cost-effective implementation path to promote the large-scale implementation of CCUS.

Finally, it is necessary to establish a CCUS ecosystem of cross-industry and cross-border cooperation, so that policy makers and enterprises can have effective dialogues, unite all parties, pool resources, and achieve synergies. It can actively cooperate and exchange with relevant departments and enterprises in Europe and the United States to jointly promote the cost reduction in CCUS and the realization of carbon neutrality goals. Enterprises must pay close attention to the relevant policy trends and technological progress of CCUS within the policy framework, actively explore new business models for carbon dioxide utilization, and seize opportunities one step ahead of others in the general trend of carbon-neutral transformation. Ample power and oil and gas enterprises with a specific capital base should actively carry out CCUS pilot projects with universities and scientific research units and strive to become world-class CCUS service providers.

Author Contributions: Conceptualization, X.X.; methodology, X.C.; software, Y.Z.; validation, T.W. and D.H.; formal analysis, D.H.; investigation, X.C.; resources, X.X.; data curation, T.W.; writing—original draft preparation, D.H.; writing—review and editing, X.X. and X.C.; visualization, Y.Z.; supervision, Y.Z.; project administration, X.X.; funding acquisition, T.W. All authors have read and agreed to the published version of the manuscript.

Funding: This research was funded by the Major Projects of National Social Science Fund, grant number 20&ZD205; the Projects of National Social Science Fund, grant number 19CFX052; Research project of State Intellectual Property Office of China, grant number SS21-B-015; The National Natural Science Foundation of China, grant number 71974144; The fund of the Key Laboratory of Cities' Mitigation and Adaptation to Climate Change in Shanghai, grant number QHBHSYS201906.

Data Availability Statement: Data is contained within the article. Further requests can made to the corresponding author.

Conflicts of Interest: The authors declare no conflict of interest.

References

1. Wang, Y.; Guo, C.; Chen, X.; Jia, L.; Guo, X.; Chen, R.; Zhang, M.; Chen, Z.; Wang, H. Carbon peak and carbon neutrality in China: Goals, implementation path and prospects. *China Geol.* **2021**, *4*, 720–746. [CrossRef]
2. Tian, H.; Lin, J.; Jiang, C. The Impact of Carbon Emission Trading Policies on Enterprises' Green Technology Innovation—Evidence from Listed Companies in China. *Sustainability* **2022**, *14*, 7207. [CrossRef]
3. Addis, T.L.; Birhanu, B.S.; Italemahu, T.Z. Effectiveness of Urban Climate Change Governance in Addis Ababa City, Ethiopia. *Urban Sci.* **2022**, *6*, 64. [CrossRef]
4. Liu, Y.; Tang, L.; Liu, G. Carbon Dioxide Emissions Reduction through TI: Empirical Evidencefrom Chinese Provinces. *Int. J. Environ. Res. Public Health* **2022**, *19*, 9543. [CrossRef]
5. Huang, X.; Ai, N.; Li, L.; Jiang, Q.; Wang, Q.; Ren, J.; Wang, J. Simulation of CO_2 Capture Process in Flue Gas from Oxy-Fuel Combustion Plant and Effects of Properties of Absorbent. *Separations* **2022**, *9*, 95. [CrossRef]
6. Wijesiri, R.P.; Knowles, G.P.; Yeasmin, H.; Hoadley, A.F.A.; Chaffee, A.L. Technoeconomic evaluation of a process capturing CO_2 directly from air. *Processes* **2019**, *7*, 503. [CrossRef]
7. Sarbassov, Y.; Duan, L.; Manovic, V.; Anthony, E.J. Sulfur trioxide formation/emissions in coal-fired air- and oxy-fuel combustion processes: A review. *Greenh. Gases Sci. Technol.* **2018**, *8*, 402–428. [CrossRef]
8. Basha, O.M.; Keller, M.J.; Luebke, D.R.; Resnik, K.P.; Morsi, B.I. Development of a conceptual process for selective CO_2 capture from fuel gas streams using [hmim][Tf2N] ionic liquid as a physical solvent. *Energy Fuel* **2013**, *27*, 3905–3917. [CrossRef]

9. Madejski, P.; Chmiel, K.; Subramanian, N.; Kuś, T. Methods and Techniques for CO_2 Capture: Review of Potential Solutions and Applications in Modern Energy Technologies. *Energies* **2022**, *15*, 887. [CrossRef]
10. Wienchol, P.; Szlęk, A.; Ditaranto, M. Waste-to-energy technology integrated with carbon capture—Challenges and opportunities. *Energy* **2020**, *198*, 117352. [CrossRef]
11. Wang, C. Green Technology Innovation, Energy Consumption Structure and Sustainable Improvement of Enterprise Performance. *Sustainability* **2022**, *14*, 10168. [CrossRef]
12. Gao, X.; Zhai, K. Performance Evaluation on Intellectual Property Rights Policy System of the Renewable Energy in China. *Sustainability* **2018**, *10*, 2097. [CrossRef]
13. Yang, X.; Yu, X.; Liu, X. Obtaining a Sustainable Competitive Advantage from Patent Information: A Patent Analysis of the Graphene Industry. *Sustainability* **2018**, *10*, 4800. [CrossRef]
14. Rubashkina, Y.; Galeotti, M.; Verdolini, E. Environmental regulation and competitiveness: Empirical evidence on the Porter Hypothesis from European manufacturing sectors. *Energy Policy* **2015**, *83*, 288–300. [CrossRef]
15. Wu, J.; Wang, C.; Hong, J.; Piperopoulos, P.; Zhuo, S. Internationalization and innovation performance of emerging market enterprises: The role of host-country institutional development. *J. World Bus.* **2016**, *51*, 251–263. [CrossRef]
16. Hong, J.; Feng, B.; Wu, Y. Do government grants promote innovation efficiency in China's high-tech industries? *Technovation.* **2016**, *57*, 4–13. [CrossRef]
17. Bigliardi, B.; Ferraro, G.; Filippelli, S.; Galati, F. The past, present and future of open innovation. *Eur. J. Innov. Manag.* **2021**, *24*, 1130–1161. [CrossRef]
18. Dabić, M.; Daim, T.U.; Aralica, Z.; Bayraktaroglu, A.E. Exploring relationships among internationalization, choice for research and development approach and technology source and resulting innovation intensity: Case of a transition country Croatia. *J. High Technol. Manag. Res.* **2012**, *23*, 15–25. [CrossRef]
19. Laermann, U.; Backfisch, M. Innovation crisis in the pharmaceutical industry? A survey. *SN Bus Econ* **2021**, *1*, 164. [CrossRef]
20. Shearmur, R.; Doloreux, D.; Laperrièred, A. Is the degree of internationalization associated with the use of knowledge intensive services or with innovation? *Int. Bus. Rev.* **2015**, *24*, 457–465. [CrossRef]
21. Sumrit, D.; Anuntavoranich, P. Using DEMATEL Method to Analyze the Causal Relations on TI Capability Evaluation Factors in Thai Technology-Based Firms. *Int. Trans. J. Eng. Manag. Appl. Sci. Technol.* **2013**, *4*, 81–103. Available online: http://TuEngr.com/V04/081-103.pdf (accessed on 14 January 2013).
22. Hai, B.; Yin, X.; Xiong, J.; Chen, J. Could more innovation output bring better financial performance? The role of financial constraints. *Financ. Innov.* **2022**, *8*, 6. [CrossRef] [PubMed]
23. Feniser, C.; Burz, G.; Mocan, M.; Ivascu, L.; Gherhes, V.; Otel, C.C. The Evaluation and Application of the TRIZ Method for Increasing Eco-Innovative Levels in SMEs. *Sustainability* **2017**, *9*, 1125. [CrossRef]
24. Lee, K.; Go, D.; Park, I.; Yoon, B. Exploring Suitable Technology for Small and Medium-Sized Enterprises (SMEs) Based on a Hidden Markov Model Using Patent Information and Value Chain Analysis. *Sustainability* **2017**, *9*, 1100. [CrossRef]
25. Wang, Y.; Zhu, Z.; Liu, Z. Evaluation of TIE of petroleum companies based on BCC–Malmquist index model. *J. Pet. Explor. Prod. Technol.* **2019**, *9*, 2405–2416. [CrossRef]
26. Lanoie, P.; Laurent-Lucchetti, J.; Johnstone, N.; Ambec, S. Environmental policy, innovation and performance: New insights on the Porter hypothesis. *J. Econ. Manag. Strategy* **2011**, *20*, 803–842. [CrossRef]
27. Wong, C.W.Y.; Lai, K.; Shang, K.C.; Lu, C.S.; Leung, T.K.P. Green operations and the moderating role of environmental management capability of suppliers on manufacturing firm performance. *Int. J. Prod. Econ.* **2012**, *140*, 283–294. [CrossRef]
28. Ghisetti, C.; Rennings, K. Environmental innovations and profitability: How does it pay to be green? An empirical analysis on the German innovation survey. *J. Clean. Prod.* **2014**, *75*, 106–117. [CrossRef]
29. Zhang, L.; Ma, X.; Ock, Y.S.; Qing, L. Research on Regional Differences and Influencing Factors of Chinese Industrial Green Technology Innovation Efficiency Based on Dagum Gini Coefficient Decomposition. *Land* **2022**, *11*, 122. [CrossRef]
30. Guo, X.F.; Yang, H.A. combination of EFG-SBM and a temporally-piecewise adaptive algorithm to solve viscoelastic problems. *Eng. Anal. Bound. Elem.* **2016**, *67*, 43–52. [CrossRef]
31. Liu, J.; Liu, H.; Yao, X.L.; Liu, Y. Evaluating the sustainability impact of consolidation policy in China's coal mining industry: A data envelopment analysis. *J. Clean. Prod.* **2016**, *112*, 2969–2976. [CrossRef]
32. Wang, W.; Yu, B.; Yan, X.; Yao, X.; Liu, Y. Estimation of innovation's green performance: A range-adjusted measure approach to assess the unified efficiency of China's manufacturing industry. *J. Clean. Prod.* **2017**, *149*, 919–924. [CrossRef]
33. Rumanti, A.A.; Samadhi, T.M.A.A.; Wiratmadja, I.I.; Reynaldo, R. Conceptual model of green innovation toward knowledge sharing and open innovation in Indonesian SME. In Proceedings of the 2017 4th International Conference on Industrial Engineering and Applications (ICIEA), Nagoya, Japan, 21–23 April 2017; pp. 182–186. [CrossRef]
34. Govindan, K.; Rajendran, S.; Sarkis, J.; Murugesan, P. Multi criteria decision making approaches for green supplier evaluation and selection: A literature review. *J. Clean. Prod.* **2015**, *98*, 66–83. [CrossRef]
35. Sun, L.; Miao, C.; Yang, L. Ecological-economic efficiency evaluation of green technology innovation in strategic emerging industries based on entropy weighted TOPSIS method. *Ecol. Indic.* **2017**, *73*, 554–558. [CrossRef]
36. Guo, Y.; Xia, X.; Zhang, S.; Zhang, D. Environmental Regulation, Government R&D Funding and Green Technology Innovation: Evidence from China Provincial Data. *Sustainability* **2018**, *10*, 940. [CrossRef]

37. Lin, S.; Sun, J.; Marinova, D.; Zhao, D. Evaluation of the green technology innovation efficiency of China's manufacturing industries: DEA window analysis with ideal window width. *Technol. Anal. Strateg. Manag.* **2018**, *30*, 1166–1181. [CrossRef]
38. Lee, H.S.; Choi, Y. Environmental Performance Evaluation of the Korean Manufacturing Industry Based on Sequential DEA. *Sustainability* **2019**, *11*, 874. [CrossRef]
39. Liu, Z.; Deng, Z.; He, G.; Wang, H.; Zhang, X.; Lin, J.; Qi, Y.; Liang, X. Challenges and opportunities for carbon neutrality in China. *Nat. Rev. Earth Environ.* **2022**, *3*, 141–155. [CrossRef]
40. Hao, J.; Chen, L.; Zhang, N. A Statistical Review of Considerations on the Implementation Path of China's "Double Carbon" Goal. *Sustainability* **2022**, *14*, 11274. [CrossRef]

Disclaimer/Publisher's Note: The statements, opinions and data contained in all publications are solely those of the individual author(s) and contributor(s) and not of MDPI and/or the editor(s). MDPI and/or the editor(s) disclaim responsibility for any injury to people or property resulting from any ideas, methods, instructions or products referred to in the content.

Article

Does Environmental Regulation Promote Corporate Green Innovation? Empirical Evidence from Chinese Carbon Capture Companies

Hong Chen [1], Haowen Zhu [2], Tianchen Sun [3], Xiangyu Chen [2], Tao Wang [4,5,*] and Wenhong Li [6,*]

1. College of Political Science and Law of Jiangxi Normal University, Nanchang 330022, China
2. International School of Law and Finance, East China University of Political Science and Law, Shanghai 200042, China
3. Law School, East China University of Political Science and Law, Shanghai 200042, China
4. College of Environmental Science and Engineering, Tongji University, Shanghai 200092, China
5. UNEP-Tongji Institute of Environment for Sustainable Development, Tongji University, Shanghai 200092, China
6. Shanghai International College of Intellectual Property, Tongji University, Shanghai 200042, China
* Correspondence: a.t.wang@foxmail.com (T.W.); liwenhong@tongji.edu.cn or liwenhong1909@126.com (W.L.)

Abstract: The proposal of the "double carbon" goal of "carbon peak, carbon neutralization" highlights the determination of China's green and low-carbon development. Carbon capture is one of the essential ways to reduce carbon dioxide (CO_2) emissions and cope with climate change. Then, how to improve the green innovation capability of organizations and promote the transformation and upgrading of enterprises with green development is a practical problem that needs to be dealt with quickly. This paper uses multiple linear regression to investigate the impact of environmental regulation on corporate green innovation and explores the mediating effect of corporate environmental investment and the moderating effect of corporate digital transformation. The analysis results show that government environmental regulation can effectively enhance the green innovation of enterprises and environmental investments play an intermediary role. However, the development of environmental regulation in China is still relatively backward, and its positive incentive role needs to be further played. As a result, the government should strengthen environmental legislation while also accelerating system development, increasing corporate investment in environmental protection, and raising protection awareness among companies using digital network technology.

Keywords: carbon capture technology; environmental regulation; green innovation

Citation: Chen, H.; Zhu, H.; Sun, T.; Chen, X.; Wang, T.; Li, W. Does Environmental Regulation Promote Corporate Green Innovation? Empirical Evidence from Chinese Carbon Capture Companies. *Sustainability* 2023, 15, 1640. https://doi.org/10.3390/su15021640

Academic Editors: Zilong Liu, Meixia Shan and Yakang Jin

Received: 4 December 2022
Revised: 9 January 2023
Accepted: 11 January 2023
Published: 14 January 2023

Copyright: © 2023 by the authors. Licensee MDPI, Basel, Switzerland. This article is an open access article distributed under the terms and conditions of the Creative Commons Attribution (CC BY) license (https://creativecommons.org/licenses/by/4.0/).

1. Introduction

Environmental protection is currently humanity's greatest worldwide challenge. The world is entering an era of low-carbon development, and the shift to renewable energy sources is a worldwide objective. Now, global energy is evolving toward high efficiency [1], cleanliness, and diversification, and key countries are accelerating the energy transition toward low carbonization or decarbonization [2]. China has proposed carbon peak and carbon neutral targets to implement environmental protection actively, which is, on the one hand, an inherent requirement for China to achieve sustainable development and an irreplaceable grip to consolidate the construction of ecological civilization and succeed in the goal of creating a beautiful China; on the other hand, it is also the responsibility of China as a responsible power to fulfill its international commitment and promote the building of a sustainable global community. President Xi Jinping announced China's new goal of actively addressing climate change twice in less than 100 days, first at the general debate on 22 September 2020. Then, on 12 December 2020, the Climate Ambition Summit was held to mark the 5th anniversary of the signing of the Paris Agreement. This strengthened China's resolve to pursue a green and low-carbon development path and illustrated the blueprint for China's future growth. It has also shown its role as a great power in the international community and provided a solid political impetus for the execution of the Paris Agreement, the worldwide climate protection process, and the green recovery after the pandemic.

President Xi Jinping stated in a speech, and he said, "China will make more effective efforts, enforce stricter rules and regulations to achieve the goal by 2060 [3]". The solemn dedication of. Then, at the general meeting, on 12 December 2020, Xi said that "by 2030 [4], China's CO2 emissions per unit of gross domestic product will decrease by more than 65 percent in comparison to 2005" [5]. In this context, the Central Economic Work Conference convened in December 2020 identified "making effective use of carbon peaking and carbon neutrality" as one of 2021's eight most important priorities. The 4th Session of the 13th National People's Congress (NPC) in March 2021 adopted the "Outline of the 14th Five-Year Plan and 2035 Vision for National Economic and Social Development of the People's Republic of China", which emphasized "completing the target of national independent response to climate change and specifying an action plan for peaking carbon emissions by 2030". In October 2022, Xi, General Secretary of the NPC Central Committee, reiterated that "we should actively and steadily promote carbon peaking and carbon neutrality" and that "achieving carbon peaking and carbon neutrality is a change that will affect the long-term development of China and the world [6]".

China's "double carbon" aim, also known as the carbon peak and carbon neutral aim, is the reflection of China's promise of green and low-carbon development [7], which will have a substantial impact on global climate change and China's future socioeconomic growth [8]. Carbon reduction must be achieved by energy substitution, energy saving, source reduction [9], efficiency improvement [10], process transformation, recycling, and carbon capture, utilization, and storage to realize this objective and achieve green, low-carbon change [11]. In this series of procedures, carbon capture, use, and storage technology are among the most feasible and good methods to reduce carbon dioxide emissions and combat climate change [12]. Carbon capture, utilization, and storage (CCUS) stands for the industrial process of letting out CO_2 from industrial [13], energy, and other emission sources or the atmosphere and directly utilizing or storing it to reduce CO_2 emissions [14]. CCUS is a crucial technology option for global low-carbon development [15]. Carbon capture technology, which plays a vital role as the principal link [16], is still in the industrial demonstration phase in China, and there is still a great deal of space for improvement [17].

China's economic growth has been driven by the traditional model of relying primarily on conventional factor inputs and resource consumption since the reform and opening policy [18]. The country has made significant contributions to the global economy. As China is no longer inclined to high-speed development, but to pursue a higher quality of development, the current issues of an inappropriate industrial structure, poor value-added technologies, and environmental limits will become "bottlenecks" for high-quality economic development [19]. In this context, academics typically view innovation as the most critical factor in sustaining high-quality economic growth. The Party and the administration also highly value innovation's crucial role in fostering economic growth. The report of the 20th Party Congress reaffirms that innovation is the primary factor driving development [20], emphasizes innovation's important place in the context of China's modernization, and makes significant arrangements surrounding innovation-driven development [21]. Based on this, how to actualize the improvement of companies' green innovation capability [22] and promote the transformation and upgrading of firms with green development [23] is a genuine challenge that must be resolved as soon as possible.

Consequently, based on the current development situation of carbon capture technology and the carbon capture industry [24] and the critical role of carbon capture technology and the carbon capture industry in achieving the goal of "double carbon" and high-quality development [25]. It is undoubtedly of great significance for China's carbon capture industry to face up to the critical position of the improvement of green innovation ability for the development of technology and industry and to keep stable and far ahead on the road of healthy development.

Many studies have pointed out that environmental regulation is an effective way to solve the problem of promoting enterprises' green innovation capability and industrial structure upgrading [26–28]. After sorting out the existing studies on the effects of various specific environmental regulation measures on firms' green innovation performance, it was found that they are relatively mature in terms of research theory and research methodology. Li et al. (2022) selected Chinese firms as a sample to study the effects of environmental regulation on firms' innovation outcomes. It was concluded that stricter environmental regulation had a positive impact on the increase in firms' innovation output [29]. Liu and Li (2022) explored how to achieve a win–win situation for both environmental protection and economic development by focusing on the impact of environmental regulation on green innovation, and their study found that a pilot carbon emissions trading policy promoted green innovation among firms in the region [30]. A more in-depth study classifies the types of environmental regulation. Song and Han (2022) decompose environmental regulation into two types, where command-based environmental regulation has a negative impact on carbon reduction and market-based environmental regulation has the opposite [31]. Sun et al. (2023) examined the heterogeneity of these two types of environmental regulation and conclude that only enhanced environmental regulation can examined the heterogeneity of these two types of environmental regulation and concluded that only enhanced environmental regulation can achieve green development of the marine economy [32]. These inconsistent findings provide two important inspirations for empirical studies: first, it is reasonable to distinguish the types of environmental regulations, and there are obvious differences in the mechanisms of action of command-based and market-based environmental regulations on green technological innovation [33]. Theoretically, market-based environmental regulations can provide more flexible and effective incentives for innovation than command-based environmental regulations [34,35]. In the process of environmental regulation policies playing a role in improving firms' green innovation, previous studies have pointed out that firms' research and development (R&D) investment and environmental protection investment play an important role, which provides necessary insights for this study [36]. Huang et al. (2021) found through an empirical study that environmental regulation can stimulate firm innovation by enhancing firms' R&D investment, which is particularly evident in Chinese low-carbon pilot cities [37]. Ahmed et al. (2022) and Guo et al. (2021) showed that increasing national public investment in renewable energy is important to curb CO_2 emission reduction and green energy transition [38,39]. Meanwhile, there are some differences in the innovation behavior of firms due to their different property rights. By combing existing studies, we found that the results differ by the nature of firms' property rights under the same environmental regulatory constraints. Castelnovo (2022) found that patents or the sales of new products were used as indicators of innovation performance, and government subsidies had a stronger effect on the innovation capacity of state-owned enterprises (SOEs). The study found that government subsidies had a stronger effect on the innovation capacity of SOEs, regardless of whether the number of patent applications or new product sales revenue was chosen as a measure of innovation performance [40]. It has also been argued that environmental regulation can have a negative impact on SOEs' performance, and SOEs tend to invest inefficiently when they receive additional credit resources, and these can lead to a decrease in firms' technical efficiency [41]. In addition, existing research on how digital transformation will affect firms' green innovation mainly supports that advancing digital transformation can help improve resource allocation efficiency and integration efficiency and improve firms' green innovation performance [42]. Digital transformation is important for companies to achieve value enhancement by promoting an efficient flow of data elements and enhancing their innovation capabilities [43,44]. Additionally, it has been shown that digital transformation also helps to improve the supply of trade credit, thus significantly enhancing the external financing capacity of firms [45]. Few existing studies have explored the significance of digital transformation in the context of environmental regulation, so it is also worth investigating what role digital transformation plays in the path of environmental regulation's impact on green innovation in carbon capture firms.

Moreover, in general, looking at the existing research in the academic circle, the relatively mature part focuses on several relatively independent discussions on government subsidies and enterprise R&D innovation, environmental regulation and industrial structure upgrading [46], intelligence, digitalization, and enterprise green innovation capabilities [47], or is more common in the relatively common realistic context. However, in the existing literature, especially in the context of China's carbon capture listed companies that have a direct significance for the realization of the "dual carbon" goal, the role of environmental regulation as an incentive for enterprises' environmental protection investment, and the role of enterprises' environmental protection investment as an intermediary for enterprises' green innovation under the support of regional intelligence and enterprise digitalization. Thus, the discussion on the complete transmission path to promote the realization of the "double carbon" goal and the transformation and upgrading of the industrial structure and high-quality development is relatively rare.

Therefore, the novelty and contribution of this paper focus on the basis of existing research, introduces and launches from the introduction and correlation of environmental regulation and green innovation of carbon capture listed companies in China, and further evaluates the impact of environmental regulation on green innovation capability of carbon capture listed companies as a whole through theoretical analysis, reasonable variable selection and empirical research under model construction, Through the analysis of data, we further explore the mechanism of environmental regulation on the green innovation promotion function of carbon capture listed enterprises, and further exert the positive incentive effect of environmental regulation policies on improving the green innovation performance of carbon capture green enterprises in China. At the same time, further study the impact of environmental regulation on green innovation through the play path of the intermediary role of enterprise environmental protection investment, and straighten out the play channel of this intermediary role. In addition, combined with relevant data, we determine the different effects of enterprises with varying rights of property on the improvement of green innovation capability of carbon capture listed enterprises under the influence of environmental regulations, so as to explore the reasons reflected behind the data and implement policies according to the property rights of different enterprises. Furthermore, it discusses the impact of digital transformation degree on environmental regulation on the green innovation process of carbon capture listed enterprises, and studies explicitly how to fully mobilize the regulatory role of the digital transformation degree of enterprises.

Through this study, we are expected to fill the gap between environmental regulation and green innovation in China's carbon capture industry at the theoretical level under the guidance of the overall goal of sustainable development [48]. With the help of the logic and suggestions of this paper, we can better form a new and more efficient linkage between the government, enterprises, and other stakeholders, serving the government at the legislative level to introduce and implement specific policy formulation on environmental regulation of carbon capture listed enterprises [49]. It serves as a feasible path for the maximum efficiency transformation of China's carbon capture listed enterprises under the general background of environmental regulation and specific policies, and focuses on the development of carbon capture enterprises, and explores constructively how environmental regulation can better promote green innovation of China's carbon capture listed enterprises from the aspects of environmental legislation policies, artificial intelligence development, and intellectual property protection. We should take various measures to promote green innovation of carbon capture technology and transformation and upgrading of the carbon capture industry from the perspective of environmental regulation, so as to give full play to comprehensive governance efficiency, and further promote the better realization of China's "dual carbon" goals, sustainable development, and high-quality development.

The introduction introduces the relevant content of carbon capture, analyzes the problem of insufficient innovation ability faced by the current enterprise development, makes a supplement based on combining the existing literature, expounds the similarities and differences between this paper and the existing research, and finally puts forward the research vision. The second part is hypothesis development, which, respectively, expounds on the relationship between environmental regulation and green innovation of enterprises, the intermediary role of environmental protection investment, the role difference of different enterprise property rights, and the regulatory role of enterprise digital transformation. The third part is the research method. While adopting the literature and hypothesis methods, this paper uses more data analysis methods to explain the relationship between various elements by establishing models to draw more scientific conclusions. The fourth and fifth parts are mainly the result analysis and discussion. The sixth part is based on the previous part and put forward the conclusion and policy implications.

2. Hypothesis Development

2.1. Environmental Regulation and Corporate Green Innovation

Enterprises are the primary source of innovation development [50], and the enhancement of their innovation capacity is the key to enhancing their core competitiveness [51], as well as playing a crucial role in technological breakthroughs [52] and industrial upgrading of their respective fields [53]. However, due to the externality of technology research and development [54], businesses cannot reap the full benefits of research and development, resulting in market failure. In addition to enterprise efforts and accumulation, government intervention is a crucial supplementary method for addressing market failure [55], mitigating firms' R&D innovation deficiencies, and promoting industrial development [56]. As for the relationship between subsidies provided by the government, a typical manifestation of government intervention, and enterprise R&D innovation, existing research has looked at the connection between government funding of business R&D and innovation performance from a variety of angles, including the scale of government subsidies [57], government pro-subsidies (i.e., subsidies when enterprises make profits), and government anti-subsidies (i.e., subsidies when enterprises lose money). The moderating influence of government subsidies on the link between firm investment in R&D and innovation performance is examined [58], and it is concluded that government subsidies provide the most significant direct external resource support for enterprise R&D and innovation. Most academics believe that environmental regulation effectively addresses the challenges above associated with industrial structure modernization [59]. As an essential component of government social regulation [60], environmental regulation stands for the government's act of regulating the economic operations of businesses to prevent pollution by creating matching policies and actions. The literature suggests that environmental regulation can correct the negative externalities of the market and improve ecological quality while also influencing technological innovation, input, and output behavior of enterprises by imposing environmental constraints on them and rationally guiding the industrial structure toward rationalization and accelerated development, thereby creating a situation with multiple benefits. To achieve high-quality economic growth and protect the environment, it is essential to investigate suitable environmental legislation and execute environmental policies with differentiation and targeting to encourage industrial structure upgrading.

For carbon capture technology and industrial innovation, carbon capture listed enterprises [61], which plays an irreplaceable role in the scientific and technological innovation and technological progress of carbon capture technology, face obstacles such as insufficient technical maturity, high economic success, a single financing channel, limited application fields, and low profitability, as well as the upgrading of their green innovation capacity and innovation level [62]. Therefore, there is an urgent need for a more rational and practical form of environmental regulation to offer governmental assistance on their journey to green innovation [27]. Consequently, exploring and implementing better environmental regulations is necessary to carry forward green innovation of listed carbon capture firms in

China, thereby empowering the construction of a green low-carbon economy, promoting the renewal of industrial structure, and is a necessary guarantee to reach the "double carbon" goal and construct a modern economy. This paper argues that reasonable environmental regulation is an essential guarantee for China's carbon capture listed enterprises to conduct R&D activities and carbon capture technology innovation and is a critical facilitator to improve the green innovation performance of Chinese carbon capture green enterprises and proposes the following hypotheses.

Hypothesis 1 (H1): *Environmental regulation significantly promotes innovation in China's carbon capture listed companies.*

2.2. The Intermediary Role of Corporate Environmental Investment

Research indicates that R&D expenditure can foster company innovation [63]. Unquestionably, enterprise R&D investment, as the foundation of enterprise innovation capacity enhancement, is a crucial way for enterprises to achieve a competitive advantage and plays an irreplaceable role in the social innovation process [64]. China's company R&D investment has exhibited a rapid development tendency from year to year, with the 2019 amount reaching CNY 2150.41 billion, or 76.9% of the total social R&D investment. This serves as a reminder of how innovation can be promoted in carbon capture technology and industry—exploring the intermediary role played by the environmental protection investment of listed enterprises of carbon capture in China during the process of environmental regulation to promote their green innovation is, without a doubt, an unavoidable requirement for understanding environmental regulation better to advance innovation and industrial upgrading of carbon capture technology and industry in China.

As an external measure, environmental investment reveals a corporation's level of environmental protection efforts. It seeks environmental, social, and economic gains as a one-of-a-kind investment [65]. Internally, the urge to survive and compete is a motivating force for innovation within businesses. Additionally, environmental protection investments provide financial support for enterprise green technology innovation; high-quality investment in environmental protection can significantly promote the green development of enterprises; and environmental protection investments and enterprise green technology innovation have a corresponding input–output relationship that contributes to enhancing their value. Consequently, according to the chain rule, environmental protection tax can impact business green technology innovation by influencing environmental protection investment [66].

Carbon capture listed firms' increased investment in environmental protection leads to a continual increase in knowledge, technological innovation, new products, and new methods, and contributes to the accumulation of corporate resources. Resource-based theory suggests that precious and scarce resources can typically become a competitive advantage for enterprises, and when enterprises have some heterogeneous resources that are difficult to imitate or replace and can be effectively utilized, they tend to promote their success; enterprises also tend to use resources to achieve improved innovation performance and thereby enhance their innovation capability; and R&D activities are a type of heterogeneous resources [67].

Suppose the logic of environmental regulation in China's listed carbon capture enterprises and environmental investment in green innovation in China's listed carbon capture enterprises can be understood more precisely. In that case, the role of environmental investment in listed carbon capture enterprises can be identified as a regulating role in the context of green innovation in the listed carbon capture firms in China [68]. Consequently, this paper argues, based on the study above, that environmental investment mediates the relationship between environmental regulation and green innovation via the influence path that enterprises can be encouraged by environmental regulation to increase their investment in environmental protection and that the increase in environmental investment can also enable firms to promote green innovation. The following hypotheses are also suggested.

Hypothesis 2 (H2): *Environmental investments by listed carbon capture companies in China mediate between environmental regulation and corporate green innovation.*

2.3. Differences in the Role of the Nature of Property Rights of Different Enterprises

Research indicates that R&D expenditure can foster company innovation [69]. Unquestionably, enterprise R&D investment, as the foundation of enterprise innovation capacity enhancement, is a crucial way for enterprises to achieve a competitive advantage and plays an irreplaceable role in the social innovation process. China's company R&D investment has exhibited a rapid development tendency from year to year, with the 2019 amount reaching CNY 2150.41 billion, or 76.9% of the total social R&D investment. This serves as a reminder of how innovation can be promoted in carbon capture technology and industry—exploring the intermediary role played by the environmental protection investment of listed enterprises of carbon capture in China during the process of environmental regulation to promote their green innovation is, without a doubt, an unavoidable requirement for gaining a better understanding of the regulation of environment to advance the innovation and industrial upgrading of carbon capture technology and industry in China [70].

As an external measure, environmental investment reveals a corporation's level of environmental protection efforts [71]. It seeks environmental, social, and economic gains as a one-of-a-kind investment. Internally, the urge to survive and compete is a motivating force for innovation within businesses. In addition, environmental protection investments provide financial support for enterprise green technology innovation; effective environmental protection investments are conducive to enterprise green technology innovation; and environmental protection investments and enterprise green technology innovation have a corresponding input-output relationship that contributes to enhancing their value [72]. Consequently, according to the chain rule, environmental protection tax can impact business green technology innovation by influencing environmental protection investment [73].

Carbon capture listed firms' increased investment in environmental protection leads to a continual increase in knowledge, technological innovation, new products, and new methods, and contributes to the accumulation of corporate resources. Resource-based theory suggests that precious and scarce resources can typically become a competitive advantage for enterprises, and when enterprises have some heterogeneous resources that are difficult to imitate or replace and can be effectively utilized, they tend to promote their success; enterprises also tend to use resources to achieve improved innovation performance and thereby enhance their innovation capability; and R&D activities are a type of heterogeneous resources [74].

Suppose the logic of environmental regulation in China's listed carbon capture enterprises and environmental investment in green innovation in China's listed carbon capture enterprises can be understood more precisely. In that case, the role of environmental investment in listed carbon capture enterprises can be identified as mediating in the context of green innovation in listed carbon capture firms in China. Consequently, based on the study above, this paper argues that environmental investment mediates the relationship between environmental regulation and green innovation via the influence path that environmental regulation can encourage businesses to invest more in the environment, and an increase in environmental investment can also spur green innovation among enterprises. The following hypotheses are also suggested [30].

Hypothesis 3 (H3): *Environmental regulation can promote SOEs to carry out green innovation.*

2.4. Moderating Role of the Degree of Digital Transformation of Enterprises

During the past several years, the digital economy, which is built on several emerging information technologies, including the Internet and cloud computing, has been critical in achieving economic growth, industrial transformation and upgrading, and the reduction of environmental pollution [75]. The national 14th Five-Year Plan and the Vision 2035 outline suggest making the digital economy and other aspects a significant pillar for achieving carbon neutrality and expanding the digital application of production and manufacturing processes [44]. The more diverse and complicated nature of digital innovation will alter the traditional competition landscape; for instance, big data has changed how businesses manage innovation and will be characterized by iterative innovation, platform innovation, and user engagement. With the acceleration of the digitization process, digital empowerment has altered organizations' business models and management styles, influenced their transformation, and upgraded. Based on research tools such as resource orchestration theory, the existing literature has constructed a model of the mechanism of digital empowerment on enterprises' green transformation and conducted a good exploration of its intrinsic transmission mechanism, as well as initiated a systematic discussion of the issue in the context of listed enterprises, which is suggestive of the extent of enterprises' digital transformation studied in this paper in terms of its operativity.

A company's research and development (R&D) capacity for green innovation demonstrates its capacity to deploy resources to address environmental issues. Green innovation in enterprises is stimulated by digital transformation, especially the listed companies' capacity for green innovation, which is mainly proven in the following three characteristics.

First, the digital transformation of businesses might enable environmental regulation to encourage green innovation by boosting the financial capacity of companies and reducing their financing limitations [76]. Digital transformation may considerably enhance the effectiveness of corporate financing and alleviate the issue of complex and costly funding for businesses. This is because digital transformation can facilitate the flow of a large amount of data and information within an enterprise to integrate and rapidly output available information, significantly improve the transparency of enterprise information, lessen the information asymmetry between internal and external to the company, and reduce external transaction costs such as information search and contract signing for investors [77]. For example, information on the government's compensation or incentives for firms' technological innovation in reducing low-end capacity or for enterprises can be transmitted externally via digital platforms, sending a good signal to external investors. Therefore, digital transformation can aid in enhancing the finance capacity of businesses, attracting more external investment, and successfully relieving financing limitations.

Second, the digital revolution of businesses enables environmental regulation to boost corporate green innovation by enhancing external oversight and reducing internal agency conflicts [78]. The character that environmental regulation played on green innovation in carbon capture listed companies may be constrained by the high level of uncertainty and risk associated with innovation activities, as well as the fact that green innovation technologies are more specialized than traditional technology innovation, with more excellent information opacity during the research, development, and application of results. The use of digital technology can make it easier for shareholders to track key performance indicators and the most recent financial data, making the management process and business results more transparent and visualized, which can help reduce the space for opportunistic speculation of corporate managers, significantly reduce their discretion and an overall reduction in the cost of monitoring innovation activities [79].

Using the findings above as a foundation, this paper illustrates that the level of digitalization also benefits the green innovation of enterprises, and the greater the degree of digital transformation within corporations, the greater the effect of environmental regulation on corporate green innovation. The following hypotheses are also suggested. The hypothetical development path is shown in Figure 1.

Hypothesis 4 (H4): *The degree of digital transformation of firms plays a positive moderating role between environmental regulation and green innovation of firms.*

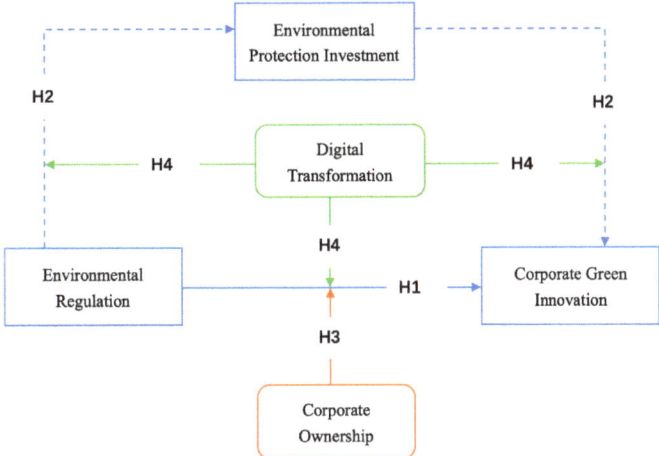

Figure 1. Hypothetical development path.

3. Research Methodology

3.1. Sample Selection and Data Sources

This paper selects A-share listed enterprises in the carbon capture sector in China from 2013 to 2020 as the research object. The data required for this paper's empirical study are obtained from the corporate annual reports of relevant enterprises, the WIND database, local government work reports, and the China Research Data Service Platform (CNRDS), and the data of patent applications are obtained from the State Intellectual Property Office. In this paper, Stata16 is used to organize the data and calculate and analyze the model.

3.2. Variable Definition

Combining the research hypothesis and theoretical model of this paper, considering this availability of data, and referring to the research results of previous scholars, the study's research methodology is as follows:

(1) Dependent variable: corporate green innovation (GRE)

Various methods can be adopted to survey green innovation in enterprises, and from the standpoint of green innovation goods, this research assesses the level of green innovation among businesses [80,81]. The innovation output indicators generally include the new product output value, revenue of recent product sales, number of corporate patent applications, number of corporate patent licenses, etc. The sample selected in this paper is the listed firms of carbon capture in China. The data disclosed in the annual reports of the listed enterprises of carbon capture in China do not compulsorily require the data of new products to be listed separately in that year. Considering the data's availability, this paper uses the single metric of innovation output that is the most straightforward and chooses the patent data of enterprises as the indicator of green innovation. The number of patent applications and the number of patents awarded are included in the patent-related statistics businesses have reported [82]. For the trade-off between these two indicators, this paper considers that the patent approval system and the preferences and efficiency of relevant institutions influence the number of patents granted. The number of patent applications can better reflect the R&D and innovation capability and level of enterprises in that year compared with the number of patent applications [83]. Drawing on the ideas of previous scholars, the article selects the number of patent applications to measure the green innovation of Chinese carbon capture of listed companies, considering the availability and applicability of data and the higher standardization of the number of patent applications.

(2) Independent variable: intensity of government environmental regulation (ER)

There are many ways to define the intensity of governmental environmental regulation. This paper measures the strength of governmental environmental regulation in terms of the importance of the environment in the work reports of prefecture-level municipal governments [84,85]. Considering the various ways and means of government environmental regulation, the vertical evolution and horizontal differences, the comparability of data and the uniformity of research standards, and the relatively stable and essential influence of local governments' work reports on governmental policies, the ratio of the number of environmental words to the frequency of local governments' work reports are chosen in this paper to measure the efforts of government environmental supervision [86,87].

(3) Mediating variable: corporate environmental protection input (INPUT)

As the central part of green innovation activities, the willingness of carbon capture listed enterprises to make environmental protection investments and the improvement of innovation capability impact green innovation ridicule [88]. Like previous studies, this paper considers capital as a variable for enterprises to make environmental protection investments, screens critical words related to environmental protection investment in construction in progress, other payables, and the management cost of the financial statements in the listed company, obtains relevant environmental protection investment data, and takes logarithms to measure the investment of enterprises spent on environmental protection [89].

(4) Moderating variable: the degree of digital transformation of enterprises (INTE)

Digitization is a complex and dynamic process that is extremely difficult to quantify, so the degree of digitization is introduced to measure the relatively static level of implementation of digitization in a company [90]. By drawing on existing research, this paper decides to use the proportion of the year-end intangible asset line items disclosed in the notes to the company's financial report that relate to digital technology to total intangible assets as a yardstick to measure the degree of digital transformation of the company [91].

(5) Control variables.

In addition, in this paper, the following control variables are presented.

① Nature of ownership (STATE): Carbon capture listed enterprises with different property rights have erratic behavior to make their decisions after receiving government subsidies, which may affect the environmental protection investment and green innovation results of carbon capture listed enterprises [92].

Carbon capture listed enterprises are classified into SOEs carbon capture listed capture listed enterprises and non-state-owned enterprises (non-SOEs) carbon capture listed enterprises in this paper, according to the nature of ownership, and assigns "0" to SOEs and "1" to non-SOEs.

② Salary payable to employees (SALA): The amount payable to employees has a particular influence on the labor efficiency and motivation of employees in listed carbon capture enterprises. If the salary payable to employees is low, the cause of employees will not be fully released in the process of promoting environmental protection investment, which is challenging to be effective during a short period, thus affecting the implementation of environmental protection investment under the environmental regulation policy and the formation of green innovation technology of enterprises, and involving the promotion of environmental regulation on the green innovation ability of listed carbon capture firms in China [93]. Conversely, if employees are rewarded and incentivized accordingly, their willingness to engage in green-based innovation can be effectively stimulated. Therefore, in this paper, the various forms of compensation given by firms to obtain the services provided by their employees and other related expenditures are used to measure the payroll payable to their employees and used as control variables.

③ Independent director ratio (DIRE): The influence that environmental regulation exerts on green innovation in carbon capture listed companies is influenced by the board composition in the process. The independent director ratio affects companies' decision-making intention on environmental investment spending under the sway of environmental regulation [94]. Therefore, in this study, one of the control variables is the ratio of independent directors to the total number of boards.

④ Operating income growth rate (RATE): The operational income growth rate indicates the company's profitability. The better profitability of the enterprise shows that the enterprise will fast forward to the maturity stage of the enterprise life cycle. From the shareholders' point of view, they want the enterprise to obtain long-term income and maintain good competitiveness. At this point, companies want to increase their market share and will continue to explore new growth areas. Previous studies have shown that a company's profitability positively influences its investment in technological innovation. Therefore, this paper uses the ratio of the current year's operating income increase to the previous year's operating income to measure the operating income growth rate as a control variable.

⑤ Gearing ratio (TDR): The gearing ratio reflects the enterprise's solvency and the firm's capital structure, measured by the ratio percentage of total liabilities to total assets. For projects with long time cycles and high risks, such as corporate environmental investments, creditors usually set restrictive terms on using the released funds. Companies with high gearing and debt service pressure will be relatively more cautious when faced with green innovation activities and invest less in environmental protection. In summary, Table 1 provides definitions and relevant descriptions of the variables used in this paper's study.

Table 1. Variable definitions.

Type	Symbol	Name	Definition
Dependent variable	GRE	Corporate Green Innovation	Annual green patent applications for enterprises
Independent variable	ER	Government environmental regulation intensity	The ratio of the frequency of environmental words to the frequency of words in the work reports of prefecture level municipal governments
Intermediate variables	INPUT	Enterprise environmental protection investment	Screen the keywords related to environmental protection investment in the notes to the financial statements of listed companies for construction in progress, other payables, and administrative expenses, obtain the relevant environmental protection investment data and take the logarithm
Adjustment variables	INTE	The degree of digital transformation of enterprises	Proportion of the portion of the year-end intangible asset line items disclosed in the notes to the company's financial report relating to digital technology to total intangible assets
Control variables	STATE	Nature of business ownership	SOEs are assigned a value of 0; non-SOEs enterprises are assigned a value of 1
	SALA	Employee payroll payable	Various forms of compensation and other related expenses are given by the enterprise to obtain the services provided by employees
	DIRE	Percentage of independent directors	Number of independent directors as a percentage of board members
	RATE	Operating income growth rate	Ratio of the increase in the enterprise's operating income for the current year to the total operating income for the previous year
	TDR	Gearing ratio	Ratio of total enterprise liabilities to total assets
	YEAR		Year fixed effects

3.3. Model Design

The research method in this paper is multiple linear regression. A variable often associated with various factors, it is more accurate and effective to use multiple independent variables to measure dependent variables than to use only one independent variable. Therefore, multiple linear regression is used in this paper instead of univariate linear regression. This paper discusses how environmental regulation directly has impact on green innovation in carbon capture firms and the regulating effect of firms' environmental investment. As shown in Equation (1), Model 1 examines the effects of environmental regulation on green innovation in carbon capture firms. As shown in Equation (2), Model 2 is a regression model to assess the effects of environmental regulation on corporate environmental investment. The variable INPUT represents a corporate environmental investment, as shown in Equation (2). Model 3 is derived from Model 1 with a new variable of corporate environmental investment. It examines how environmental investment exerts an influence on corporate green innovation.

$$GRE = \beta_0 + \beta_1 ER + \beta_2 SALA + \beta_3 DIRE + \beta_4 RATE + \beta_5 TDR + \beta_6 YEAR + \varepsilon \tag{1}$$

$$INPUT = \beta_0 + \beta_1 ER + \beta_2 SALA + \beta_3 DIRE + \beta_4 RATE + \beta_5 TDR + \beta_6 YEAR + \varepsilon \tag{2}$$

$$GRE = \beta_0 + \beta_1 ER + \beta_2 SALA + \beta_3 DIRE + \beta_4 RATE + \beta_5 TDR + \beta_6 INPUT + \beta_7 YEAR + \varepsilon \tag{3}$$

For the sake of probing the different performance of environmental regulation of green innovation for SOEs and non-SOEs, the dummy variable STATE is now introduced into the model. In Model 4, if the coefficient of the interaction term between ER and STATE is significant, this indicates that the impact of environmental regulation on green innovation is significantly different between SOEs and non-SOEs. In Model 5, if the coefficient of the interaction term between ER and STATE is significant, this indicates that the effect of environmental regulation on firms' environmental investment is significantly different between SOEs and non-SOEs. In Model 6, if the coefficient of the interaction term between INPUT and STATE is significant, this indicates that environmental regulation's influence on green innovation is significantly different between SOEs and non-SOEs.

$$GRE = \beta_0 + \beta_1 ER + \beta_2 SALA + \beta_3 DIRE + \beta_4 RATE + \beta_5 TDR + \beta_6 ER \times STATE + \beta_7 STATE + \beta_8 YEAR + \varepsilon \tag{4}$$

$$INPUT = \beta_0 + \beta_1 ER + \beta_2 SALA + \beta_3 DIRE + \beta_4 RATE + \beta_5 TDR + \beta_6 ER \times STATE + \beta_7 STATE + \beta_8 YEAR + \varepsilon \tag{5}$$

$$GRE = \beta_0 + \beta_1 ER + \beta_2 SALA + \beta_3 DIRE + \beta_4 RATE + \beta_5 TDR + \beta_6 INPUT + \beta_7 INPUT \times STATE + \beta_8 STATE + \beta_9 YEAR + \varepsilon \tag{6}$$

Next, this article studies the regulatory effect of the digital transformation of enterprises. In Model 7, if ER and INTE interaction coefficients are significant, this indicates that the digital transformation of enterprises in direct models has a significant regulatory role. In Model 8, if the cross-term between ER and INTE is significant, then the moderating effect exists in the first half of the mediated model path. If the cross-term between INPUT and INTE is significant in Model 9, this indicates that the moderating effect is present in the second half of the mediated model path.

$$GRE = \beta_0 + \beta_1 ER + \beta_2 SALA + \beta_3 DIRE + \beta_4 RATE + \beta_5 TDR + \beta_6 ER * INTE + \beta_7 INTE + \beta_8 YEAR + \varepsilon \tag{7}$$

$$INPUT = \beta_0 + \beta_1 ER + \beta_2 SALA + \beta_3 DIRE + \beta_4 RATE + \beta_5 TDR + \beta_6 ER * INTE + \beta_7 INTE + \beta_8 YEAR + \varepsilon \tag{8}$$

$$GRE = \beta_0 + \beta_1 ER + \beta_2 SALA + \beta_3 DIRE + \beta_4 RATE + \beta_5 TDR + \beta_6 INPUT + \beta_7 INPUT * INTE + \beta_8 INTE + \beta_9 YEAR + \varepsilon \tag{9}$$

4. Results

4.1. Descriptive Statistics

The data of A-share listed companies in China's carbon capture sector from 2013 to 2020 is selected in this paper. Firstly, descriptive statistics of the sample as a whole are conducted. As Table 2 shows, the sample of carbon capture listed companies' greatest and lowest patent application values differ significantly generally, however, the standard deviation exceeds the mean value, illustrating the huge gap between the green innovation performance of sample companies. The minimum value of the environmental regulation intensity of the carbon capture listed enterprises is 0.136, and the maximum value is 0.715, which indicates that a difference exists among the power of the government's environmental regulation on different carbon capture listed enterprises. At the same time, the minimum value is 0, and the maximum value is 22.122 after taking the logarithm of the investment in environmental protection by the listed carbon capture enterprises themselves, which indicates that environmental protection investment requires much financial support, with the characteristics of long cycles and high risk, which affects the enthusiasm and strategic choice of the listed carbon capture enterprises for green innovation. The highest value of the gearing ratio in the sample is 79.1%, the lowest value is 19.1%, and the average value is 49%, which indicates that the selected carbon capture listed companies have a large variability in insolvency.

Table 2. Descriptive statistics.

Variable	Obs	Mean	Std. Dev.	Min	Max
GRE	102	23.48	37.223	0	170
ER	104	0.369	0.121	0.136	0.715
INPUT	104	10.978	8.719	0	22.122
SALA	102	17.459	1.327	13.487	19.272
DIRE	102	0.349	0.032	0.273	0.429
RATE	102	0.101	0.295	−0.505	0.94
TDR	102	0.49	0.152	0.191	0.791
INTE	83	0.098	0.144	0.001	0.701
STATE	104	0.462	0.501	0	1

4.2. Correlation Analysis

According to Table 3, ER is positively correlated with GRE with a coefficient of 0.206 and a p-value of 0.038, which confirms the hypothesis that environmental regulation can enhance corporate green innovation. INPUT is positively correlated with GRE with a coefficient of 0.394 and a p-value of 0.000, which to a certain extent, can indicate that environmental investment enhances corporate green innovation.

Table 3. Pairwise correlation.

Variables	(1)	(2)	(3)	(4)	(5)	(6)	(7)	(8)
(1) GRE	1.000							
(2) ER	0.206	1.000						
	0.038							
(3) INPUT	0.394	0.111	1.000					
	0.000	0.261						
(4) SALA	0.342	−0.023	0.106	1.000				
	0.000	0.821	0.289					
(5) DIRE	−0.014	−0.032	0.174	−0.091	1.000			
	0.887	0.753	0.080	0.365				
(6) RATE	−0.039	−0.094	−0.116	−0.062	0.201	1.000		
	0.695	0.348	0.245	0.538	0.043			
(7) TDR	0.520	0.091	0.277	0.554	0.136	−0.094	1.000	
	0.000	0.364	0.005	0.000	0.172	0.346		
(8) INTE	−0.151	0.172	−0.226	0.007	−0.167	−0.111	−0.318	1.000
	0.172	0.120	0.040	0.951	0.130	0.319	0.003	

4.3. Multicollinearity Test

From the multicollinearity test in Table 4, the variance inflation factors VIF of the independent variables are below 2.4, indicating no multicollinearity between the variables.

Table 4. Multicollinearity test.

Variable	VIF	1/VIF
INPUT	1.33	0.74946
ER	1.47	0.682341
SALA	1.68	0.594215
DIRE	1.22	0.820458
RATE	1.38	0.724707
TDR	1.73	0.577725
YEAR		
2014	1.82	0.549793
2015	1.95	0.514134
2016	1.87	0.533637
2017	2.11	0.473973
2018	2.14	0.467566
2019	2.07	0.484013
2020	2.39	0.418564
Mean VIF	1.78	

4.4. Mediation Effect Test

According to Table 5, the ER coefficient in Model 1 is 84.460 and is significant ($p < 0.01$). This indicates that government environmental regulation has a facilitating effect on the green innovation of carbon capture firms. The greater the environmental supervision, the more the green innovation of an enterprise. In Model 2, the coefficient of ER is 18.866 and is significant at the ($p < 0.05$). This indicates that environmental regulation has a facilitating effect on the environmental investment of carbon capture enterprises. With the increase in government environmental regulations, the environmental protection investment of enterprises will also increase. In Model 3, the INPUT coefficient is 1.095 and is significant ($p < 0.01$). This shows that carbon capture companies' investments in environmental protection have a catalytic impact on green innovation. When enterprises increase their environmental protection investment, their green innovation will also be enhanced. The findings above show that between corporate green innovation and government environmental regulation, carbon capture businesses' environmental investments serve as a bridge. On the one hand, environmental regulation can help companies to innovate in a greener way. On the other, government environmental regulation can help businesses innovate in a more innocent way by making it easier for businesses to invest more in environmental protection. Hypothesis 1 and Hypothesis 2 are verified.

Table 5. Mediation effect test.

	GRE	INPUT	GRE
Variables	Model 1	Model 2	Model 3
INPUT			1.095 ***
			(2.74)
ER	84.460 ***	18.866 **	63.802 **
	(2.81)	(2.45)	(2.13)
SALA	0.760	−0.832	1.671
	(0.25)	(−1.08)	(0.57)
DIRE	−59.458	54.551 **	−119.189
	(−0.57)	(2.03)	(−1.15)
RATE	−0.875	−5.078	4.685
	(−0.07)	(−1.63)	(0.39)
TDR	122.981 ***	17.264 **	104.078 ***
	(4.76)	(2.61)	(4.02)
YEAR		Omission	
Constant	−76.615	−12.918	−62.470
	(−1.17)	(−0.77)	(−0.99)
Observations	102	102	102
R-squared	0.380	0.251	0.429

T-statistics in parentheses; *** $p < 0.01$, ** $p < 0.05$ (the same below).

4.5. Test for Heterogeneity of Firm Ownership

According to Table 6, the coefficient of ER*STATE is −143.437, and it passes the significant test ($p < 0.01$). This demonstrates that environmental regulation exerts more influence on SOEs' green innovation than on non-SOEs. In Model 5, ER*STATE is 22.357, which does not pass the significant test, indicating that the impact of corporate environmental investment on green innovation is not significantly different between SOEs and non-SOEs. In Model 6, the coefficient of INPUT*STATE is −2.586 and passes the significant test ($p < 0.01$). It means that environmental regulation strongly influences green innovation in SOEs and has a significantly lower impact on green innovation in non-SOEs. The above shows that the nature of ownership of carbon capture enterprises has a significant effect on environmental regulation. Under the same conditions, SOEs perform significantly better than non-SOEs in green innovation.

Therefore, Hypothesis 3 is verified.

Table 6. Test for heterogeneity of firm ownership.

Variables	GRE Model 4	INPUT Model 5	GRE Model 6
INPUT			2.090 ***
			(4.34)
ER	119.293 ***	7.734	71.037 **
	(3.44)	(0.84)	(2.49)
INPUT*STATE			−2.586 ***
			(−3.74)
STATE	−14.844 **	−3.101 *	−9.629
	(−2.26)	(−1.79)	(−1.53)
SALA	−0.878	−1.482 *	1.799
	(−0.29)	(−1.82)	(0.60)
DIRE	−46.668	43.330	−121.051
	(−0.46)	(1.62)	(−1.26)
RATE	−2.574	−4.940	1.214
	(−0.22)	(−1.62)	(0.11)
TDR	127.303 ***	18.771 ***	96.616 ***
	(5.16)	(2.88)	(3.94)
YEAR		Omission	
ER*STATE	−143.437 ***	22.357	
	(−2.69)	(1.58)	
Constant	−61.812	7.273	−72.441
	(−0.89)	(0.39)	(−1.11)
Observations	102	102	102
R-squared	0.456	0.298	0.524

T-statistics in parentheses; *** $p < 0.01$, ** $p < 0.05$, * $p < 0.1$ (the same below).

4.6. Moderating Effect Test

Table 7 shows that in Model 7, the coefficient of ER*INTE is −56.027, which fails the test of significance. Therefore, even digital transformation cannot effectively promote the positive incentive effect of environmental regulation on green innovation and fails to play a positive moderating effect. In Model 8, the coefficient of ER*INTE is −166.691 and is significant ($p < 0.01$). The data reflects that the degree of digital transformation of enterprises significantly inhibits the promotion effect of environmental regulation on enterprises' environmental investment. The influence of environmental regulation on encouraging environmental protection investment in organizations has diminished as the digital transformation of businesses has advanced, and enterprise digital transformation plays a reverse role. In Model 9, the coefficient of INPUT*INTE is −2.872, which fails the significance test. This shows that the degree of digital transformation of enterprises cannot enhance the promotion effect of enterprise environmental protection investment on enterprise green innovation and fails to play a positive moderating effect.

Table 7. Moderating effects of the degree of digital transformation of enterprises.

Variables	GRE Model 7	INPUT Model 8	GRE Model 9
INPUT			1.237 ** (2.38)
ER	123.923 *** (3.48)	21.316 *** (2.68)	86.323 ** (2.12)
INPUT*INTE			−2.872 (−0.75)
INTE	5.386 (0.16)	−0.362 (−0.05)	8.144 (0.27)
SALA	0.587 (0.15)	1.022 (1.15)	−1.731 (−0.40)
DIRE	−54.462 (−0.47)	60.595 ** (2.34)	−138.879 (−1.17)
RATE	−2.121 (−0.15)	−7.719 ** (−2.39)	10.059 (0.66)
TDR	145.635 *** (3.99)	2.169 (0.27)	145.530 *** (4.13)
YEAR		Omission	
ER*INTE	−56.027 (−0.21)	−166.691 *** (−2.79)	
Constant	−103.987 (−1.33)	−39.089 ** (−2.24)	−31.848 (−0.35)
Observations	83	83	83
R-squared	0.435	0.368	0.479

T-statistics in parentheses; *** $p < 0.01$, ** $p < 0.05$ (the same below).

The above results indicate that the degree of digital transformation only makes a substantial negative adjustment impact on the first half of the intermediary effect model, "environmental regulation-environmental inputs". Hypothesis 4 is not tested.

4.7. Robustness Test

This paper uses two approaches to robustness check the results. First, since the number of green patents granted can also reflect the green innovation of enterprises to a certain extent, the dependent variable is replaced with the number of green patents granted (GAPA) in this paper. The results of the main effects regression are shown in Table 8. There is a promoting effect of ER on green innovation, and INPUT has a mediating role between the two. This result is the same as the results above. Second, since the number of patents fits the characteristics of a discrete variable, this paper uses Poisson regression for robustness testing. The results of the main effects regression are shown in Table 9. As can be seen, this result is also the same as the results above, indicating that the results are robust.

Table 8. Substitution of variables method.

Variables	GAPA Model 1	INPUT Model 2	GAPA Model 3
INPUT			0.804 *** (2.86)
ER	76.796 *** (3.62)	18.866 ** (2.45)	61.626 *** (2.92)
SALA	0.719 (0.34)	−0.832 (−1.08)	1.389 (0.67)
DIRE	−37.906 (−0.51)	54.551 ** (2.03)	−81.770 (−1.12)
RATE	1.756 (0.20)	−5.078 (−1.63)	5.840 (0.70)
TDR	81.955 *** (4.49)	17.264 ** (2.61)	68.074 *** (3.74)
YEAR		Omission	
Constant	−64.130 (−1.39)	−12.918 (−0.77)	−53.743 (−1.21)
Observations	102	102	102
R-squared	0.394	0.251	0.446

T-statistics in parentheses; *** $p < 0.01$, ** $p < 0.05$ (the same below).

Table 9. Poisson regression method.

Variables	GAPA Model 1	INPUT Model 2	GAPA Model 3
INPUT			0.059 ***
			(15.84)
ER	1.946 ***	18.866 **	0.827 ***
	(9.67)	(2.45)	(3.81)
SALA	0.340 ***	−0.832	0.326 ***
	(10.09)	(−1.08)	(9.89)
DIRE	−1.038	54.551 **	−2.525 ***
	(−1.46)	(2.03)	(−3.69)
RATE	−0.738 ***	−5.078	−0.375 ***
	(−7.10)	(−1.63)	(−3.67)
TDR	4.778 ***	17.264 **	3.872 ***
	(24.22)	(2.61)	(18.68)
YEAR		Omission	
Constant	−6.592 ***	−12.918	−5.621 ***
	(−10.29)	(−0.77)	(−9.00)
Observations	102	102	102
R-squared	0.546	0.251	0.614

T-statistics in parentheses; *** $p < 0.01$, ** $p < 0.05$, (the same below).

5. Discussion

According to the data in Section 4.4, the environmental protection investment of carbon capture enterprises can promote green innovation. In the process of enterprises increasing environmental investment, their green innovation will also be improved. The environmental investment of carbon capture enterprises is an intermediary between government environmental regulation and enterprise green innovation [95]. On the one hand, environmental regulation can play a very direct role in improving the green innovation ability of enterprises; on the other hand, government environmental regulation can also promote green innovation by promoting enterprises to increase environmental investment.

In the process of enactment and enforcement of relevant legal provisions, the government regulates the economic activities of China's carbon capture listed enterprises to reduce pollution and correct the negative externalities of the market [96]. It influences the technological innovation, input, and output behavior of corresponding enterprises by imposing environmental constraints on China's carbon capture listed enterprises, through the innovation compensation effect and optimization of factor allocation, the industrial structure of carbon capture will be rationally guided towards rationalization and upgrading. China's carbon capture listed enterprises will increase their investment in environmental protection and will also promote the accumulation of resources of China's carbon capture listed enterprises through the emergence of technological innovation, new products, and new methods [97]. The corresponding environmental protection investment and its role in R&D activities improve the innovation capacity of enterprises.

Through the analysis of the data in Section 4.5, it can be concluded that the nature of enterprise ownership can strongly and directly affect the effectiveness of environmental regulation. Compared with non-SOEs, environmental regulations can effectively lead SOEs to carry out green transformation and upgrading. However, the impact of enterprise environmental protection investment on green innovation is not apparent between SOEs and non-SOEs. Under the same conditions, the green innovation performance of SOEs is significantly better than that of non-SOEs.

The conclusion is also consistent with the previous theoretical discussion. The extent to which enterprises with government background can improve their green innovation capability under environmental regulation is more substantial than other enterprises, to a considerable extent, because SOES have advantages in existing resources and stuff, and because SOES can more easily access funds and policy resources that are more closely related to green innovation in environmental regulation [98].

It can be concluded from the data in Section 4.6 that the degree of enterprise digital transformation plays a negative role in the role of environmental regulation in green innovation of carbon capture listed enterprises. On the one hand, the degree of enterprise digital transformation cannot enhance the promotion effect of enterprise environmental protection investment on enterprise green innovation. It cannot play a positive regulatory role [90]. On the other hand, it even plays a significant negative regulatory effect in the first half path of the intermediary effect model, "environmental regulation-environmental protection investment".

After analysis, the main reason for this phenomenon is that the existing environmental regulation mode conflicts with the more common path of corporate digital transformation. The positive regulatory role of enterprise digital transformation degree in the promotion of environmental regulation on enterprises' green innovation ability is mainly achieved through enhancing enterprises' financing ability, strengthening external supervision and alleviating internal agency conflicts, optimizing internal and external resource allocation, etc., the current digital transformation of China's carbon capture listed enterprises does not correspond to the promotion of environmental regulation on enterprises' green innovation ability, but partially led to the opposite result. The increased transparency brought about by the digital transformation has allowed investors to obtain adverse signals from environmental supervision and other negative effects. In addition, some studies have pointed out that the formation process of green innovation capability, as a manifestation of the dynamic capability concept of enterprises, may be affected by organizational inertia [99]. Organizational inertia makes the past development path of an organization fail to adjust its behavior in time according to major national policies and social environment changes, which leads to its inability to adapt to the trend and severe erosion of organizational change and innovation; in turn, it significantly inhibits the role of environmental regulation in green innovation of carbon capture listed enterprises [31]. These problems need to be further optimized and adjusted under the guidance of the government and the cooperation of enterprises, as well as comprehensively implemented under the background of digital transformation to make a positive connection between digital transformation and green innovation of carbon capture listed enterprises.

6. Conclusions

6.1. Research Finding

This paper's findings indicate that environmental regulation is a necessary guarantee for China's carbon capture listed enterprises to conduct R&D activities and carbon capture technology innovation, as well as a significant factor in enhancing the green innovation performance of China's carbon capture green enterprises [100]. Nevertheless, compared to the vast space and market for industrial optimization and upgrading, China's environmental rules are still behind. There is ample area for adjustment and optimization from full play. Therefore, continuing to play environmental supervision policies is essential for my country's carbon capture to capture the green innovation of listed companies and continue transforming and upgrading carbon capture technology and industry. In the way environmental regulation promotes green innovation in China's carbon capture enterprises, the internal environmental protection investment of carbon capture enterprises plays an intermediary role. Environmental supervision can encourage enterprises to increase environmental protection investment and promote the green innovation of firms. Under the premise of investigating the implementation of applicable environmental rules, it is also essential to consider how to actively direct the environmental protection investments of publicly traded carbon capture companies. In addition, the various property rights of listed carbon capture firms in China may influence the amount to which environmental rules affect the green innovation of listed carbon capture enterprises [101]. The heterogeneity analysis indicates that substantial differences exist in the efficacy of green innovation across firms with different property rights and geographical locations, meaning that non-state enterprises should prioritize resource accumulation and enhance resource utilization efficiency. This study did not conclude that the digital transformation of enterprises in environmental supervision concluded that the active regulation role of carbon capture listed enterprises in environmental management; however, it provides essential insights for us to concentrate on the corporate governance role of digital transformation so that carbon capture listed enterprises can successfully leverage the digital platform to take more significant advantage of environmental regulation.

6.2. Policy Implications

The first thing the government must do is boost the environmental supervision system and enhance environmental supervision. Additionally, it is important to continue to improve our capacity for environmental oversight. The government should establish strict environmental standards from the legislative level, fundamentally restrict pollution-intensive production methods, and promote the rationalization and heightening of the structure of the carbon capture industry; at the same time, the government should increase investment and subsidies for the carbon capture industry, i.e., the listed enterprises of carbon capture, and strengthen the support for the upgrade of carbon capture technology [102]. Simultaneously, the government should increase investment and subsidies for the carbon capture industry, i.e., listed carbon capture enterprises, increase support and investment in environmental research institutions, and establish special industrial development funds for the introduction of green technologies and the upgrading of low-carbon environmental protection equipment to compensate for the short-term cost increase in such innovative enterprises that caused by green innovation [103].

Secondly, enterprises should increase their environmental consciousness. Improve their sense of responsibility and positive environmental image. Provide talent assistance, adopt more flexible and effective innovative reward policies, and expand efforts to attract and educate talent while encouraging businesses to equip and introduce professional talent in pollution monitoring and environmental accounting [104,105]. Strengthen the collaboration between listed companies, universities, and scientific research institutions in carbon capture, and promote the formation of an innovation consortium led by leading companies, supported by universities and institutes, and in which all innovation subjects collaborate to accelerate the development and commercialization of clean production technology, monitoring and testing technology, and green products [106].

For firms to obtain investment in environmental protection, the proportion of matching subsidies will increase their willingness to make investments contributing to environmental protection. Developing comprehensive subsidy application procedures and requirements and strict supervision of the use of funds will ensure that enterprises use them for innovation in green technology [107,108]. Expand the financing channels for corporate environmental investment, minimize financing risks and financing constraints, lower the cost of financing for corporate environmental investment, and assist businesses in overcoming the cash flow shortage that constrains corporate environmental investment [109]. Simultaneously, the active use of economic and incentive-based policies, the development of incentives and voluntary policies to strengthen the "inherent constraints" of social and economic development, to encourage carbon capture listed companies to explore newly available resources, the result of new materials utilization, clean production technologies, and environmental pollution control technologies, to steer the transformation of carbon capture technology and industry. The company will encourage listed carbon capture enterprises to explore new resources, develop further material utilization, clean production technology, and the technology of pollution limitation, guide the transformation and upgrading of carbon capture technology and industry, and promote high-quality economic growth in China [110].

Finally, the government should continually improve digital hardware facilities and network system construction, adapt, and optimize the original management model, and continually solidify digital transformation's technological and managerial basis. Carbon capture listed enterprises should actively incorporate more subjects into the organization's innovation network to realize the effective change from value co-creation to value multi-creation, thereby promoting the construction of the enterprise ecosystem; carbon capture listed enterprises should also concentrate on the cultivation of green innovation capabilities to reduce corporate transformation and upgrading caused by lack of green innovation capabilities [111]. Through learning green knowledge and effectively sharing green resources, it is vital to establish an organization-wide culture of continuous learning, expand the degree of data exchange, and increase the innovative capacity of enterprise subjects. The government should use the digital platform to demonstrate and communicate the positive influence of environmental regulations on firm transformation and upgrading, as well as revenue growth, so that the digital medium can fulfill its mission and fulfill the expectation of enhancing the character that environmental regulations play in fostering green innovation capacity of listed firms with carbon capture [112].

6.3. Limitations and Future Research

The principal shortcomings of this paper are highlighted by the following: due to the characteristics of the carbon capture industry, to carry forward the green innovation of carbon capture listed firms, the government has implemented various environmental regulation policies. However, the diversity of policies on environmental regulation and the fact that many of them are not required to be presented in the government work report make it difficult to collect certain information and data comprehensively, which may have some impact on the carbon capture industry. Second, there is more than one way for environmental legislation to affect the green innovation of carbon capture listed companies. Due to the limitation of time and resources, this paper solely focuses on the path of corporate environmental investment to impact the green innovation of carbon capture of listed companies, eliminating other variables that may affect the green and innovative performance of listed companies. In the future, we will further investigate ways to improve the precision of the critical variables and do our best to enhance the results' dependability and generalizability. In the future, research should investigate how government environmental regulation enhances the effects of relevant policies to generate strong incentives to achieve the goals of the corresponding industrial policy, the influence of investment on environment protection in the aspect of green innovation of carbon capture of listed enterprises, and other internal and external factors on carbon capture listed companies green innovation influence.

Author Contributions: Conceptualization, H.C.; methodology, X.C.; software, W.L.; validation, T.W. and H.Z.; formal analysis, H.Z. and T.S.; investigation, X.C.; resources, H.C.; data curation, T.W.; writing—original draft preparation, H.Z.; writing—review and editing, H.C. and X.C.; visualization, W.L. and T.S.; supervision, W.L. and T.S.; project administration, H.C.; funding acquisition, T.W. All authors have read and agreed to the published version of the manuscript.

Funding: This research was funded by the Major Projects of the National Social Science Fund, grant number 20&ZD205; the Young Talent Research Program of China Association for Science and Technology (2022-316); the National Natural Science Foundation of China, grant number 71974144; the fund of the Key Laboratory of Cities' Mitigation and Adaptation to Climate Change in Shanghai, grant number QHBHSYS201906.

Institutional Review Board Statement: Not applicable.

Informed Consent Statement: Not applicable.

Data Availability Statement: Data are contained within the article. Further requests can be made to the corresponding author.

Conflicts of Interest: The authors declare no conflict of interest.

References

1. Longa, F.D.; Fragkos, P.; Nogueira, L.P.; van der Zwaan, B. System-level effects of increased energy efficiency in global low-carbon scenarios: A model comparison. *Comput. Ind. Eng.* **2022**, *167*, 108029. [CrossRef]
2. Wang, P.; Zhao, S.; Dai, T.; Peng, K.; Zhang, Q.; Li, J.; Chen, W. Regional disparities in steel production and restrictions to progress on global decarbonization: A cross-national analysis. *Renew. Sustain. Energy Rev.* **2022**, *161*, 112367. [CrossRef]
3. Sun, L.; Cui, H.; Ge, Q. Will China achieve its 2060 carbon neutral commitment from the provincal perspective. *Adv. Clim. Change Res.* **2022**, *13*, 169–178. [CrossRef]
4. Zhang, L.; Liu, X.; Chen, W. Characteristic analysis of conwentional pole and consequence pole IPMSM for electric vehicle application. *Energy Rep.* **2022**, *8*, 259–269. [CrossRef]
5. Zuo, Z.; Guo, H.; Cheng, J. An LSTM-STRIPAT model analysis of China's 2030 CO_2 emissiona peak. *Carbon Manag.* **2020**, *11*, 577–592. [CrossRef]
6. Wei, Y.; Chen, K.; Wang, X. Policy and Management of Caibon Peaking and Carbon Neutrality: A Literature Review. *Engineering* **2022**, *14*, 52–63. [CrossRef]
7. Wu, G.; Niu, D. A study of carbon peaking and carbon neutral pathways in Chinas power sector under a 1.5 °C temperature contral target. *Environ. Sci. Pollut. Res.* **2022**, *29*, 85062–85080. [CrossRef]
8. Caminade, C.; Kovat, S.; Lloyd, S.J. Impact of climate change on global malaria distribution. *Proc. Natl. Acad. Sci. USA* **2017**, *111*, 3286–3291. [CrossRef]
9. Ismailos, C.; Touchie, M.F. Achieving a low carbon housing sticks: An analysis of low-rise residential carbon reduction measures for new construction in Ontario. *Build. Environ.* **2017**, *126*, 176–183. [CrossRef]
10. Oikonomou, V.; Becchis, F.; Russolillo, D. Energy saving and energy efficiency concepts for policy making. *Energy Policy* **2009**, *37*, 4787–4796. [CrossRef]
11. Thunman, H.; Vilches, T.B.; Nguyen, H.N.T. Circular use of plastics-transformation of existing petrochemical clusters into thermochemical recycling plants with 100% plastics recovery. *Sustain. Mater. Technol.* **2009**, *22*, e00124. [CrossRef]
12. Ilbahar, E.; Colak, M. A combined methodology based on z-fuzzy numbers for sustainability assessmentof hydrogen energy storage systems. *Int. J. Hydrogen Energy* **2022**, *47*, 15528–15546. [CrossRef]
13. Zhang, T.; Zhang, W. CO_2 ingection deformation monitoring based on UAV and in SAR technology: A case study of Shizhuang, Shanxi proviance, China. *Remote Sens.* **2022**, *14*, 237. [CrossRef]
14. Shao, J.; Xiao, L. Observation of Field-Emission dependence on stored energy. *Phys. Rev. Lett.* **2015**, *115*, 264802. [CrossRef] [PubMed]
15. Shukla, P.R.; Dhar, S. Renewable energy and low carbon economy transition in India. *J. Renew. Sustain. Energy* **2010**, *2*, 031005. [CrossRef]
16. Engel, D.; Jones, E. Development of a risk-based comparison methodology of carbon capture technologies. *Greenh. Gases Sci. Technol.* **2014**, *4*, 316–330. [CrossRef]
17. Zeng, X.; Shao, R. Industrial demonstration plant for the gasification of herb residue bed two-stage process. *Bioresour. Technol.* **2016**, *206*, 93–98. [CrossRef]
18. Alanne, K.; Saari, A. Estimating the environmental burdens of residential energy supply systems through material input and emission factors. *Build Environ.* **2008**, *43*, 1734–1748. [CrossRef]
19. Budizianowski, W.M. Valud-added carbon management technologies for low CO_2 intensive carbon-based energy vectors. *Energy* **2012**, *41*, 280–297. [CrossRef]
20. Lou, R.; Zhou, N.; Li, Z. Spatial-Temporal Evolution and Sustainable Type Division of Fishery Science and Technology Innovation Efficiency in China. *Sustainability* **2022**, *14*, 7277. [CrossRef]
21. Xiao, W.; Kong, H. The impact of innovation-driven strategy on high-quality economic development: Evidence from China. *Sustainabality* **2022**, *14*, 4212. [CrossRef]
22. Li, G.; Wang, X. How green technological innovation ability influences enterprise competitiveness. *Technol. Soc.* **2019**, *59*, 101136. [CrossRef]
23. Ye, F.; Quan, Y.; He, Y.; Lin, X. The impact of government preferences and environmental regulations on green development of China's marine economy. *Environ. Impact Assess. Rev.* **2021**, *87*, 106522. [CrossRef]
24. Li, J.; Tharakan, P.; Liang, X. Technological, economic and financial prospects of carbon dioxide capture in the cement industry. *Energy Policy* **2013**, *61*, 1377–1387. [CrossRef]
25. Leonzio, G.; Fennell, P.S.; Shah, N. Analysis of technologies for carbon dioxide capture from the air. *Appl. Sci.* **2022**, *12*, 8321. [CrossRef]
26. Guo, M.; Wang, H.; Kuai, Y. Environmental regulation and green innovation: Evidence from heavily polluting firms in China. *Finance Res. Lett.* **2022**, 103624, in press. [CrossRef]
27. Ouyang, Y.; Ye, F.; Tan, K. The effect of strategic synergy between local and neighborland environmental regulations on green innovation efficiency: The perspective of industrial transfer. *J. Clean. Prod.* **2022**, *380*, 134933. [CrossRef]
28. Yu, H.; Wang, J.; Hou, J.; Yu, B.; Pan, Y. The effect of economic growth pressure on green technology innovation: Do environmental regulation, government support, and financial development matter? *J. Environ. Manag.* **2023**, *330*, 117172. [CrossRef]
29. Li, X.; Du, K.; Ouyang X; Liu, L. Does more stringent environmental regulation induce firms' innovation? Evidence from the 11th Five-year plan in China. *Energy Econ.* **2022**, *112*, 106110. [CrossRef]

30. Liu, M.; Li, Y. Environmental regulation and green innovation: Evidence from China's carbon emissions trading policy. *Finance Res. Lett.* **2022**, *48*, 103051. [CrossRef]
31. Song, W.; Han, X. Heterogeneous two-sided effects of different types of environmental regulations on carbon productivity in China. *Sci. Total Environ.* **2022**, *841*, 156769. [CrossRef] [PubMed]
32. Sun, J.; Zhai, N.; Miao, J.; Mu, H.; Li, W. How do heterogeneous environmental regulations affect the sustainable development of marine green economy? Empirical evidence from China's coastal areas. *Ocean Coast. Manag.* **2023**, *232*, 106448. [CrossRef]
33. Du, W.; Li, M.; Wang, Z. The impact of environmental regulation on firms' energy-environment efficiency: Concurrent discussion of policy tool heterogeneity. *Ecol. Indic.* **2022**, *143*, 109327. [CrossRef]
34. Tian, Y.; Feng, C. The internal-structural effects of different types of environmental regulations on China's green total-factor productivity. *Energy Econ.* **2022**, *113*, 106246. [CrossRef]
35. Li, M.; Gao, X. Implementation of enterprises' green technology innovation under market-based environmental regulation: An evolutionary game approach. *J. Environ. Manag.* **2022**, *308*, 114570. [CrossRef]
36. Bhuiyan, M.B.U.; Huang, H.J.; Villiers, C. Determinants of environmental investment: Evidence from Europe. *J. Clean. Prod.* **2021**, *292*, 125990. [CrossRef]
37. Huang, J.; Zhao, J.; Cao, J. Environmental regulation and corporate R&D investment—Evidence from a quasi-natural experiment. *Int. Rev. Econ. Financ.* **2021**, *72*, 154–174. [CrossRef]
38. Ahmed, Z.; Ahmad, M.; Murshed, M.; Shah, M.I.; Mahmood, H.; Abbas, S. How do green energy technology investments, technological innovation, and trade globalization enhance green energy supply and stimulate environmental sustainability in the G7 countries? *Gondwana Res.* **2022**, *112*, 105–115. [CrossRef]
39. Guo, J.; Zhou, Y.; Ali, S.; Shahzad, U.; Cui, L. Exploring the role of green innovation and investment in energy for environmental quality: An empirical appraisal from provincial data of China. *J. Environ. Manag.* **2021**, *292*, 112779. [CrossRef]
40. Castelnovo, P. Innovation in private and state-owned enterprises: A cross-industry analysis of patenting activity. *Struct. Change Econ. Dyn.* **2022**, *62*, 98–113. [CrossRef]
41. Yang, Z.; Shao, S.; Yang, L. Unintended consequences of carbon regulation on the performance of SOEs in China: The role of technical efficiency. *Energy Econ.* **2021**, *94*, 105072. [CrossRef]
42. Niu, Y.; Wen, W.; Wang, S.; Li, S. Breaking barriers to innovation: The power of digital transformation. *Financ. Res. Lett.* **2023**, *51*, 103457. [CrossRef]
43. Zhang, Z.; Jin, J.; Li, S.; Zhang, Y. Digital transformation of incumbent firms from the perspective of portfolios of innovation. *Technol. Soc.* **2023**, *72*, 102149. [CrossRef]
44. Feliciano-Cestero, M.M.; Ameen, N.; Kotabe, M.; Paul, J.; Signoret, M. Is digital transformation threatened? A systematic literature review of the factors influencing firms' digital transformation and internationalization. *J. Bus. Res.* **2023**, *157*, 113546. [CrossRef]
45. Malodia, S.; Mishra, M.; Fait, M.; Papa, A.; Dezi, L. To digit or to head? Designing digital transformation journey of SMEs among digital self-efficacy and professional leadership. *J. Bus. Res.* **2023**, *157*, 113547. [CrossRef]
46. Du, K.; Cheng, Y.; Yao, X. Environmental regulation, green technology innovation, and industrial structure upgrading: The road to the green transformation of Chinese cities. *Energy Econ.* **2021**, *98*, 105247. [CrossRef]
47. Tian, H.; Li, Y.; Zhang, Y. Digital and intelligent empowerment: Can big data capability drive green process innovation of manufacturing enterprises? *J. Clean. Prod.* **2022**, *377*, 134261. [CrossRef]
48. Lin, B.; Zhang, A. Can government environmental regulation promote low-carbon development in heavy polluting industries? Evidence from China's new environmental protection law. *Environ. Impact Assess. Rev.* **2023**, *99*, 106991. [CrossRef]
49. Feng, C.; Zhu, R.; Wei, G.; Dong, K.; Dong, J. Typical case of carbon capture and utilization in Chinese iron and steel enterprises: CO_2 emission analysis. *J. Clean. Prod.* **2022**, *363*, 132528. [CrossRef]
50. Song, Y.; Zhang, K.; Li, X. A novel multi-objective mutation flower pollination algorithm for the optimization of industrial enterprise R&D investment allocation. *Appl. Soft Comput.* **2021**, *109*, 107530. [CrossRef]
51. Du, Y.; Huang, R. Research on the core competitiveness of pharmaceutical listed companies based on fuzzy comprehensive evaluation. *J. Intell. Fuzzy Syst.* **2020**, *38*, 6971–6978. [CrossRef]
52. Capponi, G.; Martinelli, A.; Nuvolari, A. Breakthrough innovation and where to find them. *Res. Policy* **2021**, *51*, 104376. [CrossRef]
53. Wang, L.; Chen, F.; Knell, M. Pattern of technology upgrading—The case of biotechnology in China. *Asian J. Technol. Innov.* **2019**, *27*, 152–171. [CrossRef]
54. Lee, P.C. Investgating the knowledge spillover and externality of technology standards based on patent dada. *IEEE Trans. Eng. Manag.* **2021**, *68*, 1027–1041. [CrossRef]
55. Ball, C.; Kittler, M. Removing environmental market failure through support mechanisms: Insight from green star-ups in the British, French and German energy sectors. *Small Bus. Econ.* **2019**, *52*, 831–844. [CrossRef]
56. Zhu, Q.; Geng, Y.; Lai, K. Barriers to promoting Eco-industrial parks development in China: Perspectives from senior officials at national industrial parks. *J. Ind. Ecol.* **2015**, *19*, 457–467. [CrossRef]
57. Huang, Z.; Liao, G.; Li, Z. Loaning scale and government subsidy for promoting green innovation. *Technol. Forecast. Soc. Change* **2019**, *144*, 148–156. [CrossRef]
58. Yu, F.; Wang, L.; Li, X. The effects of government subsidies on new energy vehicle enterprises: The moderating role of intelligent transformation. *Energy Policy* **2020**, *141*, 111463. [CrossRef]

59. Xin, C.; Hao, X.; Cheng, L. Do environmental administrative penalties affect audit fees? Result from multiple econometric models. *Sustainability* **2022**, *14*, 4268. [CrossRef]
60. Wang, X.; Ge, J.; Han, A. Market impacts of environmental regulation on the production of rare earths: A computable general equilibrium analysis for Chain. *J. Clean. Prod.* **2017**, *154*, 614–620. [CrossRef]
61. Sun, X.; Zheng, Y.; Wang, B. The effect of Chinas pilot low-carbon city initiative on enterprise labor structure. *Front. Energy Res.* **2021**, *9*, 821677. [CrossRef]
62. Kainiemi, L.; Eloneva, S.; Toikka, A.; Levänen, J.; Järvinen, M. Opportunities and obstacles for CO_2 mineralization: CO_2 mineralization specific frames in the interviews of Finnish carbon capture and storage (CCS) experts. *J. Clean. Prod.* **2015**, *94*, 352–358. [CrossRef]
63. Shen, Y.; Ruan, S. Accounting Conservatism, R&D Manipulation, and Corporate Innovation: Evidence from China. *Sustainability* **2022**, *14*, 9048. [CrossRef]
64. Pan, X.; Pan, X.; Wu, X.; Jiang, L.; Guo, S.; Feng, X. Research on the heterogeneous impact of carbon emission reduction policy on R&D investment intensity: From the perspective of enterprise's ownership structure. *J. Clean. Prod.* **2021**, *328*, 129532. [CrossRef]
65. Yang, L.; Qin, H.; Xia, W.; Gan, Q.; Li, L.; Su, J.; Yu, X. Resource slack, environmental management maturity and enterprise environmental protection investment: An enterprise life cycle adjustment perspective. *J. Clean. Prod.* **2021**, *309*, 127339. [CrossRef]
66. Wu, W.; Liu, Q.; Wu, C.; Tsai, S. An empirical study on government direct environmental regulation and heterogeneous innovation investment. *J. Clean. Prod.* **2020**, *254*, 120079. [CrossRef]
67. Zhang, J.; Chen, Z.; Altuntaş, M. Tracing volatility in natural resources, green finance and investment in energy resources: Fresh evidence from China. *Resour. Policy* **2022**, *79*, 102946. [CrossRef]
68. Huang, S.; Lin, H.; Zhu, N. The influence of the policy of replacing environmental protection fees with taxes on enterprise green innovation-evidence from China's heavily polluting industries. *Sustainability* **2022**, *14*, 6850. [CrossRef]
69. Xu, X.; Chen, X.; Xu, Y.; Wang, T.; Zhang, Y. Improving the Innovative Performance of Renewable Energy Enterprises in China: Effects of Subsidy Policy and Intellectual Property Legislation. *Sustainability* **2022**, *14*, 8169. [CrossRef]
70. Chen, C.; Gu, J.; Luo, R. Corporate innovation and R&D expenditure disclosures. *Technol. Forecast. Soc. Change* **2022**, *174*, 121230. [CrossRef]
71. Li, Y.; Zhu, D. Share pledging and corporate environmental investment. *Financ. Res. Lett.* **2022**, *50*, 103348. [CrossRef]
72. Wei, F.; Zhou, L. Multiple large shareholders and corporate environmental protection investment: Evidence from the Chinese listed companies. *China J. Account. Res.* **2020**, *13*, 387–404. [CrossRef]
73. Long, F.; Lin, F.; Ge, C. Impact of China's environmental protection tax on corporate performance: Empirical data from heavily polluting industries. *Environ. Impact Assess. Rev.* **2022**, *97*, 106892. [CrossRef]
74. He, M.; Zhu, X.; Li, H. How does carbon emissions trading scheme affect steel enterprises' pollution control performance? A quasi natural experiment from China. *Sci. Total Environ.* **2023**, *858*, 159871. [CrossRef]
75. Cheng, Y.; Zhang, Y.; Wang, J.; Jiang, J. The impact of the urban digital economy on China's carbon intensity: Spatial spillover and mediating effect. *Resour. Conserv. Recycl.* **2023**, *189*, 106762. [CrossRef]
76. Hepburn, C.; Qi, Y.; Stern, N.; Ward, B.; Xie, C.; Zenghelis, D. Towards carbon neutrality and China's 14th Five-Year Plan: Clean energy transition, sustainable urban development, and investment priorities. *Environ. Sci. Ecotechnology* **2021**, *8*, 100130. [CrossRef] [PubMed]
77. Lin, B.; Ma, R. How does digital finance influence green technology innovation in China? Evidence from the financing constraints perspective. *J. Environ. Manag.* **2022**, *320*, 115833. [CrossRef]
78. Feng, S.; Chong, Y.; Yu, H.; Ye, X.; Li, G. Digital financial development and ecological footprint: Evidence from green-biased technology innovation and environmental inclusion. *J. Clean. Prod.* **2022**, *380*, 135069. [CrossRef]
79. Ma, D.; Zhu, Q. Innovation in emerging economies: Research on the digital economy driving high-quality green development. *J. Bus. Res.* **2022**, *145*, 801–813. [CrossRef]
80. Liu, S.; Wang, Y. Green innovation effect of pilot zones for green finance reform: Evidence of quasi natural experiment. *Technol. Forecast. Soc. Change* **2023**, *186*, 122079. [CrossRef]
81. Cui, S.; Wang, Y.; Zhu, Z.; Zhu, Z.; Yu, C. The impact of heterogeneous environmental regulation on the energy eco-efficiency of China's energy-mineral cities. *J. Clean. Prod.* **2022**, *350*, 131553. [CrossRef]
82. Liao, Z. Is environmental innovation conductive to corporate financing? The moderating role of advertising expenditures. *Bus. Strategy Environ.* **2020**, *29*, 954–961. [CrossRef]
83. Feng, G.; Niu, P.; Wang, J.; Liu, J. Capital market liberalization and green innovation for sustainability: Evidence from China. *Econ. Anal. Policy* **2022**, *75*, 610–623. [CrossRef]
84. Zor, S. Conservation or revolution? The sustainable transition of textile and apparel firms under the environmental regulation: Evidence from China. *J. Clean. Prod.* **2023**, *382*, 135339. [CrossRef]
85. Li, F.; Wang, Z.; Huang, L. Economic growth target and environmental regulation intensity: Evidence from 284 cities in China. *Environ. Sci. Pollut. Res.* **2022**, *29*, 10235–10249. [CrossRef]
86. Lodhia, S.; Jacobs, K.; Park, Y.J. Driving Public Sector Environmental Reporting. *Public Manag. Rev.* **2012**, *14*, 631–647. [CrossRef]
87. Zhang, S.; Qin, G.; Wang, L.; Cheng, B.; Tian, Y. Evolutionary game research between the government environmental regulation intensities and the pollution emissions of papermaking enterprises. *Discrete Dynam Nat. Soc.* **2021**, *2021*, 7337290. [CrossRef]

88. Huang, L.; Lei, Z. How environmental regulation affect corporate green investment: Evidence from China. *J. Clean. Prod.* **2021**, *279*, 123560. [CrossRef]
89. Wang, L.; Chen, C.; Zhu, B. Earnings pressure, external supervision, and corporate environmental protection investment: Comparison between heavy-polluting and non-heavy-polluting industries. *J. Clean. Prod.* **2023**, *385*, 135648. [CrossRef]
90. Li, G.; Jin, Y.; Gao, X. Digital transformation and pollution emission of enterprises: Evidence from China's micro-enterprises. *Energy Rep.* **2023**, *9*, 552–567. [CrossRef]
91. Zhang, Y.; Yang, G.; Zhang, D.; Wang, T. Investigation on recognition method of acoustic emission signal of the compressor valve based on the deep learning method. *Energy Rep.* **2021**, *7*, 62–71. [CrossRef]
92. Wang, T.; Wen, C.Y.; Seng, J.L. The association between the mandatory adoption of XBRL and the performance of listed state-owned enterprises and non-state-owned enterprises in China. *Inf. Manag.* **2014**, *51*, 336–346. [CrossRef]
93. Lu, C.; Niu, Y. Do companies compare employees' salaries? Evidence from stated-owned enterprise group. *China J. Account. Res.* **2022**, *15*, 100252. [CrossRef]
94. Xing, J.; Zhang, Y.; Xiong, X. Social capital, independent director connectedness, and stock price crash risk. *Int. Rev. Econ. Financ.* **2023**, *83*, 786–804. [CrossRef]
95. Du, L.; Lin, W.; Du, J.; Jin, M.; Fan, M. Can vertical environmental regulation induce enterprise green innovation? A new perspective from automatic air quality monitoring station in China. *J. Environ. Manag.* **2022**, *317*, 115349. [CrossRef] [PubMed]
96. Du, Z.; Xu, C.; Lin, B. Does the Emission Trading Scheme achieve the dual dividend of reducing pollution and improving energy efficiency? Micro evidence from China. *J. Environ. Manag.* **2022**, *323*, 116202. [CrossRef] [PubMed]
97. Li, Y.; Zhao, K.; Zhang, F. Identification of key influencing factors to Chinese coal power enterprises transition in the context of carbon neutrality: A modified fuzzy Dematel approach. *Energy* **2023**, *263*, 125427. [CrossRef]
98. Li, J.; Dong, K.; Wang, K.; Dong, X. How does natural resource dependence influence carbon emissions? The role of environmental regulation. *Resour. Policy* **2023**, *80*, 103268. [CrossRef]
99. Zeng, J.; Chen, X.; Liu, Y.; Cui, R.; Zhao, P. How does the enterprise green innovation ecosystem collaborative evolve? Evidence from China. *J. Clean. Prod.* **2022**, *375*, 134181. [CrossRef]
100. Liu, M.; Shan, Y.; Li, Y. Study on the effect of carbon trading regulation on green innovation and heterogeneity analysis from China. *Energy Policy* **2022**, *171*, 113290. [CrossRef]
101. Yu, Z.; Shen, Y.; Jiang, S. The effects of corporate governance uncertainty on state-owned enterprises' green innovation in China: Perspective from the participation of non-state-owned shareholders. *Energy Econ.* **2022**, *115*, 106402. [CrossRef]
102. Wei, N.; Liu, S.; Jiao, Z.; Li, X. A possible contribution of carbon capture, geological utilization, and storage in the Chinese crude steel industry for carbon neutrality. *J. Clean. Prod.* **2022**, *374*, 133793. [CrossRef]
103. Wang, L.; Dilanchiev, A.; Haseeb, M. The environmental regulation and policy assessment effect on the road to green recovery transformation. *Econ. Anal. Policy* **2022**, *76*, 914–929. [CrossRef]
104. Pang, C.; Zhou, J.; Ji, X. The Effects of Chinese Consumers' Brand Green Stereotypes on Purchasing Intention toward Upcycled Clothing. *Sustainability* **2022**, *14*, 16826. [CrossRef]
105. Xu, X.; Zhang, W.; Wang, T.; Xi, Y.; Du, H. Impact of subsidies on innovations of environmental protection and circular economy in China. *J. Environ. Manag.* **2021**, *289*, 112385. [CrossRef] [PubMed]
106. Lin, C.P.; Tsai, Y.H.; Chiu, C.K.; Liu, C.P. Forecasting the purchase intention of IT product: Key roles of trust and environmental consciousness for IT firms. *Technol. Forecast. Soc. Change* **2015**, *99*, 148–155. [CrossRef]
107. Cheng, P.; Wang, X.; Choi, B.; Huan, X. Green Finance, International Technology Spillover and Green Technology Innovation: A New Perspective of Regional Innovation Capability. *Sustainability* **2023**, *15*, 1112. [CrossRef]
108. Xu, X.; Cui, X.; Chen, X.; Zhou, Y. Impact of government subsidies on the innovation performance of the photovoltaic industry: Based on the moderating effect of carbon trading prices. *Energy Policy* **2022**, *170*, 113216. [CrossRef]
109. Chen, Y.; Han, X.; Lv, S.; Song, B.; Zhang, X.; Li, H. The Influencing Factors of Pro-Environmental Behaviors of Farmer Households Participating in Understory Economy: Evidence from China. *Sustainability* **2023**, *15*, 688. [CrossRef]
110. Wang, L.; Long, Y.; Li, C. Research on the impact mechanism of heterogeneous environmental regulation on enterprise green technology innovation. *J. Environ. Manag.* **2022**, *322*, 116127. [CrossRef]
111. Cao, H.; Zhang, L.; Qi, Y.; Yang, Z.; Li, X. Government auditing and environmental governance: Evidence from China's auditing system reform. *Environ. Impact Assess. Rev.* **2022**, *93*, 106705. [CrossRef]
112. Sun, Y.; Sun, H. Green Innovation Strategy and Ambidextrous Green Innovation: The Mediating Effects of Green Supply Chain Integration. *Sustainability* **2021**, *13*, 4876. [CrossRef]

Disclaimer/Publisher's Note: The statements, opinions and data contained in all publications are solely those of the individual author(s) and contributor(s) and not of MDPI and/or the editor(s). MDPI and/or the editor(s) disclaim responsibility for any injury to people or property resulting from any ideas, methods, instructions or products referred to in the content.

Article

Carbon Dioxide Adsorption over Activated Carbons Produced from Molasses Using H_2SO_4, H_3PO_4, HCl, NaOH, and KOH as Activating Agents

Karolina Kiełbasa [1], Şahin Bayar [2], Esin Apaydin Varol [2], Joanna Sreńscek-Nazzal [1], Monika Bosacka [3], Piotr Miądlicki [1], Jarosław Serafin [4,*], Rafał J. Wróbel [1,*] and Beata Michalkiewicz [1]

[1] Faculty of Chemical Technology and Engineering, Department of Catalytic and Sorbent Materials Engineering, West Pomeranian University of Technology in Szczecin, Piastów Ave. 42, 71-065 Szczecin, Poland
[2] Faculty of Engineering, Deptarment of Chemical Engineering, Eskisehir Technical University, Eskisehir 26555, Turkey
[3] Faculty of Chemical Technology and Engineering, Department of Inorganic and Analytical Chemistry, West Pomeranian University of Technology in Szczecin, Piastów Ave. 42, 71-065 Szczecin, Poland
[4] Department of Inorganic and Organic Chemistry, University of Barcelona, Martí i Franquès, 1-11, 08028 Barcelona, Spain
* Correspondence: jaroslaw.serafin@qi.ub.es (J.S.); rafal.wrobel@zut.edu.pl (R.J.W.)

Abstract: Cost-effective activated carbons for CO_2 adsorption were developed from molasses using H_2SO_4, H_3PO_4, HCl, NaOH, and KOH as activating agents. At the temperature of 0 °C and a pressure of 1 bar, CO_2 adsorption equal to 5.18 mmol/g was achieved over activated carbon obtained by KOH activation. The excellent CO_2 adsorption of M-KOH can be attributed to its high microporosity. However, activated carbon prepared using HCl showed quite high CO_2 adsorption while having very low microporosity. The absence of acid species on the surface promotes CO_2 adsorption over M-HCl. The pore size ranges that are important for CO_2 adsorption at different temperatures were estimated. The higher the adsorption temperature, the more crucial smaller pores were. For 1 bar pressure and temperatures of 0, 10, 20, and 30 °C, the most important were pores equal and below: 0.733, 0.733, 0.679, and 0.536 nm, respectively.

Keywords: CO_2 adsorption; activated carbon; molasses

1. Introduction

The ubiquitous global climate crisis and an improved readiness of the industrial sectors to reach a global net zero by 2050 have been concentrating on reducing greenhouse gases, especially CO_2, which represents about 77% of all greenhouse gases [1,2]. Thus, there are developing fields of research focusing on reducing CO_2 emissions, of which Carbon Capture, Utilisation, and Storage (CCUS) deserve special attention [3,4]. Although several promising industrial processes have already been demonstrated in various stages of development, none of the technologies can provide an economically viable and complete CCUS.

Therefore, it must be admitted that it additionally drives the ever-increasing request for energy-efficient adsorbent materials and methods for mitigating the energy footprint of chemical processes. Especially, activated carbons are considered very promising materials as CO_2 sorbents as they meet most of the requirements: chemical and thermal stability, low energy consumption for regeneration, hydrophobicity, and stability during regeneration [5]. Remarkably, if biomass or waste is a carbon precursor, such sorbents are also inexpensive, and most biomass-based activated carbons show high adsorption capacity and selectivity [6]. The use of activated carbons as sorbents of various gases, pollutants of water, or other liquids has been described widely [7]. Moreover, these activated carbons can also be used as catalysts [8,9].

The structure and porosity of activated carbon depend on the carbon source and can be controlled by the conditions of the synthesis such as temperature, activating agent (type and quantity), duration of activation or carbonization, etc. Knowledge of how different variables influence physicochemical properties is essential and allows the design of activated carbons for desired applications, including CO_2 adsorption. The majority of the authors demonstrate only the influence of carbonization temperature on the properties of carbon materials [10–12]. On the other hand, the significance of the activating agent quantity is rarely described. The most common activating agents used in activated carbon production are H_3PO_4, H_2SO_4, HNO_3, $ZnCl_2$, NaOH, and KOH [13]. Other activators, such as K_2CO_3, $CaCl_2$, H_2O_2, and CH_3NO, were used quite seldom [14].

The present study demonstrates the synthesis of activated carbon for CO_2 adsorption from molasses using five different activating agents. Three acids: H_2SO_4, H_3PO_4, HCl, and two bases: NaOH and KOH, were used as activating agents to provide more information on the role of chemicals in the activation process during activated carbon production from biomass. Beet molasses, a sugar industry waste, was applied as the carbon source. This is a very unusual carbon precursor because it is liquid. Apart from our team's research [15], there is only one report of obtaining activated carbon from molasses [16]. Legrouri et al. [16] produced three activated carbons from molasses for methylene blue adsorption using CO_2 and H_2SO_4 as activating agents. Our team applied activated carbon from molasses activated by a saturated solution of KOH for CH_4 adsorption [15] and activated by KOH powder for CO_2 adsorption [17,18].

The novelty of this work is the use of molasses as a carbon precursor for CO_2 adsorption combined with H_3PO_4, HCl, and NaOH as activating agents. According to our knowledge, such investigations have not been described up to now. The physicochemical properties of the activated carbons were characterized by powdered X-ray diffraction (XRD), Raman spectroscopy, scanning electron microscopy (SEM), N_2 adsorption analysis, Fourier transform infrared reflection (FTIR), and thermogravimetry. The influence of textural properties and surface chemistry on CO_2 adsorption was discussed.

2. Results

The graphitic structure and purity of activated carbons were characterized using XRD. The diffraction patterns of the activated carbons are given in Figure 1a. Activated carbons produced using acids showed two broad asymmetric peaks at 2θ, about 23 and 43°, which correspond to the plane (002) and (100/101) of the graphite structure (JCPDS 41-1487) associated with the stacking height-thickness of the layer packets (Lc) and longitudinal dimension, so-called aromatic sheets (La), respectively.

Broad peaks indicated a highly disordered carbon structure and a predominantly amorphous arrangement. Activated carbons obtained using bases showed a more disordered structure. The XRD spectra of M-KOH and M-NaOH contained only one low-intense signal. It was particularly interesting that M-KOH had a peak at $2\theta \approx 43°$ and M-NaOH had a peak at $2\theta \approx 23°$. This means that the interaction of KOH and NaOH with carbon precursors was different. During activation by NaOH, very small aromatic sheets were produced, whilst during activation by KOH, the number of the layers in the packets was supposed to be very small.

The structural parameters obtained from XRD measurements are listed in Table 1. The interlayer spacings ($d_{(002)}$) of the produced activated carbons are higher than that of graphite which has an interlaying spacing of 0.335 nm. For activated carbons obtained using acidic agents, the interlayer spacings were similar to each other and ranged from 0.363 to 0.368 nm. The value of $d_{(002)}$ for M-NaOH was considerably higher (0.376 nm). It was impossible to determine $d_{(002)}$ for M-KOH due to the absence of the peak at $2\theta \approx 23°$. For the same reason, the values of Lc and the number of layers in the packets (N) could not be determined.

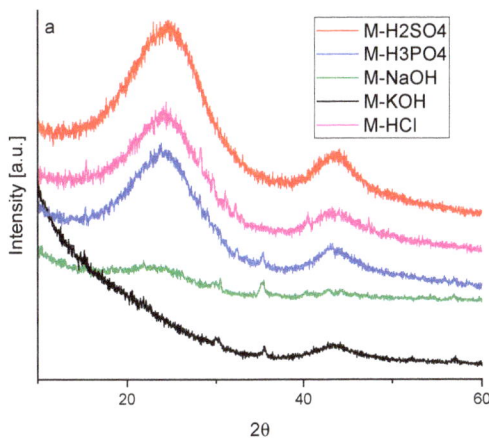
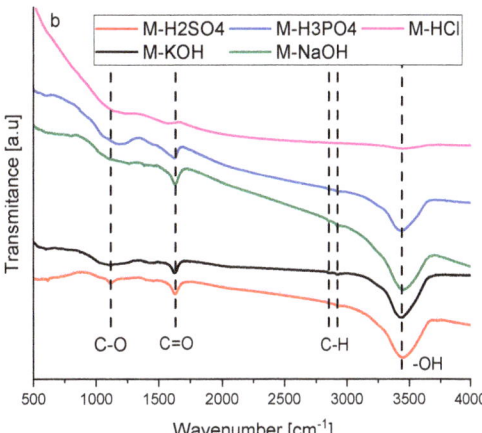

Figure 1. XRD diffraction patterns (a) and FTIR spectra (b) of the activated carbons produced from molasses.

Table 1. Structural parameters of activated carbons obtained from XRD, and I_G/I_D ratios calculated from Raman measurements.

Activated Carbon	$d_{(002)}$ [nm]	La [nm]	Lc [nm]	N	I_G/I_D
M-H$_2$SO$_4$	0.363	3.785	0.919	2.53	0.692
M-H$_3$PO$_4$	0.368	3.878	1.013	2.75	0.722
M-HCl	0.366	3.403	0.964	2.63	0.711
M-KOH		3.875			0.686
M-NaOH	0.376		1.212	3.22	0.686

Considering that the number of layers in the packets for the other samples ranged from two to three, it can be postulated that there are no parallel aromatic sheets in the structure of M-KOH. The M-KOH was built of individual layers arranged in a completely disordered manner. The dimensions of so-called aromatic sheets (La) were different for activated carbon obtained using different activation agents and ranged from 3.403 to 3.878 nm. The value of La for M-NaOH was not listed in Table 1 due to the absence of a peak at 2θ ≈ 43°. It can be assumed that during the interaction of molasses with NaOH, very small aromatic sheets were produced.

Raman spectroscopy is a common method to evaluate the defects and crystallographic disorders in carbon. In Figure S1 Supplementary Material, Raman spectra with a range of Raman shifts from 500 to 2500 cm^{-1} were presented. There were two broad overlapping peaks. The first one, centred near 1330 cm^{-1} is due to the disordered portion of the carbon–D band. The second one, centred near 1600 cm^{-1} is due to ordered graphitic crystallites of the carbon (sp^2 bonding carbon atoms)–G band. The intensities of D signals were higher than the G ones. All the spectra were normalized to the D band intensity (Figure S2). The values of the G peak maxima in Figure S2 were equal to the values of the intensity ratio of the G and D bands (I_G/I_D). The ratios of the G band to D band intensities were compiled in Table 1.

The intensity ratio of the G and D bands allows for estimating the degree of defects. The lower I_G/I_D ratio means more defects. The presence of the G band in the Raman spectra indicates that some graphene sheets can be present. The lower the intensity ratio, the smaller the size of the graphitic sheet and the higher the disorder of the graphene sheets. The values of I_G/I_D ratios for activated carbon samples were quite similar and ranged from 0.686 to 0.722. The smallest values were obtained for M-KOH and M-NaOH. The structure of the M-KOH was highly disordered and probably only single graphene sheets existed, arranged concerning each other in a random manner. The dimensions of graphene sheets in the M-NaOH were so small that the XRD signal at $2\theta \approx 43°$ was absent. The XRD results were consistent with Raman's findings.

Based on XRD and Raman investigations, it was found that all the samples are amorphous and disordered carbon materials, consisting of small aromatic carbon sheets stacked in 2 or 3 packets or even singular sheets.

The FTIR spectra are presented in Figure 1b. The bands in the range of 500–4000 cm^{-1} indicated the presence of carboxyl groups. The peak at 1630 cm^{-1} was attributed to C=O stretching vibrations [19]. The peak at 1120 cm^{-1} is characteristic of the coupled C–O stretching frequency and OH bending modes in the carboxylic group and the C–O stretching modes of ethers [20]. Two small peaks at around 2850 and 2920 cm^{-1} indicated the presence of symmetric and asymmetric C–H stretching vibrations, respectively, typical for hydrocarbons [21]. The wide and broad band cantered at 3444 cm^{-1} was attributed to the O–H stretching mode of H in hydroxyl groups. The hydroxyl groups can indicate the presence of phenolic or ether groups or water adsorbed on the surface [22]. For activated carbon activated by HCl, no transmittance bands were observed. The HCl treatment removed all the functional groups from the carbon surface. After treatment by basics, the C–O functional groups were nearly removed. The band characteristic of C–O was considerably broadened for M-H3PO4. The broad band at around 1000–1300 cm^{-1} was also typical for the phosphorous and phosphocarbonaceous compounds. It can be ascribed to the stretching mode of the hydrogen-bonded P=O and C–O stretching mode of the P–O–C and P=OOH groups [21]. The highest intensity of the peak at 1120 cm^{-1} being characteristic for C–O was observed for M-H2SO4 because of the overlapping bands of S=O stretching vibrations of the sulphates and sulfoxides. Moreover, a small S–O stretching vibration band was also found at 622 cm^{-1} [23].

Figure 2 shows the DTA–TG curves of activated carbons from room temperature to 1000 °C. The DTA curves of all samples show an effect with an onset of approx. 55 °C. This effect is related to the first stage of weight loss due to the removal of moisture adsorbed on the surface of activated carbons, which is also visible in the TG curves of all five samples [24]. No other thermal effects were recorded on the DTA curves of activated carbons M-HCl, M-KOH, and M-NaOH. The TG curves of the samples were quite similar. Their weight had been falling with a slow gradient with the increase in the temperature during the range of 30–900 °C, and the total weight loss ranged from 17.2 to 23.5 wt.%, depending on the sample. The weight loss in this range may be related to the decomposition of residual organic materials [25,26] due to the volatilization reactions of non-carbon functional groups. In the temperature range from 900 to 1000 °C, mass rapidly decreases by approx. 10 wt.%. The weight loss stage starts at 900 °C and can be attributed to the initiation of a progressive decomposition of the carbon [27].

On the DTA curve of the M-H2SO4 sample, apart from the effect, which is associated with the evaporation of adsorbed water, a second small endothermic effect with the beginning of approx. 250 °C was recorded. The second degree of weight loss is associated with this effect. At a temperature of 400 °C, the total weight loss is 20.5% by weight. This weight loss may be related to the removal of absorbed sulphur oxides associated with the use of concentrated sulphuric acid [23]. The presence of sulphur groups was shown in the FTIR spectrum (Figure 1b).

On the DTA curve of sample M-H3PO4, two endothermic effects were also recorded. The first effect at the onset of around 55 °C, as mentioned above, is attributed to moisture. The weight loss related to this effect recorded on the TG curve amounts to 16.3 wt.%. In the temperature range from 155 to 800 °C, there is a slight weight change. The second, slight endothermic effect, beginning at 900 °C, recorded on the DTA curve of sample M-H3PO4, is closely related to the second stage of weight loss recorded on the TG curve. At temperatures >800 °C, different P-containing compounds, and elemental phosphorus (P$_4$) could be formed as a result of the reduction in phosphorus compounds previously bound to the carbonaceous solid residue [28,29]. Thus, when fewer phosphate-like structures are released, more phosphorus remains in the solid residue can be released at higher temperatures to produce the latter effect. The total weight loss of the sample M-H$_3$PO$_4$ recorded on the TG curve is 32.7 wt.%. The presence of phosphorus-containing groups was shown in the FTIR spectrum (Figure 1b).

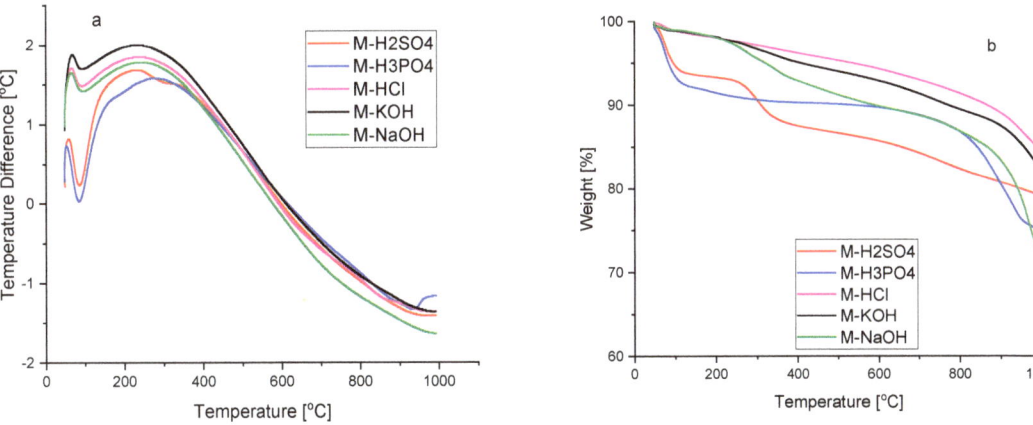

Figure 2. The DTA (**a**) and TG curves (**b**) of the activated carbons produced from molasses.

SEM pictures of the activated carbons revealed the textures of the obtained materials (Figures 3 and S3). The surface of activated carbon produced using bases looked as if they were made of corrugated petals. The surface of activated carbon produced by means of acids looked different. The picture showed a more massive, yet undulating surface.

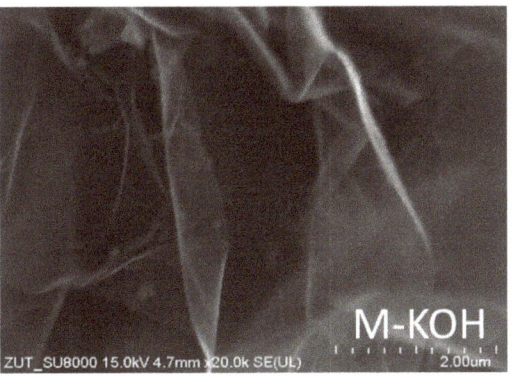 

Figure 3. SEM pictures of the activated carbon produced from molasses.

Figure 4a shows the nitrogen adsorption isotherms at a temperature of 77 K. All the isotherms were characterized by rapid growth at low-pressure P/P_0, which points that the activated carbons obtained from molasses are microporous materials. The hysteresis loops for most of the samples were very narrow, sometimes even invisible without proper magnification. The presence of the hysteresis loop indicates that mesopores also accompany the micropores. This phenomenon was the most established for activated carbon activated by sulphuric acid. Thus, it might be concluded that this modification agent contributed to the formation of the mesopores. The most rapid growth of micropores was recorded for material activated by potassium hydroxide.

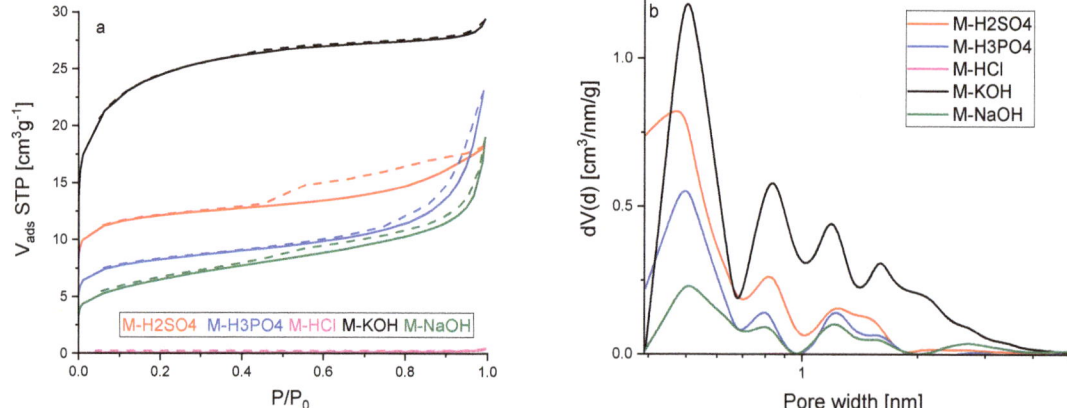

Figure 4. Adsorption–desorption isotherms of nitrogen (**a**) and micropore pore size distribution calculated by the DFT method (**b**) for activated carbons from molasses.

According to the IUPAC classification, the three nitrogen adsorption isotherms presented in Figure 4a, except M-KOH and M-HCl, represent a composite of Types Ia and IV. The isotherm of activated carbon M-KOH was Type Ib. The ideal Ia and Ib isotherms are characterized by the relatively quick achievement of the adsorption limit, which results in the course of the adsorption isotherm being parallel to the P/P_0 axis. The isotherms presented in Figure 4a were parallel to the P/P_0 axis in most ranges of P/P_0. Detailed studies on macropores have not been carried out due to the fact that these pores are not crucial for CO_2 adsorption. They are mainly used to transport the adsorbent from the outer surface to micro and mesopores. The isotherms were reversible, but for some of them, hysteresis was observed. The hysteresis was caused by capillary condensation that took place in mesopores. The sparse hysteresis loops were identified as H3 type (for M-H_3PO_4 and M-NaOH), which are observed with aggregates of plate-like particles giving rise to slit-shaped pores, as well as H4 type (for M-H_2SO_4 and M-NaOH) that can be correlated to narrow slit-like pores. The isotherm of M-HCl is placed on the x-axis in Figure 4a because the N_2 adsorption was very low. The micropore size distribution calculated by the DFT method is presented in Figure 4b. For all the materials, pores of about 0.5 nm in diameter were dominant. The highest pore volume of the size of the pores about was 0.5 nm for M-KOH. The pore volume of these pores decreased in the following order: M-H_2SO_4, M-H_3PO_4, and M-NaOH. The second maxima were observed for a pore size of about 0.8 nm. For M-KOH, two clearly developed peaks were observed with maxima at 1.2 and 1.6 nm. For the other materials, the wide peak was observed in regions 1.1–1.7 nm. It is clearly seen that activated carbon produced using KOH exhibited the highest micropore volume. The second-highest micropore volume was found for H_2SO_4.

Table 2 shows the textural properties of activated carbons obtained from molasses. The highest surface area, total pore volume, and micropore volume were achieved for M-KOH. The highest mesopore volume was obtained for M-NaOH.

Table 2. Textural properties of activated carbons obtained from molasses.

Activated Carbon	S_{BET} [m²/g]	V_p [cm³/g]	V_{mi} [cm³/g]	V_{ms} [m²/g]
M-H_2SO_4	1016	0.636	0.315	0.307
M-H_3PO_4	681	0.801	0.188	0.356
M-HCl	6	0.015	0.001	0.009
M-KOH	1970	1.020	0.635	0.362
M-NaOH	505	0.660	0.106	0.480

The values of textural properties were in good agreement with the conclusions drawn from Figure 4. The analysis of Tables 1 and 2 indicated that the activators worked in different ways. Activated carbons produced using acids had a more ordered structure than those produced using basic, but a disordered structure was not a guarantee of high porosity.

The CO_2 adsorption was investigated at temperatures of 0 °C, 10 °C, 20 °C, and 30 °C, up to a pressure of 1 bar. The adsorption results were presented in Figure 5a and S4. The highest CO_2 adsorption was observed for activated carbons activated by KOH and H_2SO_4. The lowest values were obtained for M-HCl and M-NaOH. Some authors indicate that CO_2 adsorption is correlated with textural properties, especially micropore volume [30–32]. Our results also confirmed their findings, except for M-HCl. A detailed analysis of the relationship between CO_2 adsorption capacity at temperatures of 0 °C, 10 °C, 20 °C, and 30 °C and the pore volume of pores smaller than a specific pore size was also performed (Figures S5–S8). The results of the best fit are presented in Table 3. The CO_2 adsorption at a temperature of 0 to 10 °C is mainly caused by the micropores equal to or less than 0.733 nm in diameter. Further temperature increases resulted in a reduction in the pore diameter, which determines high adsorption. Similar results were presented by Deng et al. [10]. At a temperature of 20 °C and 30 °C, the CO_2 adsorption was mainly caused by the micropores having a diameter of equal or less than 0.536 nm.

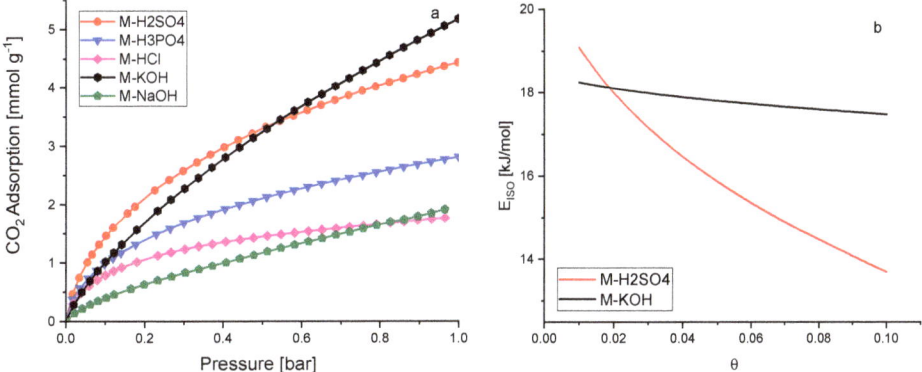

Figure 5. CO_2 adsorption at a temperature of 0 °C over activated carbons from molasses (**a**), the isosteric heat of adsorption for M-KOH and M-H_2SO_4 (**b**).

Table 3. The pore size (diameter) important for CO_2 adsorption at different temperatures.

Temperature	Pore Size	R^2
[°C]	[nm]	
0	0.733	0.990
10	0.733	0.994
20	0.536	0.998
30	0.536	0.993

Very interesting results were obtained for M-HCl. This material was not porous. The micropore volume was equal to 0.001 cm^3/g, pores equal to 0.733 nm and smaller were not detected, and the CO_2 adsorption was nearly the same as for M-NaOH, while the micropore volume was a hundred times higher. It can be assumed that the reason is the absence of acid functional groups on the surface of the carbon materials. Shafeeyan et al. [33] showed that removing acidic groups from the carbon surface increased CO_2 adsorption. To eliminate oxygen-containing acidic groups, heat treatment was applied by many authors. Such investigations were summarised in [33]. The possibility of producing activated carbon without oxygen-containing acidic groups using HCl has not been described up to now. The highest CO$_2$ adsorption at 0 °C and 1 bar was equal to 5.18 mmol/g (M-KOH). This is quite a high result compared to the others that claimed high-performance CO$_2$ adsorption, for example, 3.31 mmol/g at 0 °C and 1 bar. [34] and 3.2–5.3 mmol/g at 0 °C and 1 bar [35]. The CO$_2$ adsorption decreased with the increase in temperature, meaning that the nature of CO$_2$ sorption in over activated carbon produced from molasses is physical regardless of the activating agent.

Experimental data of carbon dioxide adsorption over the best sorbents, M-KOH and M-H$_2$SO$_4$, were modelled with Langmuir, Freundlich, Langmuir–Freundlich (Sips), Toth, Fritz–Schlunder, Radke–Prausnitz equations. The least-squares method was utilized as an error function. The best fittings of the experimental data were obtained for the Toth model (1).

Toth Equation (1):

$$q = \frac{q_m bp}{(1 + (bp)^n)^{\frac{1}{n}}} \; [\text{mmol/g}] \quad (1)$$

where

q_m—the maximum adsorption capacity [mmol/g]
b—the Toth constant [bar^{-1}]
n—the heterogeneity factor

The listed parameters above depend on the temperature according to the equations:

$$q_m = q_{m0} \cdot \exp\left[\chi\left(1 - \frac{T}{T_0}\right)\right] \quad (2)$$

$$b = b_0 \exp\left[\frac{Q}{RT_0}\left(\frac{T_0}{T} - 1\right)\right] \quad (3)$$

$$n = n_0 + \alpha\left(1 - \frac{T_0}{T}\right) \quad (4)$$

where q_{m0}, χ, Q, b_0, n_0, and α are constants and T_0 is the reference temperature [K], (273 K here).

The calculated values of constants q_m, b, and n in the Toth Equation (1) for M-KOH and M-H$_2$SO$_4$ at different temperatures were presented in Table S1. According to Equations (2)–(4), these parameters depend on the temperature. Considering Equations (2)–(4) the plots: ln(q_m) versus T, ln(b) versus 1/T, and n versus 1/T were sketched for M-KOH (Figure S9) and M-H$_2$SO$_4$ (Figure S10). On the basis of these plots, the parameters of q_{m0}, χ, Q, b_0, n_0, and α from Equations (2)–(4) were estimated and presented in Table S3. Using Equations (2)–(4) and Table S3, the CO$_2$ adsorption over M-KOH and M-H$_2$SO$_4$ can be calculated at any temperature and pressure.

The isosteric heat of adsorption, i.e., the differential enthalpy of adsorption at constant coverage (θ) was calculated based on the Clausius–Clapeyron Equation (S1), after converting this equation to linear form (S2). The isosteric heat of adsorption was calculated by plotting ln(p)$_θ$ vs. 1/T (Figures S11 and S12). The values of the pressures for given surface coverage were calculated using the Toth Equation (S1) and the parameters are listed in Table S2. Finally, the isosteric heat of adsorption as a function of surface coverage was obtained for M-KOH and M-H$_2$SO$_4$ (Figure 5b). The isosteric heat of adsorption decreased with the surface coverage of the activated carbon. The shape of the curves confirmed the physical nature of CO$_2$ adsorption over M-KOH and M-H$_2$SO$_4$. A very important difference was also observed. The isosteric heat of adsorption is nearly the same up to a surface coverage of 0.1 for M-KOH, whereas for M-H$_2$SO$_4$ the isosteric heat of adsorption decreases very fast. The reason can be explained by the difference in porosity. The CO$_2$ molecules at the initial stage of adsorption first penetrate the smallest micropores, resulting in a strong interaction between CO$_2$ and the carbon surface, and hence a high isosteric heat at lower coverage was achieved. With an increase in CO$_2$ adsorption, the wider pores became involved, and the CO$_2$-adsorbent surface interactions became weaker.

The molecules of CO_2 covered not only the surface but also filled all the volumes of the pore. The higher the diameters of the pores, the weaker the isosteric heat of adsorption. The micropore volume for M-KOH was twice as high as than for M-H_2SO_4. The surface coverage was equal to 0.1. The inner surface of the smallest micropores was covered in the case of M-KOH, so the isosteric heat of adsorption was nearly constant. For M-H_2SO_4, the micropores were filled faster and mesopore filling took place. A similar purpose for the decrease in isosteric heat of adsorption was postulated by Abdulsalam et al. [36]. In Figure 5b, the isosteric heat of adsorption ranged from 18 to 17 kJ/mol for M-KOH and from 19 to 14 kJ/mol for M-H_2SO_4. Such low values of the isosteric heat of adsorption indicate that the desorption is very easy, and sorbents can be used repeatedly. The values of the isosteric heat of adsorption were lower than usually presented (22–31 kJ/mol [37], about 38.9 kJ/mol [38], and 28–18 kJ/mol [39]). The meaning of the Q parameter in the Equation (3) is the value of the isosteric heat of adsorption when the degree of coverage is approaching 0. The values of Q are presented in Table S3 (19 kJ/mol for M-KOH and 25 kJ/mol for M-H_2SO_4) and were consistent with the values of isosteric heat of adsorption presented in Figure 5b.

3. Materials and Methods

3.1. Materials

Molasses was purchased from the sugar factory in Kluczewo, Poland (National Food Inc., Trenton, NJ, USA).

Chemical activation of beet molasses was carried out with the use of five activating agents: HCl (Chempur 35–38%), H_2SO_4 (Stanlab 96%), H_3PO_4 (Stanlab 85%), KOH (Chempur), and NaOH (Stanlab). All reagents were used without any further purification. Carbon dioxide (99.99%) and nitrogen (99.99%) were purchased from Messer Polska.

3.2. Methods

Liquid molasses was weighed and then an activating agent was added in such an amount that the mass ratio of molasses to activator was 1:1. Then, the material was vigorously mixed until the raw material was clearly saturated with a saturated aqueous solution of the activating agent and left at ambient temperature for 3 h. After this time, the impregnated material was placed in a laboratory dryer (12 h, 105 °C). The carbonaceous precursor impregnated in this way was carbonized. A physical activation process was conducted in a tubular reactor kept for 1 h in an electrical furnace at a temperature of 800 °C, and the temperature was increased to 10 °C per minute to a chosen value. The process was carried out in a nitrogen atmosphere (a flow rate equal to 14.4 dm^3/h). The activation process parameters such as time, N_2 flow rate, and heating rate of the furnace in all the experiments were identical. Next, the derived activated carbon containing the decomposition products of activating agents was rinsed with deionized water to attain a neutral reaction. When the sample was evaporated, the activated carbon was flooded with a 1 mol/dm^3 HCl solution and left behind for 20 h. In the following stage, carbons were rinsed with deionized water until complete removal of chloride ions. Then the samples were dried at a temperature of 105 °C for 20 h. The activated carbons were denoted as M-X, where X represents the activators: KOH, NaOH, HCl, H_2SO_4, and H_3PO_4.

Analysis of the phase composition of carbons was performed by means of a PANalytical Empyrean X-ray diffractometer (XRD) equipped with a $Cu_{K\alpha}$ lamp. The obtained diffraction patterns were analysed by a comparison of the location and intensity of reflexes on the obtained diffraction patterns with the standard diffraction patterns contained in the ICDD PDF4+2015 database based on X'Pert HighScore computer software.

Raman spectroscopy was utilized to determine the structure of the carbon framework of prepared carbon materials. The analyses were performed using an apparatus equipped with a CCD detector (Renishaw InVia). The samples were induced by a laser with a wavelength of 785 nm. The spectrum was obtained in a range of Raman scattering from 800 cm^{-1} to 2000 cm^{-1}. After normalization of the G peak maximum to 1, the intensity and location of the G and D peaks were assigned, and the ratio of these intensities was calculated. The ratio of G and D band intensities is generally recognized in the literature as the method for the determination of the order of the graphene layers and graphitic structure in carbon materials.

FTIR spectra of the samples were obtained by the Nicolet 6700 FT-IR spectrometer for the identification of functional groups between 500 and 4000 cm^{-1} using transmission mode. The samples were prepared using the KBr pressed-disk technique, with 1% inclusion of the material to be analysed.

To test the thermal properties of the obtained activated carbons, the samples were examined using a thermogravimetric analyser—SDT 650 DISCOVERY series (TA Instruments, New Castle, DE, USA). The tests were carried out in the temperature range from 20–25 °C to 1000 °C. The tested samples were placed in alumina crucibles, and their weight was adjusted to about 20 mg. All measurements were carried out under argon.

The morphology of the obtained activated carbons was examined using the Hitachi SU 8200 scanning electron microscope with field emission (FE-SEM).

To determine the textural properties of modified activated carbons, the low-temperature adsorption isotherms of N2 (−196 °C) were determined for the above-mentioned carbon samples by means of Sorption Surface Area and Pore Size Analyzer (ASAP 2460, Micrometrics, Novcross, GA, USA). The control and data acquisition were enabled by the ASAP software. To remove the pollutants before the adsorption measurements, the carbon samples were calcined at a temperature of 250 °C for 12 h with a heating rate of 1 °C/min under the conditions of reduced pressure.

From N$_2$ sorption isotherms, the following parameters characterizing the porous structure have been determined: surface area SBET calculated from the BET equation, total pore volume V_p determined based on the maximum adsorption of nitrogen for a value of P/P_0 about 0.99, micropore volume V_{mi} determined by the DFT method (Density Functional Theory), mesopores volume V_{ms} determined by the BJH method (Barrett—Joyner—Halenda).

The studies of CO_2 adsorption for the activated carbon were carried out at the temperatures of 0 °C, 10 °C, 20 °C, and 30 °C under the pressure of 1 bar utilizing the ASAP apparatus. In order to control the measurement temperature, the sample was located in a water bath equipped with a Peltier-cooled solid-state detector. Prior to the CO_2 adsorption measurements, the tested materials were outgassed at a temperature of 250 °C for 12 h.

4. Conclusions

CO_2 adsorption at temperatures of 0 °C, 10 °C, 20 °C, and 30 °C was investigated with over activated carbons produced from molasses using H_2SO_4, H_3PO_4, HCl, NaOH, and KOH as the activating agents. All five activated carbons were obtained under the same conditions, only the activating agent was changed. The highest CO_2 adsorption at the temperature of 0 °C and pressure of 1 bar was obtained over M-KOH (5.18 mmol/g) and M-H$_2$SO$_4$ (4.44 mmol/g) indicating that these materials are promising sorbents. The achieved high adsorption is even more promising because, by using molasses and KOH or H$_2$SO$_4$, much greater adsorption can be achieved as a result of changes in other production parameters (temperature, activating agent quantity, time of activation or carbonization, etc.).

It was stated that high microporosity promotes high CO_2 adsorption but probably the absence of acid groups on the surface is also of great importance. Interestingly, non-porous carbon material (M-HCl) showed quite good CO_2 adsorption results. The role and importance of HCl as a carbon-activating agent have to be developed.

The pore size ranges important for CO_2 adsorption at different temperatures were estimated. The higher the temperature, the smaller the pores were crucial. For 1 bar pressure and temperatures of 0 °C, 10 °C, 20 °C, and 30 °C the most important were pores equal and below: 0.733, 0.733, 0.536, 0.536 nm, respectively.

The experimental data were validated with Langmuir, Freundlich, Langmuir–Freundlich (Sips), Toth, Fritz–Schlunder, and Radke–Prausnitz isotherm models. The Toth equation was found to be the best fitting and gave the lowest least-squares error.

Comparing the curves of functions $E_{iso} = f(\theta)$ for M-KOH and M-H$_2$SO$_4$ resulted in the isosteric heat of adsorption strongly depending on the porosity. For the activated carbons with the highest micropore volume, the decrease in the isosteric heat of adsorption was slow along with the increase in surface coverage.

Based on the isosteric heat of adsorption data and the decrement of the CO_2 adsorption with temperature, it could be stated that CO_2 sorption in over activated carbon produced from molasses was physical.

The mathematical description of CO_2 adsorption characteristics (adsorption isotherms and isosteric heat of adsorption) is very important for designing an effective CO_2 adsorption system.

Supplementary Materials: The following supporting information can be downloaded at: https://www.mdpi.com/article/10.3390/molecules27217467/s1, Figure S1: Raman spectra of the activated carbons produced from molasses; Figure S2: Raman spectra of the activated carbon produced from molasses after the smoothing, baseline subtracting and normalizing to the D band intensity; Figure S3: SEM pictures of the activated carbon produced from molasses; Figure S4: CO_2 adsorption over activated carbons produced from molasses at the temperatures of 10 °C (a), 20 °C (b), 30 °C (c); Figure S5: CO_2 adsorption at 1 bar and 0 °C as a function of the volume of pores which are equal to and below given diameter; Figure S6: CO_2 adsorption at 1 bar and 10 °C as a function of the volume of pores which are equal to and below given diameter; Figure S7: CO_2 adsorption at 1 bar and 20 °C as a function of the volume of pores which are equal to and below given diameter; Figure S8: CO_2 adsorption at 1 bar and 30 °C as a function of the volume of pores which are equal to and below given diameter; Table S1: CO_2 adsorption at different temperatures at the pressure of 1 bar; Table S2: Parameters of the Toth model for CO_2 adsorption at different temperatures; Figure S9: The plots: (a) ln(qm) versus T, (b) ln(b) versus 1/T, and (c) n versus 1/T applied to the calculation of the Toth parameters in Equations (1)–(4) for M-KOH; Figure S10: The plots: (a) ln(qm) versus T, (b) ln(b) versus 1/T, and (c) n versus 1/T applied to the calculation of the Toth parameters in Equations (1)–(4) for M-H$_2$SO$_4$; Table S3: The parameters of the Toth temperature depended on equations for M-KOH and M-H$_2$SO$_4$ activated carbons; Figure S11: The plot of the function ln(p) vs. 1/T for different surface coverage for M-KOH; Figure S12: The plot of the function ln(p) vs. 1/T for different surface coverage for M-H$_2$SO$_4$.

Author Contributions: Conceptualization, K.K. and B.M.; methodology, K.K. and B.M.; validation, K.K. and B.M.; investigation, K.K., Ş.B., E.A.V., M.B., J.S.-N., J.S., P.M., R.J.W. and B.M.; data curation, K.K., Ş.B., E.A.V., M.B., J.S.-N., J.S., P.M. and B.M.; writing—original draft preparation, K.K., Ş.B., E.A.V., J.S.-N. and B.M.; writing—review and editing, K.K. and B.M.; visualization, K.K., Ş.B., E.A.V., J.S.-N. and B.M.; supervision, B.M. All authors have read and agreed to the published version of the manuscript.

Funding: This research received no external funding.

Institutional Review Board Statement: Not applicable.

Informed Consent Statement: Not applicable.

Data Availability Statement: Not applicable.

Conflicts of Interest: The authors declare no conflict of interest.

References

1. Chen, S.; Liu, J.; Zhang, Q.; Teng, F.; McLellan, B.C. A critical review on deployment planning and risk analysis of carbon capture, utilization, and storage (CCUS) toward carbon neutrality. *Renew. Sustain. Energy Rev.* **2022**, *167*, 112537. [CrossRef]
2. Nyambura, M.G.; Mugera, G.W.; Felicia, P.L.; Gathura, N.P. Carbonation of brine impacted fractionated coal fly ash: Implications for CO2 sequestration. *J. Environ. Manag.* **2011**, *92*, 655–664. [CrossRef] [PubMed]
3. Chiang, P.-C.; Pan, S.-Y. Post-combustion Carbon Capture, Storage, and Utilization. In *Carbon Dioxide Mineralization and Utilization*; Springer: Singapore, 2017; pp. 9–34.
4. Aminu, M.D.; Nabavi, S.A.; Rochelle, C.A.; Manovic, V. A review of developments in carbon dioxide storage. *Appl. Energy* **2017**, *208*, 1389–1419. [CrossRef]

5. Ma, C.; Bai, J.; Demir, M.; Hu, X.; Liu, S.; Wang, L. Water chestnut shell-derived N/S-doped porous carbons and their applications in CO2 adsorption and supercapacitor. *Fuel* **2022**, *326*, 125119. [CrossRef]
6. Sayari, A.; Belmabkhout, Y.; Serna-Guerrero, R. Flue gas treatment via CO_2 adsorption. *Chem. Eng. J.* **2011**, *171*, 760–774. [CrossRef]
7. Wang, X.; Cheng, H.; Ye, G.; Fan, J.; Yao, F.; Wang, Y.; Jiao, Y.; Zhu, W.; Huang, H.; Ye, D. Key factors and primary modification methods of activated carbon and their application in adsorption of carbon-based gases: A review. *Chemosphere* **2021**, *287*, 131995. [CrossRef] [PubMed]
8. Michalkiewicz, B.; Sreńscek-Nazzal, J.; Tabero, P.; Grzmil, B.; Narkiewicz, U. Selective methane oxidation to formaldehyde using polymorphic T-, M-, and H-forms of niobium(V) oxide as catalysts. *Chem. Pap.* **2008**, *62*, 106–113. [CrossRef]
9. Glonek, K.; Wróblewska, A.; Makuch, E.; Ulejczyk, B.; Krawczyk, K.; Wróbel, R.J.; Koren, Z.C.; Michalkiewicz, B. Oxidation of limonene using activated carbon modified in dielectric barrier discharge plasma. *Appl. Surf. Sci.* **2017**, *420*, 873–881. [CrossRef]
10. Deng, S.; Wei, H.; Chen, T.; Wang, B.; Huang, J.; Yu, G. Superior CO_2 adsorption on pine nut shell-derived activated carbons and the effective micropores at different temperatures. *Chem. Eng. J.* **2014**, *253*, 46–54. [CrossRef]
11. Li, D.; Zhou, J.; Wang, Y.; Tian, Y.; Wei, L.; Zhang, Z.; Qiao, Y.; Li, J. Effects of activation temperature on densities and volumetric CO_2 adsorption performance of alkali-activated carbons. *Fuel* **2019**, *238*, 232–239. [CrossRef]
12. Zhu, X.-L.; Wang, P.-Y.; Peng, C.; Yang, J.; Yan, X.-B. Activated carbon produced from paulownia sawdust for high-performance CO_2 sorbents. *Chin. Chem. Lett.* **2014**, *25*, 929–932. [CrossRef]
13. González-García, P. Activated carbon from lignocellulosics precursors: A review of the synthesis methods, characterization techniques and applications. *Renew. Sustain. Energy Rev.* **2018**, *82*, 1393–1414. [CrossRef]
14. Aghel, B.; Behaein, S.; Alobaid, F. CO_2 capture from biogas by biomass-based adsorbents: A review. *Fuel* **2022**, *328*, 125276. [CrossRef]
15. Sreńscek-Nazzal, J.; Kamińska, W.; Michalkiewicz, B.; Koren, Z.C. Production, characterization and methane storage potential of KOH-activated carbon from sugarcane molasses. *Ind. Crops Prod.* **2013**, *47*, 153–159. [CrossRef]
16. Legrouri, K.; Khouya, E.; Ezzine, M.; Hannache, H.; Denoyel, R.; Pallier, R.; Naslain, R. Production of activated carbon from a new precursor molasses by activation with sulphuric acid. *J. Hazard. Mater.* **2005**, *118*, 259–263. [CrossRef]
17. Kiełbasa, K.; Kamińska, A.; Niedoba, O.; Michalkiewicz, B. CO_2 Adsorption on Activated Carbons Prepared from Molasses: A Comparison of Two and Three Parametric Models. *Materials* **2021**, *14*, 7458. [CrossRef]
18. Młodzik, J.; Sreńscek-Nazzal, J.; Narkiewicz, U.; Morawski, A.; Wróbel, R.; Michalkiewicz, B. Activated Carbons from Molasses as CO_2 Sorbents. *Acta Phys. Pol. A* **2016**, *129*, 402–404. [CrossRef]
19. Ali, R.; Aslam, Z.; Shawabkeh, R.A.; Asghar, A.; Hussein, I.A. BET, FTIR, and RAMAN characterizations of activated carbon from wasteoil fly ash. *Turk. J. Chem.* **2020**, *44*, 279–295. [CrossRef]
20. O'Reilly, J.; Mosher, R. Functional groups in carbon black by FTIR spectroscopy. *Carbon* **1983**, *21*, 47–51. [CrossRef]
21. Puziy, A.; Poddubnaya, O.; Martínez-Alonso, A.; Suárez-García, F.; Tascón, J. Synthetic carbons activated with phosphoric acid. *Carbon* **2002**, *40*, 1493–1505. [CrossRef]
22. Yakout, S.; El-Deen, G.S. Characterization of activated carbon prepared by phosphoric acid activation of olive stones. *Arab. J. Chem.* **2016**, *9*, S1155–S1162. [CrossRef]
23. Gerçel, Ö.; Özcan, A.; Gerçel, H.F. Preparation of activated carbon from a renewable bio-plant of Euphorbia rigida by H_2SO_4 activation and its adsorption behavior in aqueous solutions. *Appl. Surf. Sci.* **2007**, *253*, 4843–4852. [CrossRef]
24. Brennan, J.K.; Bandosz, T.J.; Thomson, K.T.; Gubbins, K.E. Water in porous carbons. *Colloids Surfaces A Physicochem. Eng. Asp.* **2001**, *187–188*, 539–568. [CrossRef]
25. Singh, R.K.; Patil, T.; Pandey, D.; Tekade, S.P.; Sawarkar, A.N. Co-pyrolysis of petroleum coke and banana leaves biomass: Kinetics, reaction mechanism, and thermodynamic analysis. *J. Environ. Manag.* **2021**, *301*, 113854. [CrossRef] [PubMed]
26. Palakurthy, S.; Azeem, P.A.; Reddy, K.V. Sol–gel synthesis of soda lime silica-based bioceramics using biomass as renewable sources. *J. Korean Ceram. Soc.* **2022**, *59*, 76–85. [CrossRef]
27. Sadeek, S.A.; Mohammed, E.A.; Shaban, M.; Kana, M.T.A.; Negm, N.A. Synthesis, characterization and catalytic performances of activated carbon-doped transition metals during biofuel production from waste cooking oils. *J. Mol. Liq.* **2020**, *306*, 112749. [CrossRef]
28. Suárez-García, F.; Villar-Rodil, S.; Blanco, C.G.; Martínez-Alonso, A.; Tascón, J.M.D. Effect of Phosphoric Acid on Chemical Transformations during Nomex Pyrolysis. *Chem. Mater.* **2004**, *16*, 2639–2647. [CrossRef]
29. Castro-Muñiz, A.; Suárez-García, F.; Martínez-Alonso, A.; Tascón, J.M. Activated carbon fibers with a high content of surface functional groups by phosphoric acid activation of PPTA. *J. Colloid Interface Sci.* **2011**, *361*, 307–315. [CrossRef]
30. Hong, S.-M.; Jang, E.; Dysart, A.D.; Pol, V.G.; Lee, K.B. CO_2 Capture in the Sustainable Wheat-Derived Activated Microporous Carbon Compartments. *Sci. Rep.* **2016**, *6*, 34590. [CrossRef]
31. Wei, H.; Deng, S.; Hu, B.; Chen, Z.; Wang, B.; Huang, J.; Yu, G. Granular Bamboo-Derived Activated Carbon for High CO_2 Adsorption: The Dominant Role of Narrow Micropores. *ChemSusChem* **2012**, *5*, 2354–2360. [CrossRef]
32. Montagnaro, F.; Silvestre-Albero, A.; Silvestre-Albero, J.; Rodríguez-Reinoso, F.; Erto, A.; Lancia, A.; Balsamo, M. Post-combustion CO_2 adsorption on activated carbons with different textural properties. *Microporous Mesoporous Mater.* **2015**, *209*, 157–164. [CrossRef]

33. Shafeeyan, M.S.; Daud, W.M.A.W.; Houshmand, A.; Shamiri, A. A review on surface modification of activated carbon for carbon dioxide adsorption. *J. Anal. Appl. Pyrolysis* **2010**, *89*, 143–151. [CrossRef]
34. Yuan, X.; Xiao, J.; Yılmaz, M.; Zhang, T.C.; Yuan, S. N, P Co-doped porous biochar derived from cornstalk for high performance CO_2 adsorption and electrochemical energy storage. *Sep. Purif. Technol.* **2022**, *299*, 121719. [CrossRef]
35. Spessato, L.; Duarte, V.A.; Fonseca, J.M.; Arroyo, P.A.; Almeida, V.C. Nitrogen-doped activated carbons with high performances for CO2 adsorption. *J. CO_2 Util.* **2022**, *61*, 102013. [CrossRef]
36. Abdulsalam, J.; Mulopo, J.; Bada, S.O.; Oboirien, B. Equilibria and Isosteric Heat of Adsorption of Methane on Activated Carbons Derived from South African Coal Discards. *ACS Omega* **2020**, *5*, 32530–32539. [CrossRef]
37. Zhou, Y.; Tan, P.; He, Z.; Zhang, C.; Fang, Q.; Chen, G. CO_2 adsorption performance of nitrogen-doped porous carbon derived from licorice residue by hydrothermal treatment. *Fuel* **2022**, *311*, 122507. [CrossRef]
38. Rehman, A.; Nazir, G.; Rhee, K.Y.; Park, S.-J. Valorization of orange peel waste to tunable heteroatom-doped hydrochar-derived miroporous carbons for selective CO_2 adsorption and separation. *Sci. Total Environ.* **2022**, *849*, 157805. [CrossRef]
39. Sharma, M.; Snyder, M.A. Facile synthesis of flower-like carbon microspheres for carbon dioxide capture. *Microporous Mesoporous Mater.* **2022**, *335*, 111801. [CrossRef]

MDPI AG
Grosspeteranlage 5
4052 Basel
Switzerland
Tel.: +41 61 683 77 34

MDPI Books Editorial Office
E-mail: books@mdpi.com
www.mdpi.com/books

Disclaimer/Publisher's Note: The statements, opinions and data contained in all publications are solely those of the individual author(s) and contributor(s) and not of MDPI and/or the editor(s). MDPI and/or the editor(s) disclaim responsibility for any injury to people or property resulting from any ideas, methods, instructions or products referred to in the content.

www.ingramcontent.com/pod-product-compliance
Lightning Source LLC
LaVergne TN
LVHW070457100526
838202LV00014B/1739